Harvesting Food from Weeds

Scrivener Publishing
100 Cummings Center, Suite 541J
Beverly, MA 01915-6106

Bioprocessing in Food Science

Series Editor: Anil Panghal, PhD

Scope: Bioprocessing in Food Science will comprise a series of volumes covering the entirety of food science, unit operations in food processing, nutrition, food chemistry, microbiology, biotechnology, physics and engineering during harvesting, processing, packaging, food safety, and storage and supply chain of food. The main objectives of this series are to disseminate knowledge pertaining to recent technologies developed in the field of food science and food process engineering to students, researchers and industry people. This will enable them to make crucial decisions regarding adoption, implementation, economics and constraints of the different technologies.

As the demand of healthy food is increasing in the current global scenario, so manufacturers are searching for new possibilities for occupying a major share in a rapidly changing food market. Compiled reports and knowledge on bioprocessing and food products is a must for industry people. In the current scenario, academia, researchers and food industries are working in a scattered manner and different technologies developed at each level are not implemented for the benefits of different stake holders. However, the advancements in bioprocesses are required at all levels for betterment of food industries and consumers.

The volumes in this series will be comprehensive compilations of all the research that has been carried out so far, their practical applications and the future scope of research and development in the food bioprocessing industry. The novel technologies employed for processing different types of foods, encompassing the background, principles, classification, applications, equipment, effect on foods, legislative issue, technology implementation, constraints, and food and human safety concerns will be covered in this series in an orderly fashion. These volumes will comprehensively meet the knowledge requirements for the curriculum of undergraduate, postgraduate and research students for learning the concepts of bioprocessing in food engineering. Undergraduate, post graduate students and academicians, researchers in academics and in the industry, large- and small-scale manufacturers, national research laboratories, all working in the field of food science, agri-processing and food biotechnology will benefit.

Publishers at Scrivener
Martin Scrivener (martin@scrivenerpublishing.com)
Phillip Carmical (pcarmical@scrivenerpublishing.com)

Harvesting Food from Weeds

Edited by
Prerna Gupta
Navnidhi Chhikara
and
Anil Panghal

Scrivener Publishing

WILEY

This edition first published 2023 by John Wiley & Sons, Inc., 111 River Street, Hoboken, NJ 07030, USA and Scrivener Publishing LLC, 100 Cummings Center, Suite 541J, Beverly, MA 01915, USA

For more information about Scrivener publications please visit www.scrivenerpublishing.com.

Wiley Global Headquarters
111 River Street, Hoboken, NJ 07030, USA

For details of our global editorial offices, customer services, and more information about Wiley products visit us at www.wiley.com.

Library of Congress Cataloging-in-Publication Data

ISBN 978-1-119-79197-3

Cover image: Dandelion Greens and Jam | Elena Elisseeva | Dreamstime.com
Cover design by Kris Hackerott

Set in size of 11pt and Minion Pro by Manila Typesetting Company, Makati, Philippines

Contents

Preface

With the world's growing population, the demand and supply of food rich in nutrients are also growing. The food industries are working to meet this demand and at the same time keep the nutrients intact. The series, "Bioprocessing in Food Science," is an attempt to address the recent developments in food sciences. As the demand of healthy food is increasing all over the world, manufacturers are searching for new possibilities to occupy a major share in a rapidly changing food market. Presently, underutilized crops have come to play a major role in fulfilling this demand. And weeds are no less than these. Research workers from various parts of the world have also been devoting attention by working in such fields.

This book contains collective information about the phytochemical potential of various weeds and how to extract and separate them by using newer technologies like spray drying, encapsulation, and and how further to utilize them as functional foods of nutraceutical significance. This will be a new approach towards exploring the nutritional profile of weedy plants that are generally considered as waste. The biggest advantage of this book is that it will change the perspective of people toward weedy plants which can be further utilized to enhance the nutritional and functional properties of food products.

Harvesting Food from Weeds is primarily focused on different aspects in food nutrition, food science, and related subjects for individuals in the food industry, research, academia and students and assumes some knowledge of basic weeds used as food either whole, in parts, or in by-products. Traditionally various weeds are used in treatments of digestive problems, burns, blood purification, piles, arthritics, leukoderma, epilepsy, salivation, fever, cough, abdominal pain, peptic ulcer, hepatic disorder, and many others. The medicinal significance is due to the presence of phytochemicals like flavonoids, iso-flavonoids, and polyphenols. Researchers around the globe will be able to use this information as a guide in establishing the direction of future research in food processing, considering various aspects related to weeds usage in food. The main reason for writing this book now is to disseminate the wealth of knowledge on weeds nutrition and its relation to the food industry. It is envisioned for scientists, technologists, and engineers working in the area of food processing, nutrition, environmental engineering, process equipment design and product development, and students of food nutrition, food science & technology, health science, and engineering.

International peers having both the academic and professional expertise in the field extended their knowledge in the form of chapters of this book. The book illustrates a very clear and concise discussion on weeds usage in the food industry.

Today, with people's busy schedule, the demand for ready-to-serve food products is also growing. To meet this need, we have made this book by treating the subject in all its essential aspects in a simple and interesting style. We are confident that this book is going to receive an overwhelming response from the industry. Each of the chapters is supported with references owing to be an invaluable resource for the reader. Self-explanatory illustrations and tables of each chapter are an added advantage to understand the technology process and its outcome easily. We have also rationalized the index, which we decided was excessive and contained too many esoteric or trivial entries.

We are indebted to our numerous colleagues for their quality chapters and helping us to complete the book. We also thank the authorities of Lovely Professional University, Punjab, Guru Jambheshwar University of Science and Technology, and Chaudhary Charan Singh Haryana Agricultural University, Hisar for their support. Finally, we also express indebtedness and thankfulness to Scrivener Publishing and Wiley for their unfailing guidance and assistance provided in finalization and publishing the book.

Dr. Prerna Gupta
Dr. Navnidhi Chhikara
Dr. Anil Panghal

1

Chenopodium Species

Priyanka Kundu and Prerna Gupta*

Department of Food Technology and Nutrition, Lovely Professional University, (LPU), Punjab, India

Abstract

Plant-based food has been chosen for eating purposes because of its medicinal and health-promoting benefits. Components, like antioxidants, phenols, alkaloids, and flavonoids, present in them exhibit functions as anthelmintic, laxative, antiviral, antifungal, anti-inflammatory, anti-allergic, antiseptic, anti-pruritic, and so on. The presence of various minerals, vitamins high-quality proteins, carbohydrates, and lipids helps in performing various functions in the human body. *Chenopodiaceae* is an underutilized weedy crop having purposeful perspective apart from crucial nutritional benefits, it possesses nutraceutical and functional properties. The leaves of *chenopodium* species (*Chenopodium opulifolium shraeder*, *Chenopodium murale* L, *Chenopodium album* L., etc.) plants are generally used in various food products to make them more nutritious and healthier because of the presence of minerals, vitamins, fiber, and essential fatty acids. Among all the species, *C. album* L. is widely famous and consumed in the form of food items like bathua saag, parantha, dosa, and chutney throughout India. Traditionally bathua is used in treatments of digestive problems, burns, blood purification, piles, arthritics, leukoderma, epilepsy, salivation, fever, cough and abdominal pain, peptic ulcer, hepatic disorder, and so on. The medicinal significance is due to the presence of phytochemicals, like flavonoids, iso-flavonoids, and polyphenols.

Keywords: *Chenopodiaceae*, *Chenopodium album*, extruded product, plant foods, phytochemicals

**Corresponding author*: prerna.gskuastj@gmail.com

Prerna Gupta, Navnidhi Chhikara and Anil Panghal (eds.) Harvesting Food from Weeds, (1–20)

1.1 Introduction

Weeds are generally considered unwanted plants embattled for abolition as they grow in between the main crop, compete for nutrition, and do not require further treatment for better growth [1]. It is generally said that "one year of seeding corresponds to seven years of weeding" which means if we sow the weed once, it can be fed to people for seven years [2]. However, they are considered useful to humankind because of their high nutritional, functional, and phytochemical value, as well as medicinal properties [3, 4]. Unfortunately, a major part of our population is not aware of the nutritional benefits of these underutilized crops, especially people staying in urban areas because they are totally dependent on processed foods. It is very important to increase the level of knowledge regarding underutilized crops so that they can make it a part of their daily dietary habits. The incorporation of these underutilized crops into a food can make them functional foods by increasing the nutritional and phytochemical composition of the foods [4]. One fine example of this is *C. album*, which is commonly called bathua and belongs to the family *Chenopodiaceae* (Figures 1.1 and 1.2). It is widely growing in gram, mustard and wheat crop fields of the tropical and subtropical regions. This plant grows to a height of 0.3 to 3.3 m, with leaves that differ in size and form [5, 6]. The leaves and young shoots can be consumed like other leaf vegetables but their consumption should be limited due to presence of anti nutritional factor, i.e., oxalic acid [7]. *C. album* is an excellent source of various nutrients, phytochemicals, and also minerals, such as potassium, iron, calcium and phosphorus [6, 8]. The high nutritive value and medicinal properties that include anticancer, antifungal, antibacterial, antiseptic, antipruritic, antinociceptic make it very effective in treating the diseases of heart, blood, eye, and abdominal pain [9]. The processing and utilization of *C. album* can increase the nutritional and phytochemical composition of different food products. It is generally used in the preparation of dosa, potato curry, corn salad, bathua saag, parantha, chapati, and snacks [10–12]. Bathua's antimicrobial activity is also gaining researchers' attention with a zone of inhibition of 19.7 mm against *Staphylococcus aureus*, 18.3 mm against *Bacillus polymexia* and 16.37 mm against *Streptococcus faecalis* in a solvent extract of its leaves [14]. Bathua's methanolic extract inflorescence extract had the most antifungal activity, reducing fungal biomass generation by up to 60% [4]. The bioactive components of bathua leaves can be evaluated and used for the production of important drugs. It can be concluded

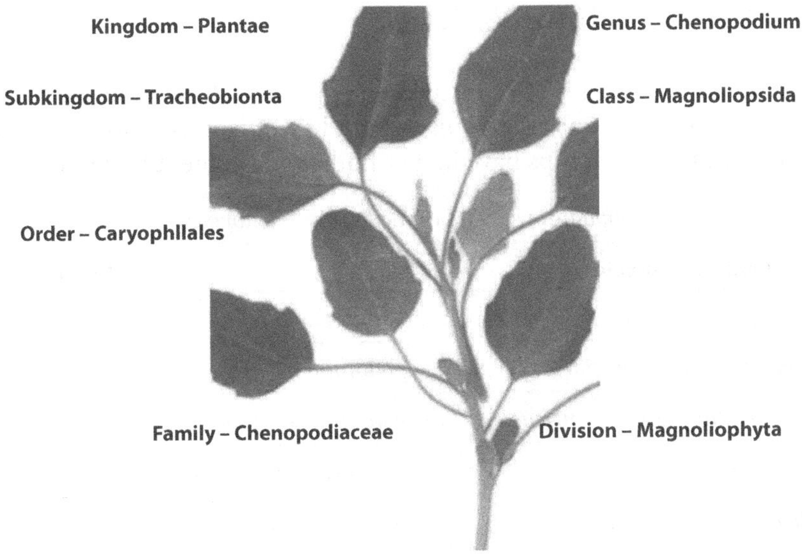

Figure 1.1 Botanical classification of *C. album.*

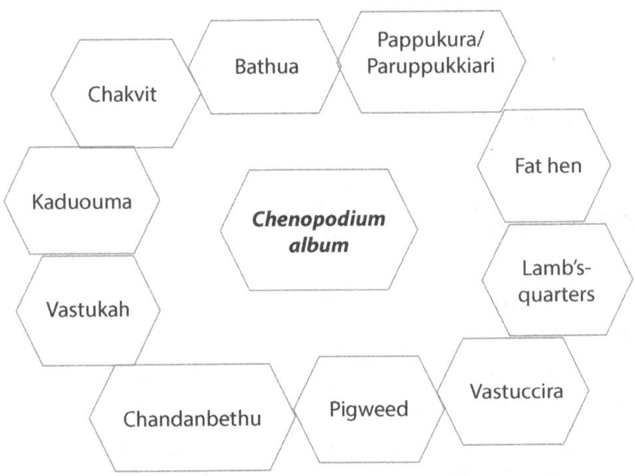

Figure 1.2 Common names of *C. album in different states* [9, 16].

that this weedy plant is actually an herb full of nutritional components [13, 15]. This plant could be a good alternative to expensive functional foods and may help in frightening against the hunger problem of India [10, 11]. In this chapter, the functional and medicinal properties of this weedy plant have been discussed.

1.2 Chenopodium Varieties

Generally, 250 species of *Chenopodium* are present throughout the world but among all these varieties *C. murale* and *C. quinoa* is mostly accepted because these weedy crops have higher nutritional content (Table 1.1) [7].

Table 1.1 Few *Chenopodium* species of the family *Chenopodiaceae.*

***Chenopodium* species**		**References**
Chenopodium album	*Chenopodium opulifolium*	[15–17]
Chenopodium ambrosioides	*Chenopodium foliosum*	
Chenopodium bonus henricus	*Chenopodium hybridum*	
Chenopodium botrys	*Chenopodium leptophyllum*	
Chenopodium chilense	*Chenopodium missouriense*	
Chenopodium ficicolium	*Chenopodium multifidum*	
Chenopodium murale	*Chenopodium pallidicaule*	[18, 19]
Chenopodium polyspermum	*Chenopodium rubrum*	
Chenopodium procerum	*Chenopodium Quinoa*	
Chenopodium urbicum	*Chenopodium vulvaria*	

1.3 Growth and Plantation

C. album is a winter annual herb with 250 species available throughout the world and 21 in India [20]. The stem of *Chenopodium* is erect with 0.3 to 3 m in height. The leaves are dark green in color with a smooth surface, variable in shape and size, and are generally found to be 15 cm long. Seeds are having 1.5 mm diameter with an acute margin and the flowers are in the form of cluster [21]. The origin of the species is from Europe. It is also found in Western Asia, America, and Mexico [22]. According to the writings of Kalm, an Agricultural economist, the Scandinavians (people living in northern Europe) used bathua as a flavoring agent during the preparation of meat. During ancient times, the Napoleon army was totally dependent on bathua seeds at the time of revenue [23]. In India Haryana, Punjab, Himachal Pradesh, Maharashtra, Rajasthan, Gujarat, Sikkim, West Bengal are the states that contribute to the production of *C. album* [24].

1.4 Health Effects

C. album is an underutilized wild plant that is loaded with vitamins and minerals (Table 1.2). Its leaves contain a large number nutrients including vitamin C, vitamin A, protein, phosphorus, calcium, potassium, iron, as well as amino acids [27]. Flavonoids, such as saponin, phenolic amide, alkaloid chinoalbicin, phenols, lignans, cinnamic acid amide, xyloside, have also been discovered to be potent phytoconstituents [28]. It can be a part of our eating habits not only because of nutrients, amino

Table 1.2 Proximate composition of *C. album* leaves.

Nutrients	Concentration	References
Moisture content	89.65%	[24]
Protein	3.7%	[25]
Fat	0.4%	
Dietary fiber	4	
Carbohydrates	2.9	[26]
Minerals	2.6	
Calcium	15 g	
Phosphorus	8 g	
Iron	4.2 gm	[23]
Magnesium	34 mg	
Potassium	452 mg	
Sodium	43 mg	[15, 24]
Thiamine	0.01 mg	
Niacin	0.6 mg	
Vitamin C	35 mg	
B-Carotene	1.47 μg	
Saponins	0.46 g	
Alkaloids	9.7 g	

acids, fiber, and vitamins but also to enhance the sensory and functional content of the meal [27]. So this plant in our everyday life has great importance to prevent health from many diseases, its antioxidant effect has active compounds, a strong natural antioxidant mixture has been reported from the leaves of bathua with its key active compounds, which are heat stable and nontoxic when consumed, its dietary invention is important in human [26, 27]. Since ancient times, it is an integral part of Ayurveda due to its health-promoting benefits. It is used to treat hepatic diseases, spleen enlargement, intestinal ulcer [24].Various bioactivities of crude and extracted compounds from the herb, such as anthelmintic, laxative, antipruritic, hypotensive, antifungal, antinociceptive properties, justified its uses in conventional medicinal. Bathua has been used since ancient times for medicinal purpose [9, 14] like to treat burns and irritation, the juice of *C. album* leaves is commonly used. The leaf juice can relieve skin inflammation and alleviate discomfort and redness. The extract of the arial portion of this plant can be used to treat arthritis by rubbing the solution on the affected part [12]. Its medicinal properties have also been addressed in several ancient books, such as Ayurveda, Charak Samhita, Atharvaveda, Sushruta Samhita from the 6th century, have also discussed about the medicinal properties of bathua [7]. Various chapters are available on medicinal properties of seeds, leaves, flowers and roots of *Chenopodium album*. The combination of bathua leaves with some cereals leads to the compensation of amino acid profiling food products, this nutritional approach can also be used for *C. album* in India [10, 11]. Its consumption should be promoted because by exploring this wild underutilized plant we can provide a new food product range to the food industry. We can provide the nutrition of green leafy vegetables to the consumers in the food of their own choice like extruded products or bakery products. It can be used as medicinal ingredient due to the presence of flavonoids and other bioactive phenolics with good antioxidant effect (Figure 1.3). Fermented foods, such as idlli, dosa, and bread, can also be made [14, 20]. *C. album* and soya that has been protein isolate can also be used to produce low fat-fried noodles, like snacks and extruded goods.

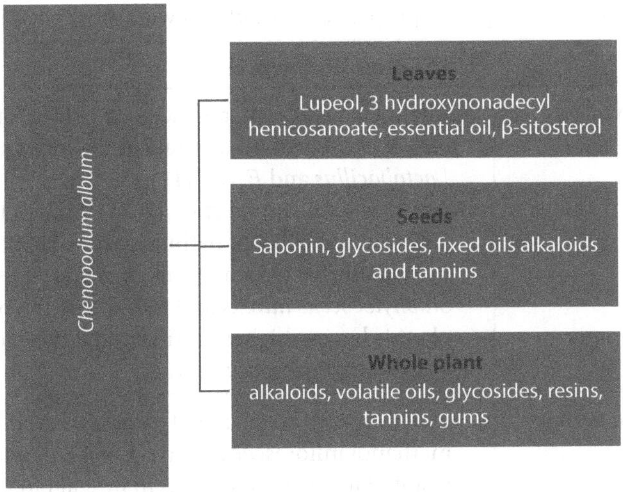

Figure 1.3 Chemical constituents of different parts of *C. album* plant [26–28].

1.5 Medicinal Values

Traditional medicine has been practiced by a significant proportion of the population for many decades, owing to the natural availability of medicinal herbs. One such common herb is *chenopodium album*, which contains alkaloids, saponins, phenols, and flavonoids as major phytochemicals [26, 27]. It specifically includes two flavonoids, like quercetin and kaempferol. It has essential oil, mineral matter, mostly potash, and a large number of albuminoids in its leaves. Due to the presence of these phytochemicals, *C. album possesses* therapeutic properties like a laxative, antirheumatic, anthelmintic, antidiarrheal, antimicrobial, antiphlogistic, antioxidant, contraceptive (Table 1.3) [26–29]. Since ancient times, people used this plant to treat various medical conditions like heart disease, anemia, kidney stone, jaundice, swelling, and burn as well [22, 25]. But with advancements in pharma industries, these traditional plants lost their value hence researchers need to focus on these plants to explore their therapeutic values (Figures 1.4 and 1.5) [28].

Table 1.3 Various therapeutic properties of *C. album* with most effective solvent.

Effects	**Utilization [References]**
Antibacterial	Methanol, chloroform, and acetone extract of *C. album* leaves exhibit antibacterial effect against *Lactobacillus* and *E. coli* [11]. Gqaza *et al* and Ahmad *et al.* [7, 31] studied the antibacterial effect of bathua leaves against human pathogenic bacteria like *Salmonella typhimurium, Staphylococcus aureus* and found that the bathua plant is having highest antibacterial property against *Staphylococcus aureus.*
Antifungal activity	The highest antifungal activity was observed in a methanol inflorescence extract of *C. album,* which resulted in a 90% reduction in fungal bio mass production [32].
Antioxidant activity	Aqueous extract of *C. album* leaves exhibits significant reducing ability and free radical scavenging effect on DPPH, hydroxyl, superoxide and hydrogen peroxide radicals when examined *in vitro* [11].
Antinociceptive effect	At doses of 100, 200, and 400 mg/kg, an ethanolic extract of the fruits of *C. album* has the ability to suppress scratching activity caused by 5-HT (5-hydroxyryptamine). 5-HT is well known for facilitating inflammatory pain of its own, as well as potentiating pain caused by other inflammatory mediators including noradrenaline and prostaglandin E. As a result, the extract's antinociceptive action may be mediated by 5-HT inhibition. In mice, the extract reduced the writing responses caused by acetic acid and an interplanetary injection of formaline [26, 27]. Investigated the effect of different solvent extract of *C. album* leaves on growth of human breast cancer cell lines and also evaluated its antioxidant activity. And found that methanolic extract of *Chenopodium* leaves showed higher anticancer activity [33–35].

(*Continued*)

Table 1.3 Various therapeutic properties of *C. album* with most effective solvent. (*Continued*)

Effects	Utilization [References]
Anti-cancer	Evaluated antitumor activity of *C. album* flavonoids. The flavonoids contained in *C. album* leaves showed antitumor activity [33, 51].
Anti- ulcer	Evaluated alcoholic extract of *C. album* leaves for its antiulcer activity. It was reported that ulcer index and gastric secretion were reduced with the use of *C. album* leaves extract [27].
Anthelmintic activity	Anthelmintic activity of *C. album* against trichostrongylid nematodes was investigated. It was analyzed that *C. album* Aqueous methanolic extracts at 3.0 g/kg on day 5 post-treatment showed a higher rate of reduction (82.2%) in eggs per gram (EPG) of feces [23, 25, 36].
Spasmolytic	Ethanol, water, chloroform, and ethyl acetate were used for the extraction process. The various extracts were examined on rabbit intestinal smooth muscles. The blunt extract demonstrated a dose-dependent improvement in smooth muscle relaxation, with the highest effect detected at 20 mg/ml (92.86%) [31].
Analgesic activity	The tail-flick procedure was used to test the blunt extract's analgesic activity in mice. From 30 to 210 minutes, a major analgesic effect was observed at a dose of 500 mg/kg [31].
Antipruritic and Antinociceptive effects	The ethanolic extract of *C. album* L. investigated for antinociceptive and antipruritic effects at doses of 100–400 mg/kg, dose-dependently inhibited scratching behavior induced by 5-HT. observed that extract of *C. album* flower possesses antipruritic and antinociceptive activities [14].

(*Continued*)

Table 1.3 Various therapeutic properties of *C. album* with most effective solvent. (*Continued*)

Effects	Utilization [References]
Anti-inflammatory activity	The occurrence of limonene, linalyl acetate, pinene, and linalool in *C. album* seed essential oil was investigated for anti-inflammatory action. It was discovered that the oil's anti-inflammatory activity is concentration-dependent. As a result, as the oil content grows, the percentage drop in ear edema rises as well [31].
Sperm immobilizing agent	*C. album* seeds were evaluated for sperm immobilizing capacity by applying the *C. album* seed decoction on the vaginal lining in rats. Results showed that a 2-mg/ml concentration of chenopodium album decoction was effective in the immobilization of spermatozoa [15].

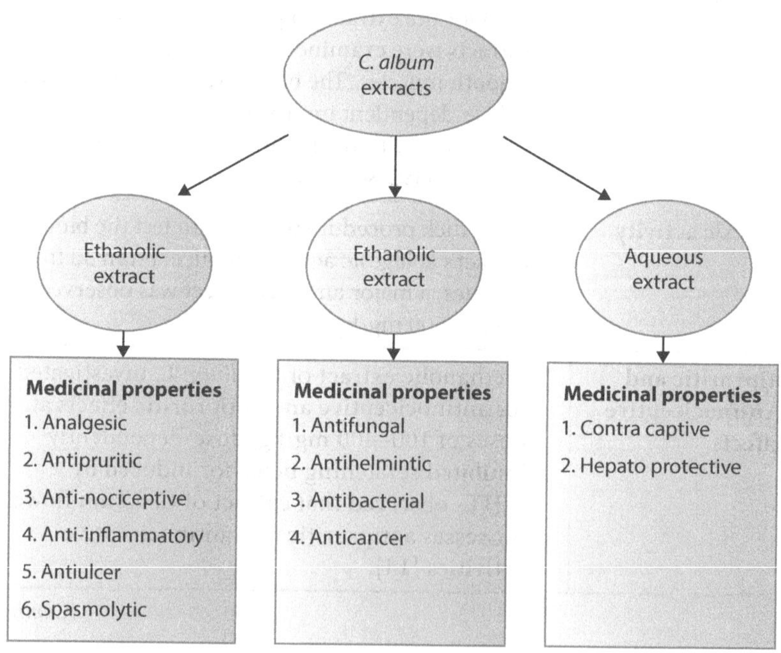

Figure 1.4 Medicinal properties of *C. album* extract with solvent showing maximum therapeutic activity [15, 27–29, 31–34, 51, 54].

1.6 Anti-Nutritional Factors

Anti-nutritional factors are chemical compounds synthesized in natural food by the normal metabolism of species. These compounds have lower nutrient absorption in the body [37–39]. However, if one continues the consumption of the same diet having a high amount of anti-nutritional factors, these toxins will reach to harmful levels in the body. These toxins not only hinder nutrient absorption but also destroy the vitamins in the food. It is important to remove or inactivate these compounds in order to protect the food values and prevent health problems (Figure 1.6). In the same way *C. album* is also having anti-nutritional factors, such as oxalates, saponins, phytates, and so on. Its anti-nutritional content is the main problem associated with this weed. It is necessary to use the appropriate techniques to reduce the anti-nutritional content to an acceptable level or minimum dose that it should not be toxic to humans [37, 40].

1.6.1 Oxalic Acid

Oxalic acid plays a major role in the formation of oxalates as combination of oxalic acid and its salts form oxalates. Many green leafy vegetables contain oxalic acid in their cell sap in three forms like magnesium salts, which are insoluble, iron, calcium, potassium, and soluble sodium salts or mixer of these two form depending on plant species [40]. Among three forms, soluble oxalates are more dangerous for human health and other minerals

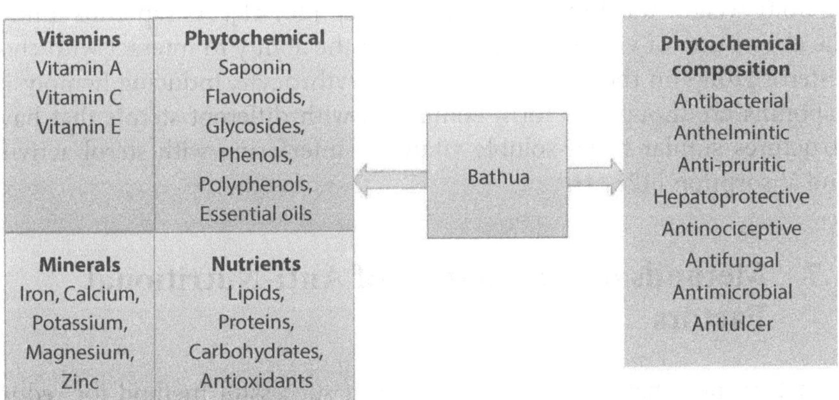

Figure 1.5 Nutritional and phytochemical composition of *C. album* plant [15, 20, 21, 27, 28, 30, 32].

preventing absorption of complexes [44]. It is important to take into consideration what amount of oxalates we are getting from our diet. If the oxalate–calcium ratio is greater than 9:4, the detrimental effect of calcium absorption is greater [42, 44].

1.6.2 Phytate

Phytic acid, also known as phytate, is a natural substance found in green leafy vegetables that serves as a major source of phosphorus storage. Phytic acid is present in plant tissues in the form of potassium, calcium, and magnesium cation salts [37, 38]. The presence of a phosphate group (negatively charged) in phytic acid chelates, and important minerals in human body hinders its absorption. 4 to 9 mg/100 g consumption of phytic acid reduces the absorption of iron by four to five times in the human body [40, 44].

1.6.3 Saponins

Saponins are nonvolatile, surface-active secondary compounds. Saponin word is derived from word sapo, meaning soap, since when shaken with water, saponin molecules form soap-like foams. They are structurally complex compounds called triterpene and steroid glycosides [38, 39]. Bitterness is the major limiting factor of saponins, as bitterness will increase with increase in the concentration of saponins. Saponins may not cause any harm to the human body when consumed in low amounts, the study has been conducted to check the bearable amount of saponins, and it was observed that the number of saponins is more than 150 mg/kg body weight was found in fetal for sheep [40, 41]. As saponins hinder the absorption of vitamins, saponins have the ability to interact with cholesterol group on the membranes of the erythrocyte, inducing hemolysis. Saponins are thought to form complexes with different sterols that have structures similar to fat-soluble vitamins, interfering with sterol activity and absorption [42–44].

1.7 Methods of Elimination of Anti-Nutritional Factors

Heat treatment has proven to be the most successful method for reducing the anti-nutritional factors found in green leafy vegetables. Blanching and cooking extract anti-nutrients by rupturing the plant cell wall and allowing soluble compounds to leach into the blanching media. Cooking and blanching techniques can effectively decrease phytic acid and oxalic acid levels [40, 45]. A study has been conducted to investigate the effect of blanching anti-nutritional factors of amaranth, bathua, fenugreek, and spinach leaves. It was reported that a combination of blanching and cooking was highly effective in reducing oxalic acid content; however, reduction in phytic acid can be achieved with 10 and 15 minutes of blanching [37, 44].

1.7.1 Drying

Antinutritional factors can be reduced with the help of drying techniques like cabinet drying, sun drying, solar drying, shade drying [45]. The impact of various drying treatments on the anti-nutritional factor of *C. album* were assessed and observed that cabinet dried *C. album* leaves shown significant decrease in antinutritional factors [46]. The impact of antinutritional factors of human health is total dependent on their consumption limit. More intake will cause serious health problems.

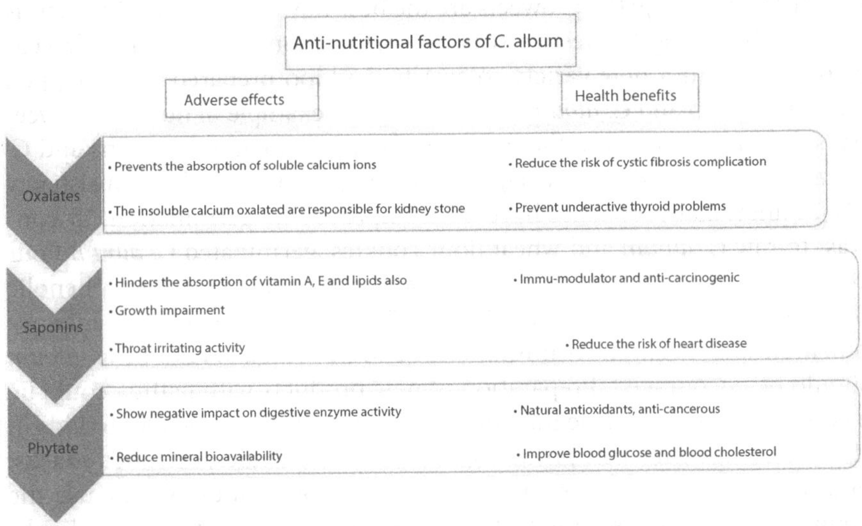

Figure 1.6 Health benefits and adverse effects of anti-nutritional factors on human health [41–45].

1.8 Traditional Food Products

Traditionally, *C. album* has been used for the preparation of various food products to enhance the flavor and nutritional value of food products.

The leaves of this plant have been incorporated into food products in different forms (powdered and liquid) and in blanched form, primarily for better retention of color and nutrients [4]. *C. album* when combined with cereals could result in nutritious food product with better susceptible response for all age groups [25].

Series of different food products prepared from bathua are: bathua dosa, [21] bathua potato curry (very popular in West Bengal), [24] bathua corn salad [21], bathua saag (a traditional dish of Punjab and Haryana. It is prepared on a stove made up of bricks and soil), bathua prantha or flat bread (chapatti). The ingredients used are wheat flour, bathua leaves, salt, oil, and water, bathua mathi (salted snack or flaky biscuit) [49, 50]. From raw and germinated *Chenopodium* flour, gluten-free cookies were prepared. Bathua pancake and boondi is prepared by using bengal gram flour and *C. album* flour [52, 53]. The physicochemical and morphological properties of prepared cookies were studied and compared with wheat flour cookies. The results clearly showed that germination increased the protein content of flour from 13.12 g/100 g to 15.45 g/100 g and reduced the fat content from 6.5 g/100 g to 4.14 g/100 g. Sweets are the best way to celebrate with friends and family during the festive season. Fortification with green leafy vegetables will add a new variety to sweets. Laddoo prepared from popped amaranth grain and *C. album* seeds are a good example of nutritious sweet balls. For its preparation, the mixture of popped amaranths grains and *C. album* will be added into the jaggery syrup up to one-third its consistency [47]. Functional properties have been improved by germination. In contrast to raw *C. album* and wheat flour cookies, germinated *C. album* flour cookies revealed 23.97/100 g highest antioxidant activity, total phenolic 671 mg/100 g, and total dietary fiber 38.77 g/100 g [49]. In order to enhance nutrition and strength in immune functions, a high intake of *C. album* should be encouraged. Preparation of new products with bathua could be a good strategy to provide the nutrition of green leafy vegetables to kids in the form of their favorite products, such as fried noodles, extruded snacks, and fermented foods (dosa, idli, bread). The food industry may use the ability of *C. album* as antioxidants. This can also provide a crucial substitute of synthetic antioxidant. A study has been done wherein cookies were prepared by supplementing *C. album* flour along with wheat flour at different levels, such as 10%, 20%, 30%, 40%, and 50%. The mixed flour was studied for physicochemical properties. The prepared cookies were examined for physical, textural, and sensorial characteristics. It was observed that the cookies prepared with 40% addition of chenopodium flour have the highest consumer acceptability. It can be inferred from the analysis that with *C. album*, functional and nutritive improvement is possible [48].

1.8.1 Conventional Food Products Supplemented with Bathua

Bathua (*C. album*) leaves were dried at 50°C to 60°C for 3 to 4 hours until 6% to 7% moisture was reached. These dehydrated leaves were integrated into two traditional foods, green gram dal and parantha, at 3% to 15% level. Results have shown that dehydrated leaves are abundant in ash, protein, and carbohydrate. The incorporation of 7% to 5% of chenopodium dehydrated leaves into green gram dal and parantha was highly acceptable. Enriched parantha (4255.66±0.6 µg/100 g) and green gram dal (984±.1.8 µg/100 g) had higher carotene content compared to control [47, 48, 52].

1.9 Future Scope

Weedy plants are not just waste but these plants can provide a wide range of various food products in affordable prices. As *C. album* is highly nutritious, it can give full nutrition to the poor population also the only thing that needs more attention is proper processing technique, which can deliver all the nutrient in the final product as well. Incorporation of bathua in fresh form or in dried form in food products like bathua parantha, bathua dosa, bathua laadoo, bathua cookies will give new taste and variety of foods. Researchers are not focusing on new approach as discussed in this chapter various extracts of this plant are showing therapeutic properties. This extract can be used to produce herbal medicines with techniques like nanoencapsulation or microencapsulation, which are new possibilities for future and can be proven as novel techniques for our new generation with cheapest source of nutrition, i.e., weeds. To maintain good physical and mental health, everybody requires a balanced diet consisting of proteins, vitamins, carbohydrates, meat, calcium. Increasing awareness of health, quality of life and nutritional insecurity leads to a shift in food consumption patterns involving more vegetables in both fresh and value-added forms.

1.10 Conclusion

Green leafy vegetables and weedy plants supply essential nutrients for human health. Vitamins, dietary fiber, amino acids are among them but lack of awareness is the key factor that people deny eating traditional foods, as well as the weedy crops. Increased awareness among the people about this crop can help in adding this plant into their daily dietary intake in different form viz simple and processed form depending on the taste of individuals. But in certain parts of the world, industrial exploitation of *C. album* is still a long way off. Bathua should be prescribed and must be taken as a part of our regular diet based on the numerous health benefits mentioned in this segment, as well as multiple past confirmed research studies, as its liberal usage is healthy and various health benefits can be derived from this natural herb. The mechanical, dietary, and medicinal properties of bathua can be further exploited in the production of healthier products, according to the studies on bathua described above.

References

1. Rao, A.N. and Chauhan, B.S., *Weeds and weed management in India-A review*, 2015.
2. Walia, U.S., *A handbook of weed management*, Kalyani Publisher, Punjab, India, 2014.
3. Grivetti, L.E., Frentzel, C.J., Ginsberg, K.E., Howell, K.L., Ogle, B.M., Bush foods and edible weeds of agriculture: Perspectives on dietary use of wild plants in Africa, their role in maintaining human nutritional status and implications for agricultural development, in: *Health and Disease in Tropical Africa: Geographical and Medical Viewpoints*, pp. 51–81, Harwood Academic Publishers, Akhtar R. London, 1987.
4. Patel, K., Neeharika, B., Suneetha, J., Kumari, B.A., Neeraja, B., Prabhakar, V., Maheswari, K.U., Effect of blanching on anti-nutritional factors of bathua leaves, *The Pharma Innovation*, 7, 4, 2018.
5. Afolayan, J. and Jimoh, F.O., Nutritional quality of some wild leafy vegetables in South Africa. *Int. J. Food Sci. Nutr.*, 60, 5, 424–431, 2009.
6. Hussain, J., Khan, A.L., Rehman, N.U., Hamayun, M., Shah, T., Nisar, M. *et al.*, Proximate and nutrient analysis of selected vegetable species: A case study of Karak region, Pakistan. *Afr. J. Biotechnol.*, 8, 12, 2725–2729, 2009.
7. Gqaza, B.M., Njume, C., Goduka, N.I., George, G., Nutritional assessment of Chenopodium album L. (Imbikicane) young shoots and mature plant-leaves consumed in the Eastern. *International Proceedings of Chemical, Biological and Environmental Engineering (IPCBEE) 2013*, 53, 97–102, 2011.
8. Pandey, M.K., Kumar, A., Singh, R., Tripathi, M., Scientific standardization of leaves of Chenopodium album L. *J. Pharmacogn. Phytochem.*, 5, 5, 1, 2016.
9. Agrawal Mona, Y., Agrawal Yogesh, P., Shamkuwar Prashant, B., Phytochemical and biological activities of Chenopodium album. *Int. J. PharmTech Res.*, 6, 1, 383–391, 2014.
10. Singh, L., Yadav, N., Kumar, A.R., Gupta, A.K., Chacko, J., Parvin, K., Tripathi, U., Preparation of value added products from dehydrated bathua leaves (Chenopodium album Linn.), *Nat. Prod. Radiance*, 6, 1, 6–10, 2007.

11. Kaur, G., Development and nutritional evaluation of value added products from dehydrated bathua *(Chenopodium album)* leaves, Doctoral dissertation, Punjab Agricultural University, Ludhiana, 2017.
12. Cutillo, F., DellaGreca, M., Gionti, M., Previtera, L., Zarrelli, A., Phenols and lignans from Chenopodium album. *Phytochem. Anal.*, 17, 344–349, 2006.
13. Milind, P. and Anupam, P., Preliminary pharmacognostic evaluations and phytochemical studies on leaf of Chenopodium album (Bathua Sag). *Asian J. Exp. Biol. Sci.*, *1*, 1, 91–95, 2016.
14. Dai, Y., Ye, W.C., Wang, Z.T., Matsuda, H., Kubo, M., But, P.P.H., Antipruritic and antinociceptive effects of *Chenopodium album L.* in mice. *J. Ethnopharmacol.*, 81, 245–250, 2018.
15. Kumar, R., Mishra, A.K., Dubey, N.K., Tripathi, Y.B., Evaluation of Chenopodium ambrosioides oil as a potential source of antifungal, antiaflatoxigenic and antioxidant activity. *Int. J. Food Microbiol.*, 115, 159–164, 2006.
16. Dembitsky, V., Shkrob., I., Hanus., L.O., Ascaridole and related peroxides from the genus Chenopodium. *Biomed. Pap. Med. Fac. Univ. Palacky Olomouc Czech. Repub.*, 152, 2, 209–215, 2008.
17. Kokanova-Nedialkova, Z., Nedialkov, P., Nikolov, S., The genus Chenopodium: Phytochemistry, ethnopharmacology and pharmacology. *Pharmacogn. Rev.*, *3*, 6, 280, 2009.
18. Bajwa, A.A., Zulfiqar, U., Sadia, S., Bhowmik, P., Chauhan, B.S., A global perspective on the biology, impact and management of Chenopodium album and Chenopodium murale: Two troublesome agricultural and environmental weeds. *Environ. Sci. Pollut. Res.*, *26*, 6, 5357–5371, 2009.
19. Huang, X. and Kong, L., Study on chemical constituents of volatile oil from Chenopodium ambrosioides L. In *2nd International Conference on Nutrition and Food Sciences*, Moscow, Russia *Zhongguo Yaoke Daxue Xuebao*, 33, 3, 256–257, 2002.
20. Yadav, N., Vasudeva, N., Singh, S., Sharma, S.K., Medicinal properties of genus Chenopodium Linn, *Nat. Prod. Radiance*, 6, 2, 131–134, 27-28 July 2013.
21. Sikarwar, I., Wanjari, M., Baghel, S.S., Vashishtha, P., Review on phytopharmacological studies on Chenopodium album Linn. *Indo Am. J. Pharm. Res.*, 3, 3089–3098, 2013.
22. Laghari, A.H., Memona, S., Nelofar, A., Khan, K.M., Yasmin, A., Determination of free phenolic acids and antioxidant activity of methanolic extracts obtained from fruits and leaves of Chenopodium album. *Food Chem.*, 126, 1850–1855, 2011.
23. Sood, P., Modgil, R., Sood, M., Chuhan, P.K., Anti-nutrient profile of different Chenopodium cultivars leaves. *Ann. Food Sci. Technol.*, *13*, 68–74, 2012.
24. Singh, L., Yadav, N., Kumar, A.R., Gupta, A.K., Chacko, J., Parvin, K., Tripathi, U., Preparation of value added products from dehydrated bathua leaves (*Chenopodiumalbum Linn.*). *Nat. Prod. Radiance*, 6, 6–10, 2007.

25. Poonia, A. and Upadhayay, A., Chenopodium album Linn: Review of nutritive value and biological properties. *J. Food Sci. Technol.*, *52*, 7, 3977–3985, 2015.
26. Lavaud, C., Voutquenne, L., Bal, P., Pouny, I., Saponins from Chenopodium album. *Fitoterapia*, 71, 338–340, 2000.
27. Nigam, V. and Paarakh, P.M., Hepatoprotective activity of Chenopodium album L. against paracetamol induced liver damage. *Pharmacologyonline*, 3, 312–328, 2011.
28. Baldi, A. and Choudhary, N.K., *In vitro* antioxidant and hepatoprotective potential of Chenopodium album extract. *Int. J. Green Pharm.*, 7, 1, 50–56, 2013.
29. Javaid, A. and Amin, M., Antifungal activity of methanol and n-hexane extracts of three Chenopodium species against Macrophomina phaseolina. *Nat. Prod. Res.*, *23*, 12, 1120–1127, 2009.
30. Pande, M. and Pathak, A., Preliminary pharmacognostic evaluations and phytochemical studies on leaf of Chenopodium album (Bathua sag). *Asian J. Exp. Biol. Sci.*, 1, 1, 91–95, 2010.
31. Ahmad, M., Mohiuddin, O.A., Mehjabeen, Jahan, N., Anwar, M., Habib, S., Alam, S.M., Baig, I.A., Evaluation of spasmolytic and analgesic activity of ethanolic extract of Chenopodium album Linn and its fractions. *J. Med. Plant Res.*, 6, 4691–4697, 2012.
32. Javaid, A. and Amin, M., Antifungal activity of methanol and n-hexane extracts of three Chenopodium species against Macrophomina phaseolina. *Nat. Prod. Res.*, *23*, 12, 1120–1127, 2009.
33. Khoobchandani, M., Ojeswi, B.K., Sharma, B., Srivastava, M.M., Chenopodium album prevents progression of cell growth and enhances cell toxicity in human breast cancer cell lines. *Oxid. Med. Cell Longev.*, 2, 3, 160–5, 2009.
34. Joshi, A. and Chauhan, R.S., Evaluation of anticancer activity of Chenopodium album leaves in BHK-21 cells. *Int. J. Univers. Pharm. Bio Sci.*, 1, 2, 92–102, 2012.
35. Jabbar, A., Zamana, M.A., Iqbal, Z., Yaseen, M., Shamim, A., Anthemintic activity of Chenopodium album (L.) and Caesalpinia crista (L.) against trichostrongylid nematodes of sheep. *J. Ethnopharmacol.*, 114, 86–91, 2007.
36. Natesh, H.N., Abbey, L., Asiedu, S.K., An overview of nutritional and antinutritional factors in green leafy vegetables. *Hortic. Int. J.*, 1, 2, 00011, 2017.
37. Samtiya, M., Aluko, R.E., Dhewa, T., Plant food anti-nutritional factors and their reduction strategies: An overview. *Food Prod. Process. Nutr.*, 2, 1, 1–14, 2020.
38. Fereidoon, S., Antinutrients and phytochemicals in food. *Developed from a Symposium Sponsored by the Division of Agricultural and Food Chemistry at the 210th National Meeting of the American Chemical Society, Chicago, Illinoi. ACS Symposium Series*, p. 2, 2012, ISSN 0097-6156.
39. Shamsuddin, A.M., Anti-cancer function of phytic acid. *Int. J. Food Sci. Technol.*, 37, 7, 769–782, 2002.

40. Umaru, H.A., Adamu, R., Dahiru, D., Nadro, M.S., Levels of antinutritional factors in some wild edible fruits of Northern Nigeria. *Afr. J. Biotechnol.*, 6, 1935–1938, 2007.
41. Gupta, R.K., Gangoliya, S.S., Singh, N.K., Reduction of phytic acid and enhancement of bioavailable micronutrients in food grains. *J. Food Sci. Technol.*, *52*, 2, 676–684, 2015.
42. Fereidoon, S., Antinutrients and phytochemicals in food. *Developed from a Symposium Sponsored by the Division of Agricultural and Food Chemistry at the 210th National Meeting of the American Chemical Society, Chicago, Illinoi. ACS Symposium Series*, p. 2, 2012, ISSN 0097-6156.
43. Hamid, N.T. and Kumar, P., Anti-nutritional factors, their adverse effects and need for adequate processing to reduce them in food. *AgricInternational*, *4*, 56–60, 2017.
44. Bora, P., Anti-nutritional factors in foods and their effects. *J. Acad. Ind. Res.*, *3*, 6, 285–290, 2014.
45. Abbas, Y. and Ahmad, A., Impact of processing on nutritional and anti-nutritional factors of legumes: A review. *Ann. Food Sci. Technol.*, 19, 2, 99–215, 2018.
46. Sood, P., Physico-chemical nutritional biological quality evaluation and value addition of Chenopodium (bathua) cultivars, Doctoral dissertation, Chaudhary Sarwan Kumar Himachal Pradesh Krishi Vishvavidyalaya, Plampur, 2011.
47. Jan, R., Saxena, D.C., Singh, S., Physico-chemical, textural, sensory and antioxidant characteristics of gluten–Free cookies made from raw and germinated Chenopodium (Chenopodium album) flour. *LWT-Food Sci. Technol.*, 71, 281–287, 2016.
48. Pandey, M.K., Kumar, A., Singh, R., Tripathi, M., Scientific standardization of leaves of Chenopodium album L. *J. Pharmacogn. Phytochem.*, 5, 5, 1, 2016.
49. Yadav, S.K. and Sehgal, S., *In vitro* and *in vivo* availability of iron from bathua (Chenopodium album) and spinach (Spinacia oleracea) leaves. *J. Food Sci. Technol.*, 39, 42–46, 2002.
50. Ali, E., The chemical constituents and pharmacological effects of Chenopodium album. *Int. J. Pharm.*, 5, 10–17, 2015.
51. Pandey, S. and Gupta, R., Screening of nutritional, phytochemical, antioxidant, antibacterial activity of Chenopodium album. *J. Pharmacogn. Phytochem.*, 3, 1–9, 2014.
52. Kaur, G., Development and nutritional evaluation of value added products from dehydrated bathua *(Chenopodium album)* leaves, Doctoral dissertation, Punjab Agricultural University, Ludhiana, 2017.
53. Laghari, A.H., Memona, S., Nelofar, A., Khan, K.M., Yasmin, A., Determination of free phenolic acids and antioxidant activity of methanolic extracts obtained from fruits and leaves of Chenopodium album. *Food Chem.*, 126, 1850–1855, 2011.

54. Kokanova-Nedialkova, Z., Nedialkov, P., Nikolov, S., The genus Chenopodium: Phytochemistry, ethnopharmacology and pharmacology. *Pharmacogn. Rev.*, 3, 6, 280, 2009.

2

Herbs of Asteraceae Family: Nutritional Profile, Bioactive Compounds, and Potentials in Therapeutics

Chinaza Godswill Awuchi[1] and Sonia Morya[2]*

[1]*School of Natural and Applied Sciences, Kampala International University, Kampala, Uganda*
[2]*Department of Food Technology and Nutrition, School of Agriculture, Lovely Professional University, Phagwara, India*

Abstract

Asteraceae is a large family of flowering plants known as Angiospermae. Asteraceae provides coffee substitutes, herbal teas, sweetening agents, artichokes, sunflower seeds, leaf vegetables, and cooking oils. Plants of Asteraceae with commercial importance include *Lactuca sativa, Cynara cardunculus, Carthamus tinctorius* (safflower), *Smallanthus sonchifolius, Helianthus annuus* (sunflower), and so on. Many bioactive compounds including essential oils, polysaccharides, flavonoids, clerodane-type diterpenes, triterpenoid saponines, phenols, salicylic acid derivatives, occur in Asteraceae, and have many biological activities, such as gastroprotective, antioxidant, cytotoxic, anti-obesity, anti-inflammatory, spasmolytic, and antimicrobial effects. Diuretic activities have been linked with leiocarposide and flavonoids, while anti-inflammatory effects are attributed to phenolic compounds such as leiocarposide, hyperoside, quercitrin, and so on. Cytotoxic activities could be attributed to the terpenic compounds. The energy store in Asteraceae mostly exists as inulin. Asteraceae produces tannins, acetylenes, alkaloids, iso/chlorogenic acid, pentacyclic triterpene alcohols, sesquiterpene lactones, and so on. Most of the substances have promising potential in medicine, diet, and therapeutics.

Keywords: Asteraceae, bioactive compounds in Asteraceae plants, medicinal properties of Asteraceae, Asteraceae therapeutics, nutritional properties of Asteraceae plants, phenolic compounds in Asteraceae

**Corresponding author*: sonia.morya8911@gmail.com

Prerna Gupta, Navnidhi Chhikara and Anil Panghal (eds.) Harvesting Food from Weeds, (21–78)

2.1 Introduction

Asteraceae, also called Compositae, composite, daisy, aster, orsunflower family, is widespread and large family of Angiospermae (Magnoliophyta or flowering plants) [1, 2]. They include at least 32,000 presently recognized species in more than 1,911 genera in 13 subfamilies. Asteraceae is only rivalled by the Orchidaceae with respect to the number of species. The study of Asteraceae family is called synantherology. Almost all the plants in Asteraceae family bear flower in dense heads (pseudanthia or capitula) encircled by involucral bracts. Each capitulum could appear as single flower when seen from distance. The involucral bracts may appear like calyx, while the enlarged peripheral (outer) flowers in each capitulum may look like petals. The term Asteraceae was derived from Ancient Greek term ἀστήρ, a genus name for *Aster*, which means star, used to denote the inflorescence which resembles a star. Species of Asteraceaeare widely and diversely distributed, and are seen in every area except in extreme Arctic and the Antarctica. They are mostly abundant in the tropics and subtropics, including Southwestern China, central Asia, Southern Africa, the Levant, the Mediterranean, Eastern Brazil, and Central America.

Several plants in Asteraceae family are perennial or annual herbs; however, a huge numbers of them are also trees, vines, or shrubs. The family is widely and diversely distributed, with the species widely spread from tropical to subpolar regions, in habiting several habitats. Most of the species are commonly found in semi-arid and arid regions of subtropics and lower temperate latitude. Asteraceae could represent about 10% of autochthonous flora in several parts of the globe. Species in Asteraceae has economic importance, providing commercial products including herbal teas, coffee substitutes, sweetening agents, artichokes, sunflower seeds, leaf vegetables (such as lettuce), and cooking oils. Additionally, many genera of Asteraceae have importance in horticultural system, including heleniums, zinnias, dahlias, chrysanthemums, fleabane, various daisies, *Echinacea* (coneflowers), and pot marigold (*Calendula officinalis*). Asteraceae plants are significant in herbal (traditional) medicine, such as yarrow, *Grindelia*, and so on. Several Asteraceae are categorized as weeds in some instances, of which several of them are invasive species common in certain regions, usually introduced by humans. Typical examples are several dandelions, thistles, ragweeds, *Bidens*, and tumble weeds.

Asteraceae plays significant role in antiparasitic treatments. Quinine discovery from *Cinchona succirubra*, along with its ensuing application as antimalarial medication present a historical breakthrough in natural antiparasitic drugs.

In 2015, the Nobel Prize for Medicine/Physiology was awarded for avermectin and artemisinin discovery, which deeply reformed parasitic disease treatment all over the world [3]. Milestone for antimalarial drug development was sesquiterpene artemisinin identification from *Artemisia annua*, which has the potency to destroy multidrug-resistant *Plasmodium falciparum* strains [4, 5]. Many semi-synthetic compounds derived from artemisinin, such as water-soluble artesunate, are developed for use in clinical practices currently [5]. Figure 2.1 shows some common plants in Asteraceae family.

The bioactive compounds and nutrients in Asteraceae have been widely noted, with many nutritional and medicinal benefits. This chapter provides information on the bioactive compounds and nutritional profile of plants in Asteraceae family, as well as the medicinal and therapeutic applications of these plants. Information about plants in other families can be accessed from other chapters. The energy in Asteraceae is generally stored as inulin, instead of starch. They produce tannins, acetylenes, several alkaloids, pentacyclic triterpene alcohols, sesquiterpene lactones, and iso/chlorogenic acid. Species in Asteraceae have essential oils of terpenoids which has no iridoids. Some secondary metabolites, including terpenoids and flavonoids, are produced by Asteraceae. Most of these bioactive compounds inhibit parasitic intestinal worms and protozoanparasites, including *Leishmania*, *Trypanosoma*, and *Plasmodium*, and thus have potential in medicine [3]. The medicinal and therapeutic uses of some plants from Asteraceae family are shown in Table 2.1.

2.1.1 History, Etymology, and Taxonomy of Asteraceae

The term Asteraceae comes to scientific terminology from *Aster*, New Latin word, and the genus type, + *-aceae*, standardized suffix used for the names of plant family in modern taxonomical classification. The name of the genus comes from the word *aster*, a Classical Latin word which means "star," and came from ἀστήρ (*astḗr*), an Ancient Greek word for "star." Compositae, an earlier name given to Asteraceae, is now known as an alternative name, to mean "composite" and to refer to typical inflorescence, a special kind of pseudanthium seen in few other families of angiosperm alone. *Daisy*, a vernacular name widely used for members in Asteraceae family, was derived from *dæġes ēaġe*, an Old English word for daisy (*Bellis perennis*), which means "day's eye"; because of how the petals open and close at dawn and dusk respectively. In 1792, Paul Dietrich Giseke, a German botanist, used the first original name Compositae to describe Asteraceae [16]. Two subfamilies were traditionally acknowledged: Liguliflorae (or *Cichorioideae*) and *Asteroideae* (or Tubuliflorae). The former has indicated as being widely paraphyletic and

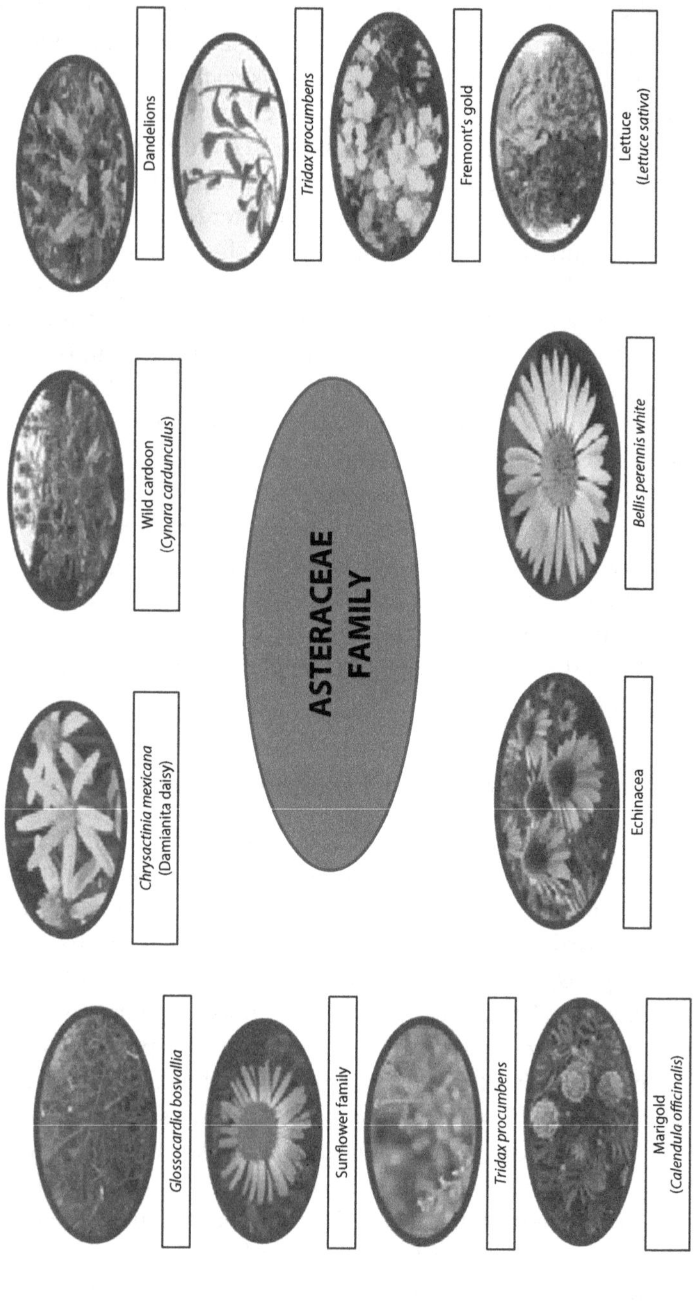

Figure 2.1 Some common plants in Asteraceae family.

Table 2.1 Medicinal and therapeutic uses of some species of Asteraceae.

Asteraceae plant	Medicinal, therapeutic, and other uses	References
Calendula officinalis L.	Its flowers are applied for treating amoebal infections and intestinal worms in pigs and pets. Cold decoctions made from the leaves are used for bloody dysentery and amoebic treatment.	[6]
Bidens pilosa L.	Fresh juices of aerial parts are applied for treating stomachache, abdominal pain, and intestinal worm infection, while the juice from whole plant and root is usedto treat malaria. Whole plant can be used for prevention and control of tick in livestock.	[7–9]
Sphaeranthus indicus L.	Paste of whole plant with salt can be used as anthelmintic. Seed powder or root portion is orally administered to destroy intestinal worms in people, especially those between 4 and 15 years.	[10]
Ageratum conyzoides (L.) L.	Herbal infusion is administered for gastrointestinal (GI) ailments including flatulence, dysentery, diarrhoea, etc. Cold decoction formulated from the aerial part is used as antimalarial treatment. It can also be used as worm medicine.	[11, 12]
Centipeda minima (L.) A. Braun & Asch.	Root decoctions are used for treating all types of fever. Leaf decoctions are commonly used for roundworm and hookworm treatment.	[13]
Blumea lacera (Burm.f.)DC.	Fresh leaf juice can be used for treating fever, including malaria and toto kill worms in children.	[14]
Caesulia axillaris Roxb.	Extracts from whole plant is administered to treat malaria and amoebic dysentery.	[15]

is currently separated into 12 subfamilies; however, the latter is still standing. Four subfamilies *Mutisioideae*, *Carduoideae*, *Cichorioideae*, and *Asteroideae* contain 99% of the diversity of species in whole family; about 3%, 11%, 14%, and 70% respectively [17–19]. Due to the complexity in morphology shown by Asteraceae family, agreements on generic circumscription have been challenging for taxonomists and botanists. Consequently, most of the genera have necessitated several revisions [20].

2.1.2 Characteristics of Herbs in Asteraceae Family

Asteraceae members are mainly herbaceous plants; however, some are trees (e.g., *Lachanodes arborea*), climbers, and shrubs. In general, due to their characteristics and their typical inflorescence, they can be easily distinguished from other plants [20], although assigning species and genera of some of them, including *Hieracium*, is difficult. Various characteristics of herbs in Asteraceae Family has been mentioned in Table 2.2.

Table 2.2 Characteristics of herbs in Asteraceae Family.

Plant part	Characteristics	References
Roots	In general, plants in Asteraceae family produce taproots, although they have fibrous root system sometimes.	[1, 20, 21]
Leaves	The leaves usually contain secretory canals having latex or resin (most common in Cichorioideae). Leaves of Asteraceae can be whorled, opposite, or alternate. The leaves could be simple, but usually incised or otherwise deeply lobed; mostly revolute or conduplicate. The margins could be toothed, lobed or entire.	[1]
Stems	Stems of Asteraceae are herbaceous, cylindrical, branched, and aerial with generally erect glandular hairs but could be prostrate to climbing. Few species in the family have rhizomes or caudices underground stems, which can be woody or fleshy based on species. The stems usually contain secretory canals having latex or resin (most common in Cichorioideae).	[20]

(*Continued*)

Table 2.2 Characteristics of herbs in Asteraceae Family. (*Continued*)

Plant part	Characteristics	References
Fruits and seeds	The fruit in plants of Asteraceae is achene-like. It is known as *cypsela* (*cypselae* plural). Although two fused carpels are present, there is only a seed and only a locule per fruit. Sometimes it may be spiny or winged since the pappus, derived from tissue of calyx, usually stays on the fruits, as in dandelion. However, in some species, pappus falls off, as in *Helianthus*.	[20, 22]
Flowers	A single flower appearance in plants of Asteraceae family is in fact a composite of smaller flowers. The composite overall appearance as single flower attracts pollinators similarly as structural appearance of individual flower in many other families of plants. Compositae, an older name for Asteraceae, was derived from the reality of what seems as single flower is in fact composite of smaller flowers. The sun rays or petals in sunflower heads are in reality distinct strap-shaped flowers known as ray flowers. The sun disk composes of circular and smaller shaped distinct flowers referred to as disc flowers.	[21]

2.1.3 Evolution of Asteraceae

Pollen grains from Antarctica Late Cretaceous are the first known fossils of Asteraceae members, and dated back to ~76 to 66 Mya and placed on the genus *Dasyphyllum,* which is extant [1]. In 2015, Barreda *et al.* [1] stated that Asteraceae crown group evolved no less than 85.9 Mya with 88 to 89 Mya stem node age. It is not yet known whether the exact reason for their remarkable success was developing highly specialized capitulum, their capability of storing energy in form of fructans (mostly inulin) instead of starch, which is beneficial in dry zones, or combination of all of them and probably other one or more factors.

2.1.4 Food and Commercial Uses of Asteraceae

Plants in Asteraceae family with commercial importance include, but not limited to, the food crops Jerusalem artichoke (*Helianthus tuberosus*), safflower (*Carthamus tinctorius*), yacón (*Smallanthussonchifolius*), sunflower (*Helianthus annuus*), globe artichoke (*Cynara scolymus*), chicory (*Cichorium*), and lettuce (*Lactuca sativa*) [23]. Some of the Asteraceae plants are commonly applied in herbal medicine and are also used in herbal teas, as well as other beverages. For example, chamomile is used as tea and has two species; the perennial Roman chamomile (*Chamaemelum nobile*) and the annual German chamomile (*Matricaria chamomilla*). Pot marigold (*Calendula*) is commercially cultivated for potpourri and herbal teas. *Echinacea* has been in use as medicinal tea in various parts of the world. *Artemisia* (wormwood genus) includes tarragon (*Artemisia dracunculus*) and absinthe (*Artemisia absinthium*). *Tagetes lucida* (winter tarragon) is cultivated for use as substitute for tarragon in climates harsh to tarragon survival.

Asteraceae have been applied in industries for various purposes. *Tagetes patula* (French Marigold) is commonly used in poultry feeds; its oil extracts are used in cigarette and cola industries [23]. Many plants of Asteraceae are copious nectar producers. As such, they play key role in evaluating populations of pollinator during their bloom [23]. Domestic sunflower (*Helianthus annuus*), knapweed (*Centaurea*), and some golden rod (*Solidago*) species are main honey plants commonly used by bee keepers and other bee farmers. Golden rod makes pollen with high protein that helps honeybees during winter [24]. Many Asteraceae have economic importance as weeds; notable of this in some parts of the world include dandelion (*Taraxacum*), ground sel (*Senecio vulgaris*), and ragwort (*Senecio jacobaea*). The genera *Tanacetum*, *Tagetes*, *Pulicaria*, and *Chrysanthemum* have species with beneficial insecticide activities [23]. Hypoallergenic latex is sourced from guayule (*Parthenium argentatum*) [23].

Several species of Asteraceae have medicinal and therapeutic properties, mostly applied as traditional antiparasitic medication [23]. The plants are usually featured in phytochemical and medical journals as the compounds of sesquiterpene lactone in them are significant in causing allergic contact dermatitis [25].

2.1.5 Medicinal and Therapeutic Uses of Asteraceae

Asteraceae plants are commonly applied for treating many diseases. At least 7000 active compounds have been reported in Asteraceae. Of these, at least 5000 have shown to have bioactivities [4]. Plants in Asteraceae family have been reported to exhibit several medicinal and therapeutic properties including antiparasitic, anti-tussive, antispasmodic, antalgic, anti-hemorrhagic, wound-healing, antibacterial, detoxifying, antitumor, anti-inflammatory, and antipyretic, and are also beneficial for remedying hypotension, hemorrhoids, leucorrhoea, lumbago, dysentery, dyspepsia, hepatoprotective, and flatulence [26].

Plants of Asteraceae are used in herbal treatment against several diseases, because of their excellent bioactive, therapeutic, and medicinal properties. *Ageratum conyzoides*is used for treating malaria, intestinal colic, dysentery, and diarrhoea [27]. *A. conyzoides* is popular for its high content of phytochemicals including terpenoids, sterols, benzofurans, flavonoids, coumarins, and alkaloids, with some compounds already identified, such as caffeic acid, fumaric acid, quercetin, coumarin, caryophyllene, various flavonoids, various sterols (such as stigmasterol and b-sitosterol), and friedelin [28]. *Bidens pilosa* is used widely by Amazon indigenous tribes for malaria treatment [3]. *Bidens pilosa* contain around 201 compounds comprising 8 porphyrins, 13 aromatics, 19 phenylpropanoids, 25 terpenoids, 60 flavonoids, 70 aliphatics, and 6 additional compounds [3]. Nevertheless, the relationship between some phytochemicals in *Bidens pilosa* and several bioactivities has not been completely established, and can be explored in future research [29]. *Blumea lacera* is applied in treatment of many fevers, such as malaria. *Blumea lacera* contains bioactive compounds, including triterpenes, b-sitosterol, a-amyrin, hentriacontane, lupeol, flavonoids, campesterol, coniferyl alcohol derivatives, fenchone, and hentriacontane [29, 30]. *Calendula officinalis* is used for many medicinal purposes, and is known to contain several flavonoids, such as quercetin-3-O-rutinoside, quercetin-3-O-glucoside, rutin, calendoflavobioside, calendoflavoside, calendoflaside, narcissin, isorhamnetin-3-Ob-D-glycoside, isoquercetin, and quercetin. *Calendula officinalis* also contain quinones, saponins, and coumarins, as well as several terpenoids, such as ursadiol, brein, erythrodiol, stigma sterols, sitosterols, and its derivatives; oleanolic acid glucosides; many glycosides of triterpene, such as calendula glycoside A, and so on [31].

Caesulia axillaris whole plant extracts are commonly used to cure malaria; however, no scientific-based evidence supports this usage [3, 10]. *Centipeda minima* is used for traditional medicine against parasites [13]. *Eclipta prostrata* is regularly used for malaria treatment [32, 33]. *E. prostrate* is popular for its phytoconstituents, including bioactive compounds, especially triterpenic acid, triterpene glycosides, columbin, wedelolactone, stigmasterol, cinnaroside, apigenin, luteolin-7-O-glucoside, b-amyrin, and eclipline. *Elephanto pusscaber,* like *Eclipta prostrata,* is commonly used for malaria treatment [27]. *Elephanto pusscaber* is also popular for its phytoconstituents, including bioactive compounds such as lipids like stigma sterol, lupeol, ethyloctadecanoate, ethyl-(Z)-9-octadecenoate, 12-octadecadienoate, ethyl-9, and ethyl hexadecanoate, and sesquiterpene lactones like isoscabertopin, scabertopin, isodeoxyelephantopin, deoxyelephantopin, and elescaberin [34]. *Sphaeranthus indicus* paste from whole plant with salt is administered as anthelmintic purposes [35]. Phytochemical analyses suggest that *Sphaeranthus indicus* contains sugars, amino acids, peptide alkaloids, alkaloids, sterol glycosides, sterols, isoflavone glycosides, flavonoid C-glycosides, flavone glycosides, sesquiterpene acids, sesquiterpene lactones, sesquiterpenoids, and eudesmanolides [36]. Additionally, essential oil extracts from *Sphaeranthus indicus* has bioactive compounds, such as methoxy frullanolides, geraniol, ocimene, methyl chavicol, spharerne, sphaeranthol, and sphaeranthine [37].

An ornamental plant of Asteraceae called *Tagetes erecta* is commonly used for treating many conditions including dysentery, cough, colic, abdominal pain, irregular menstruation, and anaemia [3, 10]. *Tagetes erecta* contain phytochemicals including tagetone, ocimene, b-sitosterol, b-daucosterol, 7-hydroxy sitosterol, lupeol, erythrodiol, erythrodiol-3-palmitate, quercetin, syringic acid, gallic acid, quercetagetin-7-O-glucoside, quercetagetin-7-methyl ether, and quercetagetin [38, 39]. In Ayurvedic medicine, *Tridax procumbens*is widely used due to its content of bioactive compounds, including bis-(2-ethylhexyl)-phthalate, stigmasterol, b-sitosterol, pentadecane, apigenin-7-O-b-D-glucoside, b-daucesosterol, and so on. [39]. Several *Vernonia* species are used in various herbal medicines worldwide, including *Vernonia cinereal, Vernonia Albicans,* and *Vernonia anthelmintica. V. anthelmintica* seeds are applied as anthelmintic, mostly in children [13, 35]. Fruit powder has been used for treatment of malaria fever, and amoebic dysentery (stomach ache) [40]. For filariasis treatment, *Vernonia albicans* powder of 10 to 20 g can be taken with milk (125 ml) combined with 10 g sugar candy and five to seven fruits of cardamom; with empty stomach, once in morning, for 3 months. The whole plant aqueous extracts are used for malaria treatment [41]. Mixture of *Vernonia cinerea* root paste and honey can be orally administered for malaria treatment; reports suggest the use of *Vernonia cinerea* for elephantiasis treatment [42, 43]. *Xanthium strumarium,* a weed, is a member of Asteraceae used as for its medicinal properties. *Xanthium strumarium* therapeutical and medicinal properties are, to a large extent, due to its phytoconstituents, such as polysterols, phenols, glycosides, and sesquiterpene lactones, all of which are distributed in all parts of the plant. Notable bioactive compounds in *Xanthium strumarium* include deacetyl xanthinin (xanthatin), xanthumin, xanthinin, sulphated glycoside, oleic acid, stromasterol, stearylalcohol, chlorobutanol, isohexacosane, b-D-glucoside of b-sitosterol, g-sitosterol, b-sitosterol, deacetyl xanthumin and thiazinedione, a- and g-tocopherol, caffeoylquinic acids, xanthanolides, hydroquinone, 4-oxo-bedfordia acid, xanthinosin, isoxanthanol, xanthanol, carboxyatractyloside, phytosterols, atractyloside, and xantho strumarin [44].

2.1.6 Asteraceae and their Phytoconstituents for Antiparasitic Treatment

Several secondary metabolites from plants in Asteraceae family have shown target-specific activities against parasites of Trypanosoma, Leishmania, and Plasmodium. These plants are extensively applied as medicines because of their bioactive compounds which include saponins, pyrethrins, triterpenoid, sesquiterpene lactones, diterpenoid and quinoline, terpenoids (triterpenes, diterpenes, sesquiterpenes, and monoterpenes), coumarins, phenolic acids, flavonoids, and alkaloids (pyridine and pyrrolizidine). Ever since artemisinin was discovered, many sesquiterpenes are known as being antiprotozoal. *In vitro*, parthenin (a sesquiterpene lactone) has shown effectiveness against *Plasmodium falciparum* with 1.29 mg/mL EC50 value. Parthenin has the capability to block specific targets of parasites responsible for trypanothione and glutathionyl spermidine synthesis from precursors of glutathione and cysteine in both Trypanosoma and Leishmania [45]. Dehydrozaluzanin C and Brevili A, sesquiterpene lactones, from *Munnozia maronii* and *Centipeda minima* respectively are known to have antiparasitic properties. Also, *Neuroleaena lobate* sesquiterpene lactones are commonly used for treating infections caused by Plasmodium [46]. Structural activities relationship analyses of *Neuroleaena lobate* showed that germanocrenolide sesquiterpenes, such as neurolenin B (EC50=0.62 mM) and A (EC50=0.92 mM), showed more potency than furanoheliangolides such as lobatin B (EC50=16.51 mM) and A (EC50=15.62 mM), respectively, against *Trypanosoma epimastigotes* and *Leishmania promastigotes* [46]. Oketch-Rabah *et al.* [47] found that *Vernonia brachycalyx* "macrocyclic germancranedi lactone 16, 17-dihydrobrachycalyxolide" had anti-plasmodial and anti-leishmanial properties.

Phenolic compounds are found extensively in Asteraceae family. Most of the plants in Asteraceae family have the therapeutic and medicinal properties such as inhibiting parasites, ameliorating oxidative stress, and so on. Polyphenols are known to be effective against some diseases, including metabolic diseases such as diabetes and Alzheimer. Gallic acid and compounds derived from it inhibit *Trypanosoma cruzitrypomastigotes* proliferation, *in vitro* [48]. Gallic acid esters n-propyl-gallate and ethyl-gallate were reported to have high activities, with 1.47 and 2.28 mg/mLEC50 values, respectively, probably because of improved lipophilicity. Oketch-Rabah *et al.* [47] showed *Vernonia brachycalyx* antiprotozoal activity, possibly linked to 2′epicycloisobrachy coumarinone epoxides and their stereoisomers. *In vitro*, the stereoisomers showed same activities against chloroquine-resistant (CQ-R) and chloroquine-sensitive (CQ-S) strains of *Leishmania major* and *Plasmodium falciparum*, with 39.2 mg/mL and 37.1 mg/mL for *Leishmania major*, and 0.15 mg/mL and 0.11 mg/mL EC50 values for *Plasmodium falciparum*, respectively.

Flavonoids widely occur in plants of Asteraceae. As early as 1987, *in vitro* study by Elford *et al.* [49] showed that in synergy with artemisinin, both casticin, and artemetin (methoxylated flavonones) work against *Plasmodium falciparum*. Additionally, exigua flavanones A/B, obtained from Artemisia indica, have been demonstrated to show anti-*Plasmodium falciparum* activity, *in vitro*. Pérez-Victoria *et al.* [50] reported that flavonoids may have effect on the Leishmania transport mechanisms. AP-glycoprotein-like transporter C-terminal nucleotide-binding domain in *Leishmania tropica* encoded by ltrmdr1 genes and took part in multidrug resistance (MDR) of parasitehas been over expressed in *E. coli* in form of hexa-histidine marked protein and made pure. Recombinant domain of *L. tropica* bound efficiently to various flavonoids classes with affinity in this order: glucorhamnosyl-flavone<isoflavone<flavanone<flavone. The relative affinity depends on hydroxyl group presence at C-3 and C-5 positions, and increased further by 1,1-dimethylallyl hydrophobic substituent at C-8 position.

Brandão *et al.* [51] showed that crude extracts from various Bidens species and their fractions have antimalarial properties, and stated with evidence that it was because of flavonoids and polyacetylene presence. In addition, Kumari *et al.* [52] and Tobinaga *et al.* [53] extracted (R)-1,2-dihydroxytrideca-3,5,7,9,11-pentayne (polyacetylene compound) from *Bidens pilosa* leaves, which demonstrated potent antimalarial activities against *P. falciparum*. The polyacetylene compound was tested *in vivo* with *Plasmodium berghei* NK-65-infected mice. It was reported that the polyacetylene compound decreased the red blood cells (RBCs) parasitemia. Calendula is popular for content of bioactive compounds, such as polysaccharides, carotenoids, flavonoids, triterpene saponins, triterpene alcohols, and so on. Szakiel *et al.* [54] extracted many derivatives of oleanolic acid glycosides, which they tested on *Heligmosomoides polygyrus*; oleanolic acid glycosides wormicidal activities were superior to aglycone wormicidal activities. The activity level depended on property of sugar side-chain at position C-3. Glucuronic acid (the first glucuronides sugar molecule) connected to the aglycone may be essential for antiparasitic actions of the compounds [54, 55]. *E. prostrate* has been studied for its antiparasitic properties, including anthelmintic [56, 57], antileishmanial [58, 59], and antimalarial activities [60]. Khanna *et al.* [59] extracted dasyscyphin C from *E. Prostrate* leaves and showed that it has antileishmanial activities against *Leishmania tropica, Leishmania aethiopica,* and *Leishmania major*. Zahari *et al.* [61] isolated deoxy elephantopin, a sesquiterpene lactone, from *E. scaber* and showed its anti *Trypanosomabrucei* rhodesience properties. Also, *Tridax procumbens* has shown significant antileishmanial activities against *Leishmania Mexicana* promastigotes. (3S)-16, 17-didehydrofalcarinol, oxylipin, was reported to be the active compound [62].

2.1.7 Antiparasitic Properties of Terpenoids and Flavonoids in Asteraceae

Flavonoids widely occur in Asteraceae, and are known for their chemical stability and structural diversity. Flavonoids have been largely extracted from species of Asteraceae and are applied at lower hierarchical level for taxonomic markers [63]. Flavonols and flavones are commonly distributed in Asteraceae family, such as glycosides of apigenin, flavanone derivatives, quercetin, kaempferol, luteolin, epigallocatechin and epicatechin. Only fewer reports about antigiardial activities in Asteraceae exist. Epigallocatechin, epicatechin, quercetin, and kaempferol from Cistaceae (*Helianthemum glomeratum*) aerial parts have demonstrated *in vitro* antigiardial properties against *Giardia lamblia*, with 8.06, 1.64, 8.73, and 26.47 mg/mL IC50 values, respectively [64]. Correlation of structural activity indicates that the 4-keto group and 2,3-double bond of flavones may not play a part in antiprotozoal activities, as epigallocatechin and epicatechin lack these structures, and still maintain biological activities. Similarly, unlike flavones, the epigallocatechin and epicatechin benzene diolmoiety is not coplanar with part of heterocyclic as their flavan-3-ol C-2 structure is sp3. Additionally, there have been many reports supporting that glycosylated flavonoids have antigiardial activities. Likewise, a tiliroside of C-3 glycosylated flavone [64, 65], extracted from *Hieracium glomeratum*, has antigiardial inhibitory properties with 17.36 mg/mL IC50 value.

In 2015, Klongsiriwet *et al.* [66] showed that at 250 mM, luteolin and quercetins are very effective in reducing *H. contortus* L3 larvaeen sheathment, *in vitro*. Tasdemir *et al.* [67] evaluated flavonoids antileishmanial and antitrypanosomal properties, including those of their analogues, *in vivo* and *in vitro*, and also the quantitative relationship between their activity and structure [67]. In the study, it was demonstrated that quercetin, luteolin, 3 hydroxyflavone, and fisetin are major promising antileishmanial phytochemicals against *Leishmania donovani*, with 1.0, 0.8, 0.7, and 0.6 mg/ml IC50 values, respectively. The authors reported moderate efficacy antitrypanosomal activities of these phytochemicals against *Trypanosoma cruzi* and *Trypanosoma brucei* rhodesiense, and concluded that quercetin and 7, 8-dihydroxyflavone may ameliorate infections caused by parasites in mouse, and are effective and promising antiprotozoal agents. Mead and McNair [68] evaluated antiparasitic properties of isoflavones and flavonoids against *Encephalitozoon intestinalis* and *Cryptosporidium parvum*. They found that apigenin and quercetin have properties against *E. intestinalis* at 50 and 15 mM EC50, respectively, while low activities of quercetin and luteolin were reported against *C. parvum*. Inhibition was not detected with epigallocatechin-3-gallate (EGCG) or rutin against *Encephalitozoon intestinalis* and *Cryptosporidium parvum*. Lehane and Saliba [69] studied effect of several dietary flavonoids on development of two strains of *Plasmodium falciparum*. They reported that luteolinhad 12±1 mM and 11±1 mM IC50 values for strains 7G8 and 3D7, respectively. Although luteolin prevented parasite growth progression past stage of young trophozoite, it, however, had no effect on susceptibility of parasite to antimalarial drugs artemisinin or chloroquine. Nour *et al.* carried out study which reported moderate antiparasitic properties of methoxylated flavonoids, including ageconyflavone, 5,6,7,3,4,5-hexamethoxy-flavone, 5-methoxynobiletine, 5,6,7,5-tetramethoxy-3,4-methylene dioxyflavone, and eupalestin against many parasites; *Plasmodium falciparum, Leishmania donovani, Trypanosomacruzi,* and *Trypanosoma brucei rhodesiense* [3].

Terpenoids are among the large groups of phytochemicals with at least 20,000 known compounds. Based on number of carbons, terpenoids are subdivided into sesquiterpenes, tetraterpenes, triterpenes, diterpenes, and sterols. Many triterpenes, sterols, and sesquiterpenes are extracted from plants in Asteraceae family. Sesquiterpenes in leaves of Asteraceae are subdivided into bicyclic and monocyclic. Sitosterol and stigmasterol are the common sterols found in Asteraceae family. Sesquiterpenes from the species of *Vernonia* have antiparasitic properties against *P. falciparum*. Bioactive compounds from *Vernonia amygdalina* leaves, including hydroxyvernolide, vernolide, vernodalol, and vernodalin, have potent properties with 11.4, 8.4, 4.2, and 4 mg/mL IC50 values, respectively [70]. Another bioactive compound called 16,17-dihydrobrachycalyxolide (or sesquiterpene dilactone), obtained from *V. Brachycalyx* leaves, showed anti-plasmodial properties against various multidrug-resistant *Plasmodium falciparum* strains Dd2, V1/S, 3D7, and K39 with16, 3.0, 13.7, and 4.2 mg/mLIC50 values, respectively [47]. Goffin *et al.* [71] extracted and purified tagitinin C, a sesquiterpene lactone, from *Tithonia diversifolia* ether extract and showed anti-plasmodial properties against *P. falciparum* (0.75 mg/mL IC50 value). Becker *et al.* [72] reported dehydrobrachylaenolide and urospermal A-15-O-acetate as major active compounds associated with the antiplasmodial properties against *P. falciparum* W2 and 3D7 strains. Also, Ganfon *et al.* [73] studied *Acanthospermum hispidum* antiparasitic properties *in vitro* by extracting two sesquiterpene lactones, (9 a-acetoxy-15-hydroxy-8b-(2-methylbutyry-499 loxy)-14-oxo-4,5-transacantho spermolide), and (15-acetoxy-8 b-[(2-methylbutyryloxy)]-14-oxo-4,5-cis-acanthospermolide), which showed antiplasmodial properties against 3D7 (a strain sensitive to chloroquine) with 2.23 and 2.9 m MIC50 values, respectively.

Derivatives of lupeol and squalene are most common triterpenes. Oleanolic acid, a pentacyclic triterpenoid, widely occurs in Asteraceae family and has been reported to have antileishmanial and antimalarial properties [74, 75]. In 2015, Yamamoto *et al.* [75] investigated the ursolic acid *in vivo* and *in vitro* activities against *Leishmania amazonensis*. While oleanolic acid had small effect on *L. amazonensis* promastigote at 100 mg/ml, they reported that ursolic acid destroyed *L. amazonensis* promastigotes with 6.4 mg/ml EC50, similar to miltefosine. The likely mechanism through which ursolic acid eliminated promastigotes could by programmed cell death, with high dependence on mitochondrial activities, and no dependence on caspase 3/7. Additionally, ursolic acid had no toxicity on BALB/c mice peritoneal macrophages, and it mayeradicate intracellular amastigotes, linked to the production of nitric oxide (NO). Yamamoto *et al.* [75] concluded that ursolic acid is a promising candidate for cutaneous leishmaniasis treatment.

Betulinic acid in Asteraceae is known to have therapeutic and medicinal properties. Enwerem *et al.* [76] studied betulinic acid anthelmintic properties on *C. elegans* and found strong anthelmintic property at 100 mg/mL, which can be compared to that of piperazine. Bringmann *et al.* [77] carried out *in vitro* study and reported that betulinic acid showed antimalarial properties against Plasmodium falciparum asexual erythrocytic stages. Steele *et al.* [78] also carried out a study and reported that betulinic acid inhibits *P. falciparum* (*in vitro*), while experiments *in vivo* did not decrease parasitemia (about 500 mg/mL in *P. Berghei* infected mice with malaria) and showed toxicity. Ndjakou Lenta *et al.* [79] extracted betulinic acid and tested its activities against W2 strain of *Plasmodium falciparumin vitro*. They reported that betulinic acid has high potency with 2.33 mg/mL IC50. Additionally, Nweze *et al.* [80] conducted *in vitro* study and reported that b-sitosterol had moderate antitrypanosomal properties against S427 strain of *Trypanosoma brucei*, with 12.5 mg/mL IC50.

2.1.8 Further Discussion on Antiparasitic, Therapeutic, and Medicinal Properties of Asteraceae

In 2011, Zhu *et al.* [81] analyzed "the ranking of drug-productive plant families based on the ratio of the approved drugs to reported bioactive natural products (including leads of the approved and clinical trials drugs)" and reported that among medicines obtained from plants, few top families of plant are used to produce approved drugs in high numbers. They stated that Asteraceae is fourth-highest family relevant to drug production for several approved drugs, such as choleretic, antiviral, antihepatotoxic, ant-inflammatory, antiglaucoma, anticancer, and antiparasitic agents [81]. From 7229 species of Asteraceae family, 25 clinical drugs (many with approvals and some under clinical trials at the time of the report) have been documented within 1016 drugs that are searchable [79, 80]. Several FDA-approved drugs derived from nature are fully or partly obtained from Asteraceae, including santonin, dihydroartemisinin, co-artemether, coarsucam, artesunate, artemisinin, artemether, and arteether, which areall derived from Artemisia species. Additionally, few drugs have also been tested as antiparasitics through clinical trials, including artelinic acid, and artemisone [82]. Traditional information has been valuable instrument in quest for novel medicines derived from plant [83]. It is estimated that tens to hundreds of thousands of plant species are traditionally used worldwide [84], although only small percentage of the plant species have been well studied for biological properties [85, 86]. The antitrypanosomal properties of 50% ethanol extracts from leaves of *Xanthium strumarium* were evaluated *in vivo* and *in vitro.* The ethanol extract showed trypanocidal properties against mice infected with *Trypanosoma evansi* [87]. The researchers stated that xanthinin could be the phytochemical responsible for the trypanocidal properties.

Most people are influenced by natural resources, such as plant resources within their environment, which usually provides their medicine, shelter, food, and so on [88, 89]. Saslis-Lagoudakis *et al.* [90] studied the plants in Asteraceae family and offer large-scale insight about plant bioactive compounds, which are the bedrock of traditional medicines. The authors reported that the plants are applied as traditional medicines in various parts of the world, and they also related to the groups used in producing pharmaceutical medicines. Gertrude *et al.* [91] investigated the anthelmintic properties of leaves of *Bidens pilosa* against the larvae and eggs of

Haemonchus contortus. They reported that ethanol extracts may inhibit *Haemonchus contortus* growth. Also, antileishmanial properties of leaves of *B. Pilosa* have been reported by many authors [92, 93]. The wormicidal and anthelmintic properties of leaves of *Blumea lacera* were studied against *Pheretima posthuman* and *Ascarislum bricoides* and found to be promising [94]. *Calendula officinalis* is traditionally used by for treating worm infections. Nikmehr *et al.* [95] reported that methanol extracts have anti-leishmanial properties. A wet land plant called *Caesulia axillaris* is commonly applied for malaria treatment in some parts of the world. The phytochemicals in the plant has not been studied sufficiently, except for those on the essential oils of the plant. *Tageteserecta, Sphaeranthus indicus,* and *Centipeda minima* have been in use for anthelmintic herbal medicines for treating worm infections. Yu *et al.* [96] reported antiparasitic properties of Centipedaminima crude extracts against *Plasmodium falciparum, Entamoeba histolytica,* and *Giardia intestinalis. Sphaeranthus indicus* crude extracts also have antiparasitic properties against *Setaria digitate, Entamoeba histolytica,* and *Ascaridia galli* [97]. *Tagetes erecta* aqueous and organic extracts show anthelmintic and antiparasitic properties. *Elephanto pusscaber* crude extracts have anthelmintic properties against *Pheretima posthuma. In vivo* and *in vitro* studies proved *Vernonia anthelmintica* anthelmintic bioactivities against *Haemonchus contortus* [86]. In some of these plants, further studies are required to identify the compounds responsible for the antiparasitic, medicinal, and therapeutical properties.

2.1.9 Nutritional Composition of Plants in Asteraceae Family

This section focuses on the nutritional compositions of plants in Asteraceae family. Wild leafy vegetables have gained significant interests increased in many parts of the world as they offer high levels of nutrients and several health benefits. Many species in Asteraceae family are cultivated worldwide, and several non-cultivated ones are consumed as vegetables, although information on their nutrient constituents has not been completely studied. García-Herrera *et al.* [98] reported the nutrient composition of plants from Asteraceae family. The study presented valuable information on mineral content, soluble and insoluble fiber content, available carbohydrates, fat, and proteins, which play role in advancing the knowledge of nutrient compositions of six species of Asteraceae commonly eaten in several regions of the world. The six species studied include *Silybum marianum (L.), Scolymus hispanicus L., Cichorium intybus L., Sonchus oleraceus L., Chondrilla juncea L.,* and *Taraxacum obovatum.* They reported these species have (16 to 472 mg/100 g), (375 to 1772 mg/100 g), and (2.3 to 13.4 g/ 100 g) for Ca, K, and fiber content respective. *Chondrilla juncea* contribute to intakes of fiber and micronutrients, especially Mn which has about 50% of Recommended Dietary Allowances (RDA) [98].

García-Herrera *et al.* [98] reported moisture range from 71.8 to 93.8 g/100 g in the Asteraceae plants. The moisture content is often high for greens, and may affect their microbiological stability and preservation; however, it provides valuable properties including texture and palatability. Guil-Guerrero *et al.* [99] showed values for *So. Oleraceus, So. Tenerrimus, and So. Asper* (87.1 g/100 g) (all among Sonchus species). *Taraxacum obovatum* had lower moisture compared to that of 91.5 g/100 g reported by Escudero *et al.* [100] for *Taraxacum officinale* (common dandelion). Other Asteraceae, including lettuce, have relatively slightly higher levels of moisture of 94.5% to 95% than the species studied by García-Herrera *et al.* [98] (65.9–93.8%) [98, 101]. García-Herrera *et al.* [98] reported that moisture had diluting effects on nutrient levels, as indicated through negative correlation seen on moisture and microelements, macro-elements, and proximate composition, except sodium (possibly more affected by factors like composition of soil). The facts were mostly evident for peeled S. marianum leaves, which recorded more reliable compositions, indicating higher levels of moisture (as anticipated from fleshy their edible tissue), and lower ash, fibre, carbohydrate, protein, and fat (resulting in extremely low energy, with 51.8 kJ/100 g average). They reported that the statistics indicated no direct effects collection site or the year on the plants 'chemical compositions, which could be as a result of the effects of various factors including growing status, soil composition, sun exposure, precipitation, temperature, and the interactions with animals or other plants in same environment, eventually creating the bio-variability in the plants 'chemical compositions.

The fat contents reported by García-Herrera *et al.* [98], were within the range of the data in literature for other species of Asteraceae [101, 102]. In comparison to the results of 0.5% to 0.7% wet weight from the publication of Guil-Guerrero *et al.* [99], regarding species of *Sonchus* genus, fat was marginally lower (0.20 to 0.41 g per 100 g wet weight) for samples of *So. oleraceus* in the study of García-Herrera *et al.* [98] *S. marianum* and *Sc. hispanicus* reportedly showed lowest fat [98]; Souci *et al.* [101] also reported lower fat for these Asteraceae plants than other leafy plants of Asteraceae family, including radicchio/lettuce, 0.17% to 25%. This could be as a result of the various types of plant tissues. Major metabolic reactions take place in the leaves' mesophyll, which causes different chemical compositions in the mesophyll and mid-ribs; this is in line with other studies [103, 104], showing different distributions of unsaturated and saturated fatty acids in the two species, in comparison with other leafy vegetables.

García-Herrera *et al.* [98] reported that protein ranged from 0.34 to 6.13 g/100 g for *C. juncea* and *S. hispanicus*. Their values are closely related to other values for *Taraxacum* sp. (1.31 g/100 g) and for *Sonchus* sp. (3.17 to 3.25 g per 100 g) [99, 100]. Jan *et al.* [105] reported 14.10% protein in leaves of *Ci. intybus*, while García-Herrera *et al.* [98] reported higher value (19.65%) for *C. intybus*. The values included for *So. oleraceus* and *C. juncea* were higher than those of Souci *et al.* [101] for Asteraceae leafy plants, including radicchio or lettuce (protein contents of 1.2 to 1.3 g/100 g).

For carbohydrates, *C. juncea* was reported to have the highest total available carbohydrates (TAC) (5.17% average) with values related to vegetables, including cabbage (5.6%) and broccoli (6.6%), as reported by Maynard and Hochmuth [102]. Guil-Guerrero *et al.* [99] showed 1.66% average value for various *Sonchus* species. The carbohydrates content of species of *Cichorium* were reported to be about 4 g/100 g for edible parts [102]. The carbohydrate content of *Taraxacumo bovatum* (3.34 g/100 g) reported by García-Herrera *et al.* [98] was in same range with reports of 4.94 g/100 g for *T. officinale* as reported by Escudero *et al.* [100]. García-Herrera *et al.* [98] observed positive correlations between carbohydrates in all Asteraceae species studied.

Crotalaria juncea, also called Indian hemp, brown hemp, sunn hemp, or Madras hemp, has been reported to have high fiber between 4.1% and 13.4%. Current reports show that people need to increase fiber intakes, because of its benefits on digestive system, such as intestinal function regulation, preventing colon cancer, enhanced glucosetolerance, and so on [106]. Most wild vegetables in Asteraceae family have high fiber, except few such as *Silybummarianum* (about 2.55 g per 100 g). Their values are higher than most conventional leafy vegetables in same family, including radicchio or lettuce; or those in other families, including spinach, parsley, and celery (about 4.3 g per 100 g); and also higher than most conventional vegetables often regarded as rich sources of fiber, as have been reported for artichokes (10.8 g), chives leaves (6 g), and asparagus shoots (1.5 g), all per 100 g [101, 102]. Some researchers have shown similar values of fiber in some Asteraceae plants [99, 107]. Jan *et al.* [105] reported 17.6% dry weight (dw) of fiber content for *Ci. intybus*, while García-Herrera *et al.* [98] reported 44.4% dw for the same plant, which is notably higher. Also, García-Herrera *et al.* [98] reported 7.01 g for *T. obovatum* fiber content, while Escudero *et al.* [100] reported 4.05 g for *T. officinale*, all per 100 g. Lachman *et al.* [108] showed 1.68% fiber content for leaves of *Smallanthus sonchifolia*. Except *S. marianum*, values of fiber in the Asteraceae plants studied by García-Herrera and others were between 4.38 and 9.65%, which is richer than the 4.44% in *Montia Fontana* L. reported as vegetable with rich fiber content by Tardío *et al.* [109]. Additionally, all Asteraceae species studied by a group of researchers were reported to have higher ash values of 1.54 to 3.19 than *Montia Fontana* (1.13%), lettuce and other leafy plants in Asteraceae family [101]. Asteraceae family can clearly contribute to the fiber intakes improvement in population of the world. The fiber in edible parts of *C. juncea* and *Sc. hispanicus* (per 100 g) can in most instances provide roughly half of the [106] Recommended Dietary Allowances (RDA) for adults.

Plants in Asteraceae family provide relatively low energy per 100 g. Edible plants of Asteraceae evaluated by García-Herrera *et al.* [98] were reported to have low energy values per 100 g, ranging from 125 to 226 kJ, except *S. marianum.*

Asteraceae plants have considerable amounts of ash and several minerals. *Silybummarianum Ci. intybus,* and *Sc. hispanicus* was reported to have 1.78, 1.54, and 3.19 g ash content per 100 g, respectively [98]. Related values were reported for other species, such as *Sonchus* sp. [99, 110]. Lower values of ash were reported for *Ci. intybus* and *T. officinale* [100, 105]. However, mineral contents are affected by environmental factors, including composition of soil. Potassium is significantly higher than sodium in many species of Asteraceae family, just as in most plants. Sodium content in *So. oleraceus* is about 150 mg per 100 g [98, 110] for many Asteraceae species of *Sonchus* genus. Asteraceae plants have considerable levels of Zn, Mn, Fe, Cu, and so on. Potassium and calcium in *T. obovatum* (average of 213 mg/100 g) are higher than the levels in *T. officinale* (average of 58.9 mg/100 g) [100]. Magnesium content for *T. officinale* is 39.8 mg/100 g [100]. *C. juncea, T. obovatum,* and *Sc. hispanicus* have low sodium compared to *Sisymbriumirio, Rumexcrispus, Plantagomajor,* and *Crithmummaritimum,* but have more sodium than *Verbena officinalis* [110]. *Cardariadraba* and *Amaranthusviridis* have moderate contents of macro-elements. Zinc and copper in *Ci. intybus* have average values of 6.79 and 1.49 mg/100 g dw, respectively, as reported by García-Herrera *et al.* [98] while Jan *et al.* [105] reported 10 and 2.50 mg per 100 g for same species. Compared to other wild greens, potassium and calcium contents of plants in Asteraceae family usually range from 31.4 to 472 mg per 100 g [98, 109]. Lettuce and radicchio, which are among the cultivated leafy Asteraceae plants, have Ca and K contents of 25 to 35 g and 190 to 200 g per 100 g, respectively [101].

Asteraceae vegetables are good sources of calcium (about 300 mg/100 g). 100 g of *C. juncea* and *Sc. hispanicus* contribute approximately 47% of RDA for calcium in adults [111]. Although calcium from plant sources has absorption impairment due to high presence of substances that bind to calcium, including oxalic acid, often resulting in forming calcium oxalates, which are insoluble, these antinutrients often occur in high levels in vegetables. To prevent this antinutritional effect, ratio of oxalic acid-Ca below 2.5 in food has been recommended [112]. The contents of oxalic acid of some Asteraceae plants have been studied [98], and they gave allowance for ratio of oxalic acid-Ca estimation, which was usually less than 2.5, with the exception for *S. Oleraceus and S. marianum* that were 3.28 and 7.77 respectively. Vegetables with very high contents of oxalic acid are not recommended for consumption by those with high susceptibility to oxalate renal calculi formation. *C. juncea* and *T. obovatum* low ratios of oxalic acid-calcium of 0.41 and 0.31 respectively, indicating that calcium from by them can be absorbed more easily in the gut. A 100 g of *C. juncea* leaves provide 25 to 80% of RDA for Mn [111], which are higher than the those provided by many other vegetables commonly consumed, which often range from 4 to 22% [101]. Additionally, the species provides Cu and Zn of about 100% and 48% RDA respectively. Along with *T. obovatum*, it gives high Fe of about 36% RDA [98]. The contributions have significance because the trace elements play essential roles in human metabolism and overall health. Trace elements function as cofactors to enzymes, including antioxidant enzymes, and help in ameliorating oxidative stress and preventing chronic diseases, such as cataract, asthma, cardiovascular disease, neurodegenerative diseases, and cancer, among others [113, 114]. Some plants in Asteraceae family provide more nutrients such as K, Ca, fiber, and so on than several commonly consumed vegetables from other plant families.

2.1.10 A Case Study of Nutritional Value and Bioactive Compounds in *Cynara cardunculus* L., Popular Plant of Asteraceae

The Asteraceae plant *Cynara cardunculus* L. is commonly found in some parts of the world, especially in Mediterranean basin, due to its great adaptability to several climate conditions, soil conditions, as well as abiotic stress factors, such as water paucity and high levels of salinity [115, 116]. The species are globe artichoke and cardoon. Cardoon, also called leafy cardoon, is categorized into *C. cardunculus* var. *sylvestris* (wild cardoon) and *C. cardunculus* var. *altilis* (cultivated cardoon) [115, 117].

In 2018, Petropoulos *et al.* [118] evaluated the nutritional value and bioactive compounds of species of Asteraceae. They found that edible parts have high moisture of 81.2% and low fat (0.59% fw), but good carbohydrates sources (13.76 g per 100 g fw) and energy values of 72.4 kcal per 100 g). The major sugars detected by Petropoulos *et al.* [118] include glucose, fructose, and sucrose at 0.09, 0.10, and 0.42 g per 100 gfw, respectively.

Heads of cardoon have great nutritional values and have been known to be rich carbohydrates sources. Although wild cardoon petioles and heads have been consumed as vegetable in several local cuisines, there are no sufficient reports available about their nutritional values, especially the cultivated cardoons. However, similar results on nutritional value including sugars have been reported for the spinier and less spiny ecotype. The variations seen in the ecotypes of wild cardoon may be partly because of the variations in moisture levels of the heads. Cropage, harvesting stage, and cultivation practices could also have effects on the chemical compositions of cardoon [120, 121, 199].

All parts of cardoon and many plants in Asteraceae family have considerable mineral contents. On dry weight basis, cardoon heads and leaves are good sources of Ca (2662, 2119, and 1199 mg per 100 g dry weight for leaf blades, leaf midribs, and heads respectively) and K (2400, 2800, and 3000 mg per 100 g for leaf blades, leaf midribs, and heads respectively). Leaf blades are good sources of Fe (23 mg per 100 g), while leaf midribs have Na (1,853 mg per 100 g). Cardoon seeds have Na, K, Mg, and Mn contents of 18, 653, 6.0, and 483 mg per 100 g, respectively. All parts of the plant are known to be rich in dietary minerals, although there are variations in different plant parts. Cardoon has high potassium and may be low in sodium and calcium in leaves of cultivated cardoon, although growth stage has significant effect on the mineral contents of cardoon leaves [122], just like many other plants, both in and outside Asteraceae family. Rouphael *et al.* [123] reported similar levels of Mg and K and low Ca in cardoon cultivated in floating culture. Colla *et al.* [124] demonstrated significant differences in dietary mineral contents of various cardoon genotypes cultivated in saline environment.

The fatty acids (FA) in seed oils and heads of cardoon are shown in Table 2.4 and 1.5. Seedoil has been reported to be 24.5% dry weight. At least 22 various fatty acids have been reported in cardoon heads, while at least 18 have been detected in seed oils [118]. Some of the detected FA in cardoon heads includes oleic acid (4.6%), myristic acid (5.3%), caproic acid (6.4%), stearic acid (6.7%), linoleic acid (7.8%), behenic acid (10.7%), palmitic acid (47.2%), and so on. Fatty acids composition of cardoon seed oils includes stearic acids (2.78%), palmitic acid (9.37%), oleic acids (21.11%), and linoleic acids (64.86%), and so on (see Table 2.3). Cardoon seed oils and heads significantly differ in fatty acid profiles, with polyunsaturated (PUFA) and saturated (SFA) fatty acids as most abundant fats in seed oils and heads, respectively (see Table 2.4). The differences in profile of fatty acids have substantial impacts on n-6/n-3 ratio (0.57 and 9.0 for seed oils and heads, respectively) and PUFA-SFA ratios (4.94 and 0.10 for seed oils

Table 2.3 Fatty acids contents of cardoon seed oils and heads (%).

Fatty acid %	Seed oils	Heads
C6:0	0.009 ± 0.001	6.4 ± 0.4
C8:0	Not Detected (ND)	1.60 ± 0.07
C10:0	ND	1.66 ± 0.04
C11:0	ND	0.14 ± 0.01
C12:0	ND	1.72 ± 0.01
C14:0	0.094 ± 0.003	5.32 ± 0.03
C14:1	ND	0.41 ± 0.03
C15:0	0.021 ± 0.001	0.98 ± 0.01
C16:0	9.37 ± 0.09	47.2 ± 0.6
C16:1	0.107 ± 0.003	0.55 ± 0.02
C17:0	0.072 ± 0.003	0.51 ± 0.02
C18:0	2.78 ± 0.01	6.7 ± 0.1
C18:1n9	21.11 ± 0.02	4.64 ± 0.02
C18:2n6c	64.86 ± 0.07	7.76 ± 0.07
C18:3n3	0.108 ± 0.005	0.52 ± 0.04
C20:0	0.277 ± 0.003	1.27 ± 0.01
C20:1	0.11 ± 0.03	ND
C20:3n3+C21:0	0.380 ± 0.001	ND
C20:5n3	0.082 ± 0.007	0.20 ± 0.01
C22:0	0.461 ± 0.001	10.7 ± 0.5
C22:1n9	0.008 ± 0.001	0.08 ± 0.01
C22:2CIS13	ND	0.04 ± 0.01
C22:6n3	ND	0.15 ± 0.01
C23:0	0.022 ± 0.007	ND

C6:0, caproic acid; C8:0, caprylic acid; C10:0, capric acid; C11:0, undecanoic acid; C12:0, lauric acid; C14:0, myristic acid; C14:1, myristoleic acid; C15:0, pentadecanoic acid; C16:0, palmitic acid; C16:1, palmitoleic acid; C17:0, heptadecanoic acid; C18:0, stearic acid; C18:1n9, oleic acid; C18:2n6c, linoleic acid; C18:3n3, α-linolenic acid; C20:0, arachidic acid; C20:1, eicosenoic acid; C20:3n3+C21:0cis-11,14,17, eicosatrienoic acid and heneicosanoic acid; C20:5n3, eicosapentaenoic acid; C22:0, behenic acid; C22:1n9, erucic acid; C22:2CIS13, docosadienoic acid; C22:6n3, docosahexaenoic acid; C23:0,tricosanoic acid.

Source: [118]

Table 2.4 Saturated and unsaturated fatty acids contents of cardoon seed oils and heads (%).

Type of fatty acid	Seed oils	Heads
SFA	13.23±0.08	85.7±0.2
MUFA	21.34± 0.02	5.68±0.03
PUFA	65.43±0.07	8.7±0.1
PUFA/SFA	4.94± 0.07	0.10±0.001
n-6/n-3	0.57±0.07	9.0±0.4

SFA, saturated fatty acids; MUFA, monounsaturated fatty acids; PUFA, polyunsaturated fatty acids; n6/n3, ratio of omega6/omega-3 fatty acids; nd, not detected.
Source: [118]

and heads, respectively). These values and ratios are significant to nutritional values, nutraceutical usage, and functional features of foods.

Cardoon seeds and heads are good sources of FA; the seeds contain high oil that may serve as food and for industrial applications. Maccarone *et al.* [125] reported 25.6 and 24.9% dry weight for oil yields from seeds and heads of cardoon. Maccarone *et al.* [125] and Curt *et al.* [126] reported oleic acid, linoleic acid, palmitic acid, and stearic acid as the main fatty acids in cardoon.

Fatty acids profiles are significant to nutritional values of foods. Although heads of cardoon can serve food purposes, the PUFA/SFA ratio and then-6/n-3 ratio do not necessarily show substantial health benefits, compared seed oils that have high functional properties and nutritional value. The importance of PUFA/SFA ratio and then-6/n-3 ratio for nutritional profile of foods have been noted, and cardoon seed oils have showed excellent health nutritional benefits with PUFA/SFA ratio and then-6/n-3 ratio recording values below 4.94 and 4.0, respectively, and above 0.57 and 0.45, respectively, indicating that the potentials for human consumption [118, 127, 128]. Some researchers have shown the essential roles of long chain polyunsaturated fatty acids in human diets, particularly n-3 and n-6 fatty acids, and also showed that 10:1n-6/n-3 ratios are commonly found in Western diet and are essential to the production of pro-essential aggregatory FA, including eicosanoids [129, 130]. Cardoon heads have amounts not within those of both ratios. It appears though that fatty acids profile relies on genotype, cardoon heads have high nutrient composition, and especially fatty acids profile [118]. Further, mineral composition of Cardoon is highlighted in Table 2.5.

Table 2.5 Mineral composition of cardoon (mg/100 g dry weight).

Plant part	Fe	Zn	K	Mg	Na	Ca	Mn
Leaf midribs and petioles	13.0 ± 0.6	0.15 ± 0.04	3000 ± 346	285 ± 45	1853 ± 266	2119 ± 90	3.5 ± 0.3
Heads	12.2 ± 0.2	1.17 ± 0.06	2800 ± 145	389 ± 15	240 ± 34	1199 ± 156	5.5 ± 0.3
Seeds	13 ± 3	3.95 ± 0.16	653 ± 58	483 ± 62	18 ± 4.0	1197 ± 195	6.0 ± 0.2
Leaf blades	23 ± 2	0.54 ± 0.06	2400 ± 200	191 ± 16	1027 ± 479	2662 ± 125	6.3 ± 0.2

Source: [118]

2.2 Future Prospects

Most studies on Asteraceae family worldwide focused more on chemical analysis. There are several *in vitro* studies and publications. More studies need to be carried out to study the less commonly studied plants in Asteraceae family. Further studies focusing on the bioactive compounds, nutritional compositions, and therapeutic properties of several plants in Asteraceae family are strongly recommended for better understanding of the relevance of this great plant family in nutrition, medicine, pharmacology, and food science.

2.3 Conclusion

Asteraceae is widespread and large family of Angiospermae. Several plants in Asteraceae family are perennial or annual herbs; however, a huge number of them are also trees, vines, or shrubs. The family is widely and diversely distributed, with the species widely spread from tropical to subpolar regions, inhabiting several habitats. Species in Asteraceae has economic importance, providing commercial products including herbal teas, coffee substitutes, sweetening agents, artichokes, sunflower seeds, leaf vegetables (such as lettuce), and cooking oils. Additionally, many genera of Asteraceae have importance in horticultural system, including heleniums, zinnias, dahlias, chrysanthemums, fleabane, various daisies, *Echinacea* (coneflowers), and pot marigold (*Calendula officinalis*). Asteraceae has economic importance, through providing coffee substitutes, herbal teas, sweetening agents, artichokes, sunflower seeds, leaf vegetables (e.g., lettuce), and cooking oils. Plants of Asteraceae which have commercial importance include *Lactuca sativa*, *Cynara cardunculus L.*, *Solidago virgaurea L.*, *Helianthus tuberosus*, *Carthamus tinctorius* (safflower), *Smallanthus sonchifolius*, *Helianthus annuus* (sunflower), *Cynara scolymus*, *Cichorium* (chicory), and so on. Many bioactive compounds including essential oils, polysaccharides, flavonoids, clerodane-type diterpenes, triterpenoid saponines, phenolic compounds, salicylic acid derivatives, have been found in Asteraceae, including*Cynara cardunculus L.*, *Solidago virgaurea L.*, *Helianthus tuberosus*, and so on. Studies have shown that these natural bioactive compounds have many biological activities, as gastroprotective, antioxidant, cytotoxic, anti-obesity, anti-inflammatory, spasmolytic, antimicrobial effects. Diuretic activities have been linked with leiocarposide and flavonoids, while anti-inflammatory effects are attributed to phenolic compounds such as leiocarposide, hyperoside, quercitrin, and so on in extracts from Asteraceae. Many studies suggest that the cytotoxic activities could be attributed to the terpenic compounds (sesquiterpenes, diterpenes, and triterpenes) in extracts from species in Asteraceae. The energy store in Asteraceae exists as inulin, not starch. Asteraceae produce tannins, acetylenes, alkaloids, iso/chlorogenic acid, pentacyclic triterpene alcohols, sesquiterpene lactones, and so on. They have essential oils of terpenoid, which do not have iridoids. They have essential oils of terpenoid which do not have iridoids. Plants in Asteraceae family have unique chemical, biochemical, physical, and biological properties, which are essential to both humans and animals.

References

1. Barreda, V.D., Palazzesi, L., Tellería, M.C., Olivero, E.B., Ian Raine, J., Forest, F., Early evolution of the angiosperm clade Asteraceae in the Cretaceous of Antarctica. *Proc. Natl. Acad. Sc. U.S.A.*, 112, 35, 10989–10994, 2015, doi: 10.1073/pnas.1423653112.
2. Jeffrey, C., Compositae: Introduction with key to tribes, in: *Families and Genera of Vascular Plants, vol. VIII, Flowering Plants, Eudicots, Asterales*, J.W. Kadereit and C. Jeffrey (Eds.), p. 61–87, Springer-Verlag, Berlin, 2007.
3. Panda, S.K. and Luyten, W., Antiparasitic activity in Asteraceae with special attention to ethnobotanical use by the tribes of Odisha, India. *Parasite*, 25, 10, 20182018. https://doi.org/10.1051/parasite/2018008.
4. Alvarenga, S.A., Ferreira, M.J., Emerenciano, V., Cabrol-Bass, D., Chemosystematic studies of natural compounds isolated from Asteraceae: characterization of tribes by principal component analysis. *Chemometr. Intell. Lab. Syst.*, 56, 27–37, 2001.
5. Kuhn, T. and Wang, Y., Artemisinin-an innovative cornerstone for anti-malaria therapy. *Prog. Drug Res.*, 66, 385–422, 2008.
6. Lans, C., Turner, N., Khan, T., Brauer, G., Ethno-veterinary medicines used to treat endo parasites and stomach problems in pigs and pets in British Columbia, Canada. *Vet. Parasitol.*, 148, 325–340, 2007.
7. Subhuti, D., *Bidens: A popular remedy escapes notice of Western Practitioners*, ITM, India, 2000.
8. WaboPone, J., Fossi Tankoua, O., Yondo, J., Komtangi, M.C., Mbida, M., Bilong Bilong, C.F., The *in vitro* effects of aqueous and ethanolic extracts of the leaves of Ageratum conyzoides (Asteraceae) on three life cycle stages of the parasitic nematode Heligmosomoides bakeri (Nematoda: Heligmosomatidae). *Vet. Med. Int.*, 140293, 2011.
9. Wanzala, W., Takken, W., Mukabana, W.R., Pala, A.O., Hassanali, A., Ethnoknowledge of Bukusu community on livestock tick prevention and control in Bungoma district, Western Kenya. *J. Ethnopharmacol.*, 140, 298–324, 2012.

10. Panda, A. and Misra, M.K., Ethnomedicinal survey of some wetland plants of South Orissa and their conservation. *Indian J. Tradit. Knowl.*, 10, 296–303, 2011.
11. Pattanaik, C., Reddy, C., Dhal, N., Phytomedicinal study of coastal and dune species of Orissa. *Indian J. Tradit. Knowl.*, 7, 263–268, 2008.
12. Pattnaik, A., Sarkar, R., Sharma, A., Yadav, K.K., Kumar, A., Roy, P., Mazumder, A., Karmakar, S., Sen, T., Pharmacological studies on Buchanania lanzan Spreng. A focus on wound healing with particular reference to anti-biofilm properties. *Asian Pac. J. Trop. Biomed.*, 3, 967–974, 2013.
13. Panda, S.K., Ethno-medicinal uses and screening of plants for antibacterial activity from Similipal Biosphere Reserve, Odisha, India. *J. Ethnopharmacol.*, 151, 158–175, 2014.
14. Shukla, A.N. and Rawat, A.K.S., An ethnobotanical study of medicinal plants of Rewa district, Madhya Pradesh. *Indian J. Tradit. Knowl.*, 9, 191–202, 2010.
15. Pani, M., Nahak, G., Sahu, R.K., Review on ethno-medicinal plants of Odisha for the treatment of malaria. *Int. J. Pharmacogn. Phytochem. Res.*, 7, 156–165, 2014.
16. Solbrig, O.T., Subfamilial nomenclature of compositae. *Taxon*, 12, 229–235, 1963. JSTOR 1216917.
17. Panéro, J.J., Freire, S.E., Ariza Espinar, L., Crozier, B.S., Barboza, G.E., Cantero, J.J., Resolution of deep nodes yields an improved backbone phylogeny and a new basal lineage to study early evolution of Asteraceae. *Mol. Phylogenet. Evol.*, 80, 1, 43–53, 2014, doi: 10.1016/j.ympev.2014.07.012.
18. Zhi-Xi, F., Bo-Han, J., Nie, B., Gao, T., A comprehensive generic-level phylogeny of the sunflower family: Implications for the systematics of Chinese Asteraceae. *J. Syst. Evol.*, 54, 4, 416–437, 2016, doi: 10.1111/jse.12216.
19. Funk, V.A. and Fragman-Sapir, O., 22. Gymnarrheneae (Gymnarrhenoideae), in: *Systematics, Evolution, and Biogeography of Compositae*, V.A. Funk, A. Susanna, T. Stuessy, R. Bayer (Eds.), p. 327–332, International Association for Plant Taxonomy, Vienna, 2009.
20. Judd, W.S., Campbell, C.S., Kellogg, E.A., Stevens, P.F., *Plant systematics: A phylogenetic approach*, Sinauer Associates, Sunderland, 2007.
21. Morhardt, S. and Morhardt, E., *California desert flowers: An introduction to families, genera, and species*, University of California Press, the US, 2004, ISBN-10: 0520240030.
22. McKenzie, R.J., Samuel, J., Muller, E.M., Skinner, A.K.W., Barker, N.P., Morphology of cypselae in subtribe arctotidinae (compositae–arctotideae) and its taxonomic implications. *Ann. Mo. Bot. Gard.*, 92, 4, 569–594, 2005. JSTOR 40035740. Archived from the original on 21 April 2014. Retrieved 21 April 2014.
23. Singh, R. *et al.*, Diversity of host plants of aphids (homoptera: aphididae) infesting asteraceae in India. *Int. J. Zool. Investig.*, 1, 2, 137–167, 2015.

24. The Xerces Society, Five Plants to Help Pollinators. Goldenrods are among the most important late-season pollinator plants, The Xerces Society, United States, 2016.
25. Odom, R.B., James, W.D., Berger, T.G., *Andrews' diseases of the skin: Clinical dermatology*, p. 1135, W.B. Saunders Company, United States, 2000, ISBN 0-7216-5832-6.
26. Lin, Y.L., Chang, C.C., Lee, I.J., Review on phytochemical study of Asteraceae in Taiwan (1996-2005). *J. Chin. Med.*, 1934, 135–149, 2008.
27. Pattanayak, B., Dhal, N.K., Parida, S., Ethnobotanical survey of antimalarial plants of Odisha, India. *Asian J. Sci. Technol.*, 7, 2529–2536, 2016.
28. Kamboj, A. and Saluja, A.K., Ageratum conyzoides L.: A review on its phytochemical and pharmacological profile. *Int. J. Green Pharm.*, 2, 59–68, 2008.
29. Bartolome, A.P., Villaseñor, I.M., Yang, W.-C., Yang, W.-C., *Bidens pilosa* L. (Asteraceae): Botanical properties, traditional uses, phytochemistry, and pharmacology. *Evid.-Based Complement. Alternat. Med.*, 340215, 2013.
30. Mishra, P., Kumar Mishra, S., lacera Pratibha Mishra, B.D., Irchhiaya, R., Phytochemical investigation and spectral studies of isolated flavonoid from ethanolic extract of whole plant. *J. Pharmacogn. Phytochem.*, 4, 1–4, 2015.
31. Muley, B., Khadabadi, S., Banarase, N., Phytochemical constituents and pharmacological activities of Calendula officinalis Linn (Asteraceae): A review. *Trop. J. Pharm. Res.*, 8, 455–465, 2009.
32. Pani, M., Nahak, G., Sahu, R.K., Review on ethnomedicinal plants of Odisha for the treatment of malaria. *Int. J. Pharmacogn. Phytochem. Res.*, 7, 156–165, 2014.
33. Rout, S.D. and Panda, S.K., Ethnomedicinal plant resources of Mayurbhanj district, Orissa. *Indian J. Tradit. Knowl.*, 9, 68–72, 2010.
34. Das, M. and Mukherjee, A., Elephantopus scaber L.: An overview. *Indian J. Life Sci.*, 4, 51–54, 2014.
35. Panda, S.K., Rout, S.D., Mishra, N., Panda, T., Phytotherapy and traditional knowledge of tribal communities of Mayurbhanj district, Orissa, India. *J. Pharmacogn. Phytotherapy*, 3, 101–113, 2011.
36. Ramachandran, S., Review on Sphaeranthus indicus Linn. (Koṭṭaikkarantai). *Pharmacogn. Rev.*, 7, 157–169, 2013.
37. Mahajan, N.G., Chopda, M.Z., Mahajan, R.T., A review on Sphaeranthus indicus Linn: Multipotential medicinal plant. *Int. J. Pharm. Res. Allied Sci.*, 4, 48–74, 2015.
38. Shetty, L.J., Sakr, F.M., Al-Obaidy, K., Patel, M.J., Shareef, H., A brief review on medicinal plant Tagetes erecta Linn A. *J. Appl. Pharm. Sci.*, 5, 91–95, 2015.
39. Saini, A., Kumar Soni, H., Gupta, P., A review on Tridax procumbens. *Imp. J. Interdiscip. Res.*, 2, 308–319, 2016.
40. Mohanty, N., Panda, T., Sahoo, S., Rath, S.P., Herbal folk remedies of Dhenkanal district, Odisha, India. *Int. J. Herb. Med.*, 3, 24–33, 2015.

41. Kantamreddi, V.S.S., Parida, S., Kommula, S.M., Wright, C.W., Phytotherapy used in Orissa state, India for treating malaria. *Phytother. Res.*, 23, 1638-1641, 2009.
42. Panda, S.K., Das, D., Tripthathy, N.K., Phytochemical investigation and anthelmintic activity of various root extracts of Gmelina arborea. *Asian J. Plant Sci. Res.*, 5, 54–58, 2015.
43. Toyang, N.J. and Verpoorte, R., A review of the medicinal potentials of plants of the genus Vernonia (Asteraceae). *J. Ethnopharmacol.*, 146, 681–723, 2013.
44. Kamboj, A. and Saluja, A.K., Phytopharmacological review of Xanthium strumarium L. (Cocklebur). *Int. J. Green Pharm.*, 4, 129–139, 2010.
45. Gallwitz, H., Bonse, S., Martinez-Cruz, A., Schlichting, I., Schumacher, K., Krauth-Siegel, R.L., Ajoene is an inhibitor and subversive substrate of human glutathione reductase and Trypanosoma cruzi trypanothione reductase: Crystallographic, kinetic, and spectroscopic studies. *J. Med. Chem.*, 42, 364–372, 1999.
46. François, G., Passreiter, C.M., Woerdenbag, H.J., Van Looveren, M., Antiplasmodial activities and cytotoxic effects of aqueous extracts and sesquiterpene lactones from Neurolaena lobata. *Planta Med.*, 62, 126–129, 1996.
47. Oketch-Rabah, H.A., Lemmich, E., Dossaji, S.F., Theander, T.G., Olsen, C.E., Cornett, I.C., Arsalan, K.V., Christensen, S.V., Two new antiprotozoal 5-methylcoumarins from Vernonia brachycalyx. *J. Nat. Prod.*, 60, 458–461, 1997.
48. Koide, T., Nose, M., Inoue, M., Ogihara, Y., Yabu, Y., Ohta, N., Trypanocidal effects of gallic acid and related compounds. *Planta Med.*, 64, 27–30, 1998.
49. Elford, B.C., Roberts, M.F., Phillipson, J.D., Wilson, R.J.M., Potentiation of the antimalarial activity of qinghaosu by methoxylated flavones. *Trans. R. Soc. Trop. Med. Hyg.*, 81, 434–436, 1987.
50. Pérez-Victoria, J.M., Chiquero, M.J., Conseil, G., Dayan, G., Di Pietro, A., Barron, D., Castanys, S., Gamarro, F., Correlation between the affinity of flavonoids binding to the cytosolic site of Leishmania tropica multidrug transporter and their efficiency to revert parasite resistance to daunomycin. *Biochemistry*, 38, 1736–1743, 1999.
51. Brandão, M.G., Krettli, A.U., Soares, L.S., Nery, C.G., Marinuzzi, H.C., Antimalarial activity of extracts and fractions from Bidens pilosa and other Bidens species (Asteraceae) correlated with the presence of acetylene and flavonoid compounds. *J. Ethnopharmacol.*, 57, 131–138, 1997.
52. Kumari, P., Misra, K., Sisodia, B., Faridi, U., Srivastava, S., Luqman, S., Darokar, M., Negi, A., Gupta, M., Singh, S., Kumar, J., A promising anticancer and antimalarial component from the leaves of Bidens pilosa. *Planta Med.*, 75, 59–61, 2009.
53. Tobinaga, S., Sharma, M., Aalbersberg, W., Watanabe, K., Iguchi, K., Narui, K., Sasatsu, M., Waki, S., Isolation and identification of a potent antimalarial and antibacterial polyacetylene from Bidens pilosa. *Planta Med.*, 75, 624–628, 2009.

54. Szakiel, A., Ruszkowski, D., Grudniak, A., Kurek, A., Wolska, K., Doligalska, M., Janiszowska, W., Antibacterial and antiparasitic activity of oleanolic acid and its glycosides isolated from marigold (Calendula officinalis). *Planta Med.*, 74, 1709–1715, 2008.
55. Takahashi, H.T., Novello, C.R., Ueda-Nakamura, T., Filho, B.P.D., Palazzo de Mello, J.C., Nakamura, C.V., Thiophene derivatives with antileishmanial activity isolated from aerial parts of Porophyllum ruderale (Jacq.). Cass. *Molecules*, 16, 3469–3478, 2011.
56. Bhinge, S.D., Hogade, M.G., Chavan, C., Kumbhar, M., Chature, V., *In vitro* anthelmintic activity of herb extract of Eclipta prostrate L. against Pheretima posthuma. *Asian J. Pharm. Clin. Res.*, 3, 229–230, 2010.
57. Kamaraj, C. and Rahuman, A.A., Efficacy of anthelmintic properties of medicinal plant extracts against Haemonchus contortus. *Res. Vet. Sci.*, 91, 400–404, 2011.
58. Singh, S.K., Bimal, S., Narayan, S., Jee, C., Bimal, D., Das, P., Bimal, R., Leishmania donovani: Assessment of leishmanicidal effects of herbal extracts obtained from plants in the visceral leishmaniasis endemic area of Bihar, India. *Exp. Parasitol.*, 127, 552–558, 2011.
59. Khanna, V.G., Kannabiran, K., Getti, G., Leishmanicidal activity of saponins isolated from the leaves of Eclipta prostrata and Gymnema sylvestre. *Indian J. Pharmacol.*, 41, 32–35, 2009.
60. Bapna, S., Adsule, S., Shirshat Mahendra, S., Jadhav, S., Patil, L.S., Deshmukh, R.A., Anti-malarial activity of Eclipta alba against Plasmodium berghei infection in mice. *J. Commun. Dis.*, 39, 91–104, 2007.
61. Zahari, Z., Jani, N.A., Amanah, A., Latif, M.N.A., Majid, M.I.A., Adenan, M.I., Bioassay-guided isolation of a sesquiterpene lactone of deoxyelephantopin from Elephantopus scaber Linn. active on Trypanosome brucei rhodesience. *Phytomedicine*, 21, 282–285, 2014.
62. Martio-Quintal, Z., del Rosario Garcia-Miss, M., Mut-Martin, M., Matus-Moo, A., Torres-Tapia, L.W., Peraza-Sanchez, S.R., The leishmanicidal effect of (3 S)-16, 17-didehydrofalcarinol, an oxylipin isolated from Tridax procumbens, is independent of NO production. *Phytother. Res.*, 24, 1004–1008, 2009a.
63. Martín-Quintal, Z., Moo-Puc, R., González-Salazar, F., Chan-Bacab, M.J., Torres-Tapia, L.W., Peraza-Sánchez, S.R., *In vitro* activity of Tridax procumbens against promastigotes of Leishmania mexicana. *J. Ethnopharmacol.*, 122, 463–467, 2009b.
64. Calzada, F., Meckes, M., Cedillo-Rivera, R., Antiamoebic and antigiardial activity of plant flavonoids. *Planta Med.*, 65, 78–80, 1999.
65. Meckes, M., Calzada, F., Tapia-Contreras, A., Cedillo-Rivera, R., Antiprotozoal properties of Helianthemum glomeratum. *Phytother. Res.*, 13, 102–105, 1999.
66. Klongsiriwet, C., Quijada, J., Williams, A.R., Mueller-Harvey, I., Williamson, E.M., Hoste, H., Synergistic inhibition of Haemonchus

contortus exsheathment by flavonoid monomers and condensed tannins. *Int. J. Parasitol.: Drugs Drug Resist.*, 5, 127–134, 2015.

67. Tasdemir, D., Kaiser, M., Brun, R., Yardley, V., Schmidt, T.J., Tosun, F., Ruedi, P., Antitrypanosomal and antileishmanial activities of flavonoids and their analogues: *In vitro, in vivo*, structure-activity relationship, and quantitative structure-activity relationship studies. *Antimicrob. Agents Chemother.*, 50, 1352–1364, 2006.
68. Mead, J.R. and McNair, N., Antiparasitic activity of flavonoids and isoflavones against Cryptosporidium parvum and Encephalitozoon intestinalis. *FEMS Microbiol. Lett.*, 259, 153–157, 2006.
69. Lehane, A.M. and Saliba, K.J., Common dietary flavonoids inhibit the growth of the intraerythrocytic malaria parasite. *BMC Res. Notes*, 1, 26, 2008.
70. Koshimizu, K., Ohigashi, H., Huffman, M.A., Use of Vernonia amygdalina by wild chimpanzee: Possible roles of its bitter and related constituents. *Physiol. Behav.*, 56, 1209–1216, 1994.
71. Goffin, E., da Cunha, A.P., Ziemons, E., Tits, M., Angenot, L., Frederich, M., Quantification of tagitinin C in Tithonia diversifolia by reversed-phase high-performance liquid chromatography. *Phytochem. Anal.*, 14, 378–380, 2003.
72. Becker, J.V., van der Merwe, M.M., van Brummelen, A.C., Pillay, P., Crampton, B.G., Mmutlane, E.M., Parkinson, C., van Heerden, F.R., Crouch, N.R., Smith, P.J., Mancama, D.T., Maharaj, V.J., *In vitro* anti-plasmodial activity of Dicoma anomala subsp. gerrardii (Asteraceae): Identification of its main active constituent, structure–activity relationship studies and gene expression profiling. *Malar. J.*, 10, 295, 2011.
73. Ganfon, H., Bero, J., Tchinda, A.T., Gbaguidi, F., Gbenou, J., Moudachirou, M., Frédérich, M., Quetin-Leclercq, J., Antiparasitic activities of two sesquiterpenic lactones isolated from Acanthospermum hispidum D.C. *J. Ethnopharmacol.*, 141, 411–417, 2012.
74. Musayeib, N., Mothana, R., Gamal, A., Al-Massarani, S., Maes, L., *In vitro* antiprotozoal activity of triterpenoid constituents of Kleinia odora growing in Saudi Arabia. *Molecules*, 18, 9207–9218, 2013.
75. Yamamoto, E.S., Campos, B.L.S., Jesus, J.A., Laurenti, M.D., Ribeiro, S.P., Kallás, E.G., Rafael-Fernandes, M., Santos- Gomes, G., Silva, M.S., Sessa, D.P., Lago, J.H.G., Levy, D., Passero, L.F.D., The effect of ursolic acid on leishmania (Leishmania) amazonensis is related to programed cell death and presents therapeutic potential in experimental cutaneous leishmaniasis. *PLoS One*, 10, e 0144946, 2015.
76. Enwerem, N.M., Okogun, J.I., Wambebe, C.O., Okorie, D.A., Akah, P.A., Anthelmintic activity of the stem bark extracts of Berlina grandiflora and one of its active principles, Betulinic acid. *Phytomedicine*, 8, 112–114, 2001.
77. Bringmann, G., Saeb, W., Assi, L., François, G., Sankara Narayanan, A., Peters, K., Peters, E.M., Betulinic acid: Isolation from Triphyophyllum peltatum and

Ancistrocladus heyneanus, antimalarial activity, and crystal structure of the benzyl ester. *Planta Med.*, 63, 255–257, 1997.
78. Steele, J.C.P., Warhurst, D.C., Kirby, G.C., Simmonds, M.S.J., *In vitro* and *in vivo* evaluation of betulinic acid as an antimalarial. *Phytother. Res.*, 13, 115–119, 1999.
79. Ndjakou Lenta, B., Devkota, K.P., Ngouela, S., Fekam Boyom, F., Naz, Q., Choudhary, M.I., Tsamo, E., Rosenthal, P.J., Sewald, N., Anti–plasmodial and cholinesterase inhibiting activities of some constituents of Psorospermum glaberrimum. *Chem. Pharm. Bull.*, 56, 222–226, 2008.
80. Nweze, N.E., Anene, B.M. and Asuzu, I.U., *In vitro* anti-trypanosomal activities of crude extracts, β-sitosterol and α-sulphur from Buchholzia coriacea seed. *African Journal of Biotechnology (AJB)*, 10, 69, 15626-15632, 2011.
81. Zhu, F., Qin, C., Tao, L., Liu, X., Shi, Z., Ma, X., Jia, J., Tan, Y., Cui, C., Lin, J., Tan, C., Jiang, Y., Chen, Y., Clustered patterns of species origins of nature-derived drugs and clues for future bioprospecting. *Proc. Natl. Acad. Sci.*, 108, 12943–12948, 2011.
82. Newman, D.J. and Cragg, G.M., Natural products as sources of new drugs from 1981 to 2014. *J. Nat. Prod.*, 79, 629–661, 2016.
83. Cox, P.A., Will tribal knowledge survive the millennium? *Science*, 287, 44–45, 2000.
84. McChesney, J.D., Venkataraman, S.K., Henri, J.T., Plant natural products: Back to the future or into extinction? *Phytochemistry*, 68, 2015–2022, 2007.
85. Gurib-Fakim, A., Medicinal plants: Traditions of yesterday and drugs of tomorrow. *Mol. Aspects Med.*, 27, 1–93, 2006.
86. Soejarto, D.D., Fong, H.H.S., Tan, G.T., Zhang, H.J., Ma, C.Y., Franzblau, S.G., Gyllenhaal, C., Riley, M.C., Kadushin, M.R., Pezzuto, J.M., Xuan, L.T., Hiep, N.T., Hung, N.V., Vu, B.M., Loc, P.K., Dac, L.X., Binh, L.T., Chien, N.Q., Hai, N.V., Bich, T.Q., Cuong, N.M., Southavong, B., Sydara, K., Bouamanivong, S., Ly, H.M., Thuy, T.V., Rose, W.C., Dietzman, G.R., Ethnobotany/ethnopharmacology and mass bioprospecting: Issues on intellectual property and benefit-sharing. *J. Ethnopharmacol.*, 100, 15–22, 2005.
87. Talakal, T.S., Dwivedi, S.K., Sharma, S.R., *In vitro* and *In vivo* antitrypanosomal activity of Xanthium strumarium leaves. *J. Ethnopharmacol.*, 49, 141–145, 1995.
88. Panda, S.K., Mohanta, Y.K., Padhi, L., Park, Y.-H., Mohanta, T.K., Bae, H., Large scale screening of ethnomedicinal plants for identification of potential antibacterial compounds. *Molecules*, 21, 293, 2016.
89. Panda, S.K., Padhi, L., Leyssen, P., Liu, M., Neyts, J., Luyten, W., Antimicrobial, anthelmintic and antiviral activity of plants traditionally used for treating infectious disease in the Similipal Biosphere Reserve, Odisha, India. *Front. Pharmacol.*, 8, 658, 2017.
90. Saslis-Lagoudakis, C.H., Savolainen, V., Williamson, E.M., Forest, F., Wagstaff, S.J., Baral, S.R., Watson, M.F., Pendry, C.A., Hawkins, J.A., Phylogenies reveal predictive power of traditional medicine in bioprospecting. *Proc. Natl. Acad. Sci.*, 109, 15835–15840, 2012.

91. Tayo, G.M., Payne, V.K., Poné, J.W., Claire, K.M., Jeannette, Y., Alidou, M.N., Mbida, M., Bilong, C.F.B., *In vitro* ovicidal and larvicidal activities of aqueous and ethanolic extracts of the leaves of Bidens pilosa (Asteraceae) on Heligmosomoides bakeri (Nematoda: Heligmosomatidae). *Int. J. Phytomed. Relat. Ind.*, 4, 121–125, 2012.
92. Gachet, M.S., Lecaro, J.S., Kaiser, M., Brun, R., Navarrete, H., Muñoz, R.A., Bauer, R., Schuhly, W., Assessment of anti-protozoal activity of plants traditionally used in Ecuador in the treatment of leishmaniasis. *J. Ethnopharmacol.*, 128, 184–197, 2010.
93. Moyo, P., Botha, M.E., Nondaba, S., Niemand, J., Maharaj, V.J., Eloff, J.N., Louw, A.I., Birkholtz, L., *In vitro* inhibition of Plasmodium falciparum early and late stage gametocyte viability by extracts from eight traditionally used South African plant species. *J. Ethnopharmacol.*, 185, 235–242, 2016.
94. Pattewar, A.M., Dawalbaje, A.B., Gundale, D.M., Pawar, P.B., Kavtikwar, P.G., Yerawar, P.P., Pandharkar, T.M., Patawar, V.A., Phytochemistry and anthelmintic studies on Blumea lacera. *Indo Glob. J. Pharm. Sci.*, 2, 390–396, 2012.
95. Nikmehr, B., Ghaznavi, H., Rahbar, A., Sadr, S., Mehrzadi, S., *In vitro* antileishmanial activity of methanolic extracts of Calendula officinalis flowers, Datura stramonium seeds, and Salvia officinalis leaves. *Chin. J. Nat. Med.*, 12, 423–427, 2014.
96. Yu, H.W., Wright, C.W., Cai, Y., Yang, S.L., Phillipson, J.D., Kirby, G.C., Warhurst, D.C., Antiprotozoal activities of Centipeda minima. *Phytother. Res.*, 8, 436–438, 1994.
97. Nisha, M., Kalyanasundaram, M., Paily, K.P., Abidha, Vanamail, P., Balaraman, K., *In vitro* screening of medicinal plant extracts for macrofilaricidal activity. *Parasitol. Res.*, 100, 575–579, 2006.
98. García-Herrera, P., Sánchez-Mata, M.C., Cámara, M., Fernández-Ruiz, V., Díez-Marquees, C., Molina, M., Tardío, J., Nutrient composition of six wild edible Mediterranean Asteraceae 4 plants of dietary interest. *J. Food Compos. Anal. YJFCA 2403*, 1–8, 2014. 2014 Published by Elsevier Inc.
99. Guil-Guerrero, J.L., Giménez-Giménez, A., Rodríguez-García, I., Torija-Isasa, M.E., Nutritional composition of Sonchusspecies (S. asper L., S. oleraceus L. and S. tenerrimus L.). *J. Sci. Food Agric.*, 76, 628–632, 1998a.
100. Escudero, N.L., De Arellano, M.L., Fernández, S., Albarracín, G., Mucciarelli, S., Taraxacum officinale asafood source. *Plant Foods Hum. Nutr.*, 58, 1–10, 2003.
101. Souci, S.W., Fachmann, W., Kraut, H., *Food composition and nutrition tables*, Medpharm Scientific Publishers, Stuttgart, 2008.
102. Maynard, D.N. and Hochmuth, G.J., *Knott's handbook for vegetable growers*, 5th ed, John Wiley & Sons, New York, USA, 2007.
103. Morales, P., Carvalho, A.M., Sánchez-Mata, M.C., Cámara, M., Molina, M., Ferreira, I.C.F.R., Tocopherol composition and antioxidant activity of Spanish wild vegetables. *Genet. Resour. Crop Evol.*, 59, 851–863, 2012a.
104. Morales, P., Ferreira, I.C.F.R., Carvalho, A.M., Sánchez-Mata, M.C., Cámara, M., Tardío, J., Fatty acids characterization of some Spanish wild vegetables. *Food Sci. Technol. Int.*, 18, 3, 281–290, 2012b.

105. Jan, G., Kahan, M., Ahmad, M., Iqbal, Z., Afzal, A., Afzal, M., Shah, G.M., Majid, A., Fiaz, M., Zafar, M., Waheed, A., Gul, F., Nutritional analysis, micronutrients and chlorophyll contents of Cichoriumintybus L. *J. Med. Plant Res.*, 5, 12, 2452–2456, 2011.
106. EFSA Panel on Dietetic Products, Nutrition, and Allergies, Scientific opinion on dietary reference values for carbohydrates and dietary fibre. *EFSA J.*, 8, 3, 1462, 2010.
107. Femenia, A., Robertson, J.A., Waldron, K.W., Selvendran, R.R., Cauliflower (Brassicaoleracea L.), globe Artichoke (Cinarascolymus) and chicorywitloof (Cichoriumintybus) processing by-products as sources of dietary fibre. *J. Sci. Food Agric.*, 77, 511–518, 1998.
108. Lachman, J., Fernández, E.C., Orsák, M., Yacon [Smallanthussonchifolia (Poepp.etEndl.) H. Robinson] chemical composition and use a review. *Plant Soil Environ.*, 49, 6, 283–290, 2003.
109. Tardío, J., Molina, M., Aceituno-Mata, L., Pardo-de-Santayana, M., Morales, R., Fernández-Ruiz, V., Morales, P., García, P., Cámara, M., Sánchez-Mata, M.C., Montiafontana L. (Portulacaceae), an interesting wild vegetable traditionally consumed in the Iberian Peninsula. *Genet. Resour. Crop Evol.*, 58, 1105–1118, 2011.
110. Guil-Guerrero, J.L., Giménez-Martínez, J.J., Torija-Isasa, M.E., Mineral nutrient composition of edible wildplants. *J. Food Compos. Anal.*, 11, 322–328, 1998b.
111. Trumbo, P., Schlicker, S., Yates, A.A., Poos, M., Dietary reference intakes for energy, carbohydrate, fiber, fat, fatty acids, cholesterol, protein and amino acids. *J. Am. Diet. Assoc.*, 102, 1621–1630, 2002.
112. Guil, J.L., Torija, M.E., Giménez, J.J., Rodríguez-García, I., Giménez, A., Oxalic acid and calcium determination in wild edible plants. *J. Agric. Food Chem.*, 44, 1821–1823, 1996.
113. Phillips, K.M., Tarragó-Trani, M.T., Gebhardt, S.E., Exler, J., Patterson, K.Y., Haytowitz, D.B., Pehrsson, P.R., Holden, J.M., Stability of vitamin C in frozen raw fruit and vegetable homogenates. *J. Food Compos. Anal.*, 23, 253–259, 2010.
114. McDermott, J.H., Antioxidant nutrients: Current dietary recommendations and research update. *J. Am. Pharm. Assoc.*, 40, 6, 785–799, 2000.
115. Pagnotta, M.A., Fernandez, J.A., Sonnante, G., Egea-Gilabert, C., Genetic diversity and accession structure in European Cynaracardunculus collections. *PLoS One*, 12, e0178770, 2017, doi: 10.1371/journal.pone.0178770.
116. Ceccarelli, N., Curadi, M., Picciarelli, P., Martelloni, L., Sbrana, C., Giovannetti, M., Globe artichokeasa functional food. *Med. J. Nutr. Metab.*, 3, 197–201, 2010, doi: 10.1007/s12349-010-0021-z.
117. Raccuia, S.A., Mainolfi, A., Mandolino, G., Melilli, M.G., Genetic diversity in Cynaracardunculus revealed by AFLP markers: Comparison between cultivars and wild types from Sicily*—. *Plant Breed.*, 123, 280–284, 2004, doi: 10.1111/j.1439-0523.2004.00983.x.

118. Petropoulos, S.A., Pereira, C., Tzortzakis, N., Barros, L., Ferreira, I.C.F.R., Nutritional value and bioactive compounds characterization of plant parts from Cynaracardunculus L. (Asteraceae) cultivated in Central Greece. *Front. Plant Sci.*, 9, 459, 2018, doi: 10.3389/fpls.2018.00459.
119. Lombardo, S., Pandino, G., Mauromicale, G., Knödler, M., Carle, R., Schieber, A., Influence of genotype, harvest time and plant part on polyphenolic composition of globe artichoke[Cynaracardunculus L. var. scolymus(L.) Fiori]. *Food Chem.*, 119, 1175–1181, 2010, doi: 10.1016/j.foodchem.2009.08.033.
120. Pandino, G., Lombardo, S., Mauromicale, G., Chemical and morphological characteristics of new clones and commercial varieties of globe artichoke (Cynaracardunculusvar. scolymus). *Plant Foods Hum. Nutr.*, 66, 291–297, 2011a, doi: 10.1007/s11130-011-0247-z.
121. Pandino, G., Lombardo, S., Mauromicale, G., Williamson, G., Phenolic acids and flavonoids in leaf and floral stem of cultivated and wild Cynaracardunculus L. genotypes. *Food Chem.*, 126, 417–422, 2011b, doi: 10.1016/j.foodchem.2010.11.001.
122. Borgognone, D., Cardarelli, M., Rea, E., Lucini, L., Colla, G., Salinity source-induced changes in yield, mineral composition, phenolic acids and flavonoids in leaves of artichoke and cardo on grown in floating system. *J. Sci. Food Agric.*, 94, 1231–1237, 2014, doi: 10.1002/jsfa.6403.
123. Rouphael, Y., Cardarelli, M., Lucini, L., Rea, E., Colla, G., Nutrient solution concentration affects growth, mineral composition, phenolic acids, and flavonoids in leaves of artichoke and cardoon. *Hortic. Sci.*, 47, 1424–1429, 2012, doi: 10.1002/jsfa.3641.
124. Colla, G., Rouphael, Y., Cardarelli, M., Svecova, E., Rea, E., Lucini, L., Effects of salinestress on mineral composition, phenolic acids and flavonoids in leaves of artichoke and cardoongenotypes grown in floating system. *J. Sci. Food Agric.*, 93, 1119–1127, 2013, doi: 10.1002/jsfa.5861.
125. Maccarone, E., Fallico, B., Fanella, F., Mauromicale, G., Raccuia, S.A., Possible alternative utilization of Cynaraspp. II. Chemical characterization of their grainoil. *Ind. Crops Prod.*, 10, 229–237, 1999, doi: 10.1016/S0926-6690(99)00027-8.
126. Curt, M.D., Sánchez, G., Fernández, J., The potential of Cynaracardunculus L. for seed oil production in a perennial cultivation system. *Biomass Bioenergy*, 23, 33–46, 2002, doi: 10.1016/S0961-9534(02)00030-2.
127. Simopoulos, A.P., The importance of the omega-6/omega-3 fatty acid ratio in cardiovascular disease and other chronic diseases. *Exp. Biol. Med.*, 233, 674–688, 2008, doi: 10.3181/0711-MR-311.
128. Guil, J.L., Torija, M.E., Giménez, J.J., Rodriguez, I., Identification of fatty acids in edible wild plants by gas chromatography. *J. Chromatogr. A*, 719, 229–235, 1996, doi: 10.1016/0021-9673(95)00414-9.
129. Simopoulos, A.P., Omega-3 fatty acids and antioxidants in edible wild plants. *Biol. Res.*, 37, 263–277, 2004, doi: 10.4067/S0716-97602004000200013.
130. Harnack, K., Andersen, G., Somoza, V., Nutrition & metabolism and docosahexaenoic acid as affected by the ratio of n6/n3 fatty acids. *Nutr. Metab.*, 6, 8, 2009, doi: 10.1186/1743-7075-6-8.

Appendix A: Comprehensive List of Plants in Asteraceae Family

A

Aylacophora Cabrera
Ayapanopsis R.M.King
Ayapana Spach
Axiniphyllum Benth
Avellara Blanca & C.Diaz
Austrosynotis C.Jeffrey
Austroliabum H.Rob. & Brettell
Austroeupatorium R.M.King
Austrocritonia R.M.King
Austrobrickellia R.M.King
Atrichoseris A.Gray
Atrichantha Hilliard
Atractylodes DC.
Atractylis L.
Athroisma DC
Athrixia Ker Gawl.
Athanasia L.
Astranthium Nutt. - western daisy
Asteromoea Blume
Asteriscus Mill.
Asteridea Lindl.
Aster L. - aster
Asplundianthus R.M.King
Aspilia Thouars
Ascidiogyne Cuatrec
Asanthus R.M.King - brickellbush
Artemisiopsis S.Moore
Artemisia L. - mugwort, sagewort, wormwood
Arrowsmithia DC.
Arrojadocharis Mattf.
Arrhenechthites Mattf.
Arnoseris Gaertn.
Arnoglossum Raf. - Indian plantain
Arnicastrum Greenm.
Arnica L. - arnica
Arnaldoa Cabrera
Aristeguietia R.M.King
Arida
Argyroxiphium DC - silversword
Argyrophanes Schltdl.
Argyroglottis Turcz.
Argyranthemum Webb - dill daisy
Argentipallium
Arctotis L.
Arctotheca J.C.Wendl. - capeweed
Arctogeron DC.
Arctium L. - burdock
Arctanthemum Tzvelev
Archibaccharis Heering
Arbelaezaster Cuatrec.
Apostates Lander
Aposeris Neck. ex Cass.
Apodocephala Baker
Aphyllocladus Wedd.
Aphanostephus DC - doze daisy
Aphanactis Wedd.
Apargidium Torr. & A.Gray
Apalochlamys (Cass.) Cass.
Anvillea DC.
Anura (Juz.) Tscherneva
Antithrixia DC.
Antiphiona Merxm.
Antillia R.M.King
Antheropeas Rydb. - Easter bonnets
Anthemis L. - Roman chamomile
Antennaria Gaertn. - pussytoes
Anomostephium DC.
Anisothrix O.Hoffm. ex Kuntze
Anisopappus Hook. & Arn.
Anisocoma Torr. & A.Gray
Anisochaeta DC.
Angianthus J.C.Wendl.

Angelphytum G.M.Barroso
Anemocarpa
Andryala L.
Anderbergia
Ancistrophora A.Gray
Ancistrocarphus
Ancathia DC.
Anaxeton Gaertn.
Anastraphia D.Don
Anaphalis DC - pearly everlasting
Anaphalioides (Benth.) Kirp.
Anacyclus L.
Anacantha (Iljin) Sojak
Amphoricarpos Vis.
Amphipappus - chaffbush
Amphiglossa DC
Amphiachyris - broomweed
Ampelaster - climbing aster
Amolinia R.M.King
Ammobium R.Br. ex Sims
Amellus L.
Ameghinoa Speg.
Ambrosia L. - bursage, ragweed
Amboroa Cabrera
Amblyopappus Hook
Amblyolepis DC
Amblyocarpum Fisch
Amberboa (Pers.) Less.
Amauriopsis
Amauria Benth
Alvordia Brandegee
Alomiella R.M.King
Alomia Kunth
Almutaster - alkali marsh aster
Allopterigeron Dunlop
Alloispermum Willd.
Allardia Decne.
Allagopappus Cass.
Aliella Qaiser & Lack
Alfredia Cass.
Alepidocline S.F.Blake
Aldama La Llave
Alciope DC. ex Lindl.
Alcantara Glaz. ex G.M.Barroso
Albertinia Spreng.
Alatoseta Compton
Ajaniopsis C.Shih
Ajania Poljakov
Ainsliaea DC.
Agrianthus Mart. ex DC.
Agoseris Raf. - mountain dandelion
Agnorhiza (Jeps.) W.A.Weber
Agiabampoa Rose ex O.Hoffm.
Ageratum L. - whiteweed
Ageratinastrum Mattf.
Ageratina Spach - snakeroot
Ageratella A.Gray ex S.Watson
Afroaster J.C.Manning & Goldblatt
Aetheorhiza Cass.
Aequatorium B.Nord
Aegopordon Boiss.
Aedesia O.Hoffm.
Adenothamnus D.D.Keck
Adenostyles A. Kern.
Adenostemma Pers. - medicine plant
Adenophyllum - dogweed
Adenopappus Benth
Adenoon Dalzell
Adenoglossa B.Nord.
Adenocritonia R.M.King
Adenocaulon Hook - trailplant
Adenanthellum B.Nord.
Adelostigma Steetz
Acunniana Orchard
Actinospermum Elliott
Actinoseris (Endl.) Cabrera
Actinobole Endl.
Acroptilon Cass - Russian knapweed, hardheads,
Acroclinium A.Gray
Acritopappus R.M.King
Acrisione B.Nord.

Acourtia D.Don - desert peony
Acomis F.Muell.
Acmella Rich.
Acilepis D.Don
Acilepidopsis H.Rob.
Achyrothalamus O.Hoffm.
Achyropappus Kunth
Achyrocline (Less.) DC.
Achyrachaena Schauer - blow wives
Achnopogon Maguire
Achnophora F.Muell.
Achillea L. - yarrow
Achaetogeron A.Gray
Acanthostyles R.M.King
Acanthospermum Schrank
Acantholepis Less.
Acanthodesmos C.D.
Acanthocladium F.Muell.
Acanthocephalus Kar. & Kir.
Acamptopappus (A.Gray) A.Gray
Abrotanella Cass.
Aaronsohnia Warb

B
Burkartia Crisci
Buphthalmum L.
Bryomorphe Harv.
Brintonia - mock goldenrod
Brickelliastrum R.M.King - brickellbush
Brickellia Elliott - brickellbush
Brenandendron H.Rob.
Bradburia Torr. & A.Gray
Bracteantha Anderb.
Brachythrix Wild
Brachyscome Cass.
Brachylaena R.Br.
Brachyglottis J.R.Forst. & G.Forst.
Brachyactis - rayless aster
Brachionostylum Mattf.
Brachanthemum DC.
Bothriocline Oliv. ex Benth.
Borrichia Adans. - seaside tansy
Bombycilaena (DC.) Smoljan.
Boltonia L›Her.- doll›s daisy
Boeberoides (DC.) Strother
Boeberastrum (A.Gray) Rydb.
Blumeopsis Gagnep.
Blumea DC. - false oxtongue
Blepharizonia (A.Gray) Greene
Blepharispermum DC.
Blepharipappus Hook
Blennospora A.Gray
Blennosperma Less. - stickyseed
Blanchetia DC
Blakiella Cuatrec.
Blakeanthus R.M.King
Blainvillea Cass.
Bishovia R.M.King
Bishopiella R.M.King
Bishopanthus H.Rob.
Bigelowia DC - rayless goldenrod
Bidens L. - Spanish needles
Bethencourtia
Berroa Beauverd
Berlandiera DC - greeneyes
Berkheya Ehrh.
Berardia Vill.
Benitoa D.D.Keck
Belloa J.Remy
Bellium L.
Bellis L. - daisy
Bellida Ewart
Bejaranoa R.M.King
Bedfordia DC.
Bebbia Greene - sweetbush
Basedowia E.Pritz.
Bartlettina R.M.King
Bartlettia A. Gray
Barrosoa R.M.King
Barroetea A.Gray
Barnadesia Mutis ex L.f.

Barkleyanthus H.Rob. - willow ragwort
Baltimora L. - baltimora
Balsamorhiza Hook. - balsamroot
Balduina Nutt. – honeycomb head
Bajacalia
Baileya Harv. - desert marigold
Bahianthus R.M.King
Bahia Lach. - bahia
Bafutia C.D.Adams
Baeriopsis J.T.Howell
Badilloa R.M.King
Baccharis L. - baccharis

C

Cynara L. - artichoke
Cymophora B.L.Rob.
Cymbopappus B.Nord.
Cymbonotus Cass.
Cymbolaena Smoljan.
Cylindrocline Cass.
Cyclolepis Gillies ex D.Don
Cyclachaena Fresen. ex Schltdl.
Cyathomone S.F.Blake
Cyathocline Cass.
Cyanthillium
Cyanopsis - knapweed
Cuspidia Gaertn.
Cuniculotinus
Cullumia R.Br.
Cuchumatanea Seid.
Cuatrecasasiella H.Rob.
Cuatrecasanthus
Crupina (Pers.) DC.
Crossothamnus R.M.King
Crossostephium Less.
Croptilon Raf. - scratchdaisy
Cronquistianthus R.M.King
Cronquistia R.M.King
Crocidium Hook - spring-gold
Critoniopsis Sch.Bip.
Critoniella R.M.King
Critoniadelphus R.M.King
Critonia P.Browne - thoroughwort
Crinitaria
Crepis L. - hawksbeard
Crepidiastrum Nakai
Cremanthodium Benth.
Cratystylis S.Moore
Crassocephalum Moench - ragleaf
Craspedia G.Forst.
Cousiniopsis Nevski
Cousinia Cass.
Coulterella Vasey & Rose
Cotula L. - waterbuttons
Cosmos Cas.
Corymbium L.
Correllia A.M.Powell
Coronidium Paul G. Wilson
Corethrogyne DC - sandaster
Corethamnium R.M.King
Coreopsis L. - tickseed
Coreocarpus Benth.
Conyza Less. - horseweed
Conoclinium R.M.King - mistflower
Conoclinium DC.
Conocliniopsis R.M.King
Condylopodium R.M.King
Condylidium
Complaya Strother
Commidendrum DC.
Comborhiza
Comaclinium Scheidw. & Planch.
Columbiadoria G.L.Nesom
Colobanthera Humbert
Coleostephus Cass.
Coleocoma F.Muell.
Coespeletia Cuatrec.
Cnicus L. - blessed thistle
Cnicothamnus Griseb.
Clibadium L.
Clappia A.Gray - clapdaisy

Cladochaeta DC.
Cladanthus Cass.
Cissampelopsis (DC.) Miq.
Cirsium Mill. - thistle
Cineraria L.
Cichorium L. - chicory
Ciceronia Urb.
Cicerbita Wallr.
Chuquiraga Juss.
Chucoa Cabrera
Chthonocephalus Steetz
Chrysothamnus Nutt. - rabbitbrush
Chrysopsis (Nutt.) Elliott - goldenaster
Chrysophthalmum Sch.Bip. ex Walp.
Chrysoma Nutt.
Chrysogonum L.
Chrysocoma L.
Chrysocephalum Walp.
Chrysanthemum L.
Chrysanthemoides Fabr.
Chrysanthellum Rich.
Chrysactinium (Kunth) Wedd.
Chrysactinia A.Gray
Chrysactinia A.Gray
Chronopappus DC.
Chromolepis Benth.
Chromolaena DC - thoroughwort
Chresta Vell. ex DC.
Chorisis DC.
Chondropyxis D.A.Cooke
Chondrilla L.
Chloracantha G.L.Nesom
Chlamydophora Ehrenb. ex Less.
Chionopappus Benth
Chionolaena DC.
Chimantaea Maguire, Steyerm. & Wurdack
Chiliotrichum Cass.
Chiliotrichiopsis Cabrera
Chiliophyllum Phil.
Chiliocephalum Benth.
Chiliadenus Cass.
Chevreulia Cass.
Chersodoma Phil.
Cheirolophus Cass.
Chardinia Desf.
Chaptalia Vent. - sunbonnetts
Chamomilla - chamomilla, pineapple weed
Chamaemelum Mill. - dogfennel
Chamaeleon Cass.
Chamaegeron Schrenk
Chamaechaenactis Rydb.
Chaetymenia Hook. & Arn.
Chaetospira S.F.Blake
Chaetopappa DC - least daisy
Chaetanthera Ruiz & Pav.
chaetadelpha A.Gray ex S.Watson - skeletonweed
Chaenactis DC - pincushion
Chacoa R.M.King
Ceruana Forssk.
Ceratogyne Turcz.
Cephalosorus A.Gray
Cephalorrhynchus Boiss.
Cephalopappus Nees & Mart.
Cephalipterum A.Gray
Centratherum Cass.
Centipeda Lour
Centaurothamnus Wagenitz & Dittrich
Centauropsis Bojer ex DC.
Centaurodendron Johow
Centaurea L. - knapweed, cornflower, star thistle
Celmisia Cass.
Caxamarca
Cavea W.W.Sm. & Small
Cavalcantia R.M.King
Caucasalia
Catolesia
Catatia Humbert
Catananche L.

Catamixis Thomson
Castenedia R.M.King
Castalis Cass.
Cassinia R.Br.
Carthamus L. - distaff thistle
Carterothamnus R.M.King
Carramboa Cuatrec.
Carphochaete A.Gray - bristlehead
Carphephorus Cass. - chaffhead
Carpesium L.
Carminatia Moc. ex DC.
Carlina L. - carline thistle
Carduus L. - plumeless thistle
Carduncellus Adans.
Cardopatium Juss.
Cancriniella Tzvelev
Cancrinia Kar. & Kir.
Canadanthus G.L.Nesom - mountain aster
Campuloclinium DC
Camptacra N.T.Burb.
Campovassouria R.M.King
Camchaya Gagnep.
Calyptocarpus Less.
Calycoseris A.Gray - tackstem
Calycocorsus F.W.Schmidt
Calycadenia DC - western rosin weed
Calotis R.Br.
Calotesta P.O.Karis
Calostephane Benth.
Calorezia Panero
Calopappus Meyen
Calomeria Vent.
Calocephalus R.Br.
Callistephus Cass.
Callilepis DC.
Callicephalus C.A.Mey.
Calendula L. - marigold
Calea L.
Caesulia Roxb.
Cacosmia Kunth
Cacaliopsis A.Gray
Cacalia L. - Indian plantain
Cabreriella Cuatrec.
Cabobanthus H.Rob.

D

Dyssodia Cav. - dogweed
Dysodiopsis (A.Gray) Rydb. - dogfennel
Dyscritothamnus B.L.Rob
Dyscritogyne R.M.King
Dymondia Compton
Duseniella K.Schum.
Duidaea S.F.Blake
Duhaldea DC.
Dugesia A.Gray
Dugaldia (Cass.) Cass.
Dubyaea DC.
Dubautia Gaudich.
Dresslerothamnus H.Rob.
Dracopis (synonym of *Rudbeckia* L.) - coneflower
Doronicum L. - false leopardbane
Doniophyton Wedd.
Dolomiaea DC.
Dolichothrix Hilliard & B.L.Burtt
Dolichorrhiza (Pojark.) Galushko
Dolichoglottis B.Nord.
Dolichlasium Lag.
Doellingeria Ness. - whitetop
Dittrichia Greuter
Dithyrostegia A.Gray
Disynaphia Hook. & Arn.ex DC.
Distephanus (Cass.) Cass.
Dissothrix A.Gray
Disparago Gaertn.
Dipterocypsela S.F.Blake
Dipterocome Fisch. & C.A.Mey.
Diplostephium Kunth
Diplazoptilon Y.Ling

Diodontium F.Muell.
Dinoseris Griseb.
Dimorphotheca Moench - cape marigold
Dimorphocoma F.Muell. & Tate
Dimerostemma Cass.
Dimeresia A. Gray
Digitacalia Pippen
Dieteria
Dielitzia P.S.Short
Didelta L›Her.
Dicranocarpus A.Gray
Dicoria Torr. & A.Gray - twinbugs
Dicoma Cass.
Dichromochlamys Dunlop
Dichrocephala L›Her. ex DC.
Dichaetophora A.Gray
Dicercoclados C.Jeffrey & Y.L.Chen
Diaphractanthus Humbert
Diaperia Nutt.
Dianthoseris Sch.Bip.
Diacranthera R.M.King
Dewildemania O.Hoffm.
Desmanthodium Benth.
Denekia Thunb.
Dendroseris D.Don
Dendrosenecio (Hauman ex Humbert) B.Nord.
Dendrophorbium (Cuatrec.) C.Jeffrey
Dendrocacalia (Nakai) Tuyama
Dendranthema (DC.) Des Moul. - arctic daisy
Delilia Spreng.
Delamerea S.Moore
Delairea Lem. - capeivy
Deinandra - often included in *Hemizona*
Decazesia F.Muell.
Decastylocarpus Humbert
Decachaeta DC.
Daveaua Willk. ex Mariz
Dasyphyllum Kunth
Dasycondylus R.M.King
Darwiniothamnus Harling
Damnxanthodium Strother
Damnamenia
Dahlia Cav.
Dacryotrichia Wild

E

Exomiocarpon Lawalree
Ewartiothamnus Anderb.
Ewartia Beauverd
Evax (synonym of *Filago* L.) - pygmy cudweed
Evacidium Pomel
Eutrochium - Joe-Pye weed
Euthamia (Nutt.) Elliott - goldentop
Eutetras A.Gray
Euryops (Cass.) Cass.
Eurydochus Maguire & Wurdack
Eurybia (Cass.) Cass.
Euphrosyne DC
Eupatorium L. - thoroughwort, snakeweed
Eupatoriopsis Hieron.
Eupatorina R.M.King
Eupatoriastrum Greenm.
Eumorphia DC
Euchiton Cass. - cudweed
Eucephalus Nutt. - Cascade Aster
Ethulia L.f
Espeletiopsis Cuatrec.
Espeletia Mutis ex Humb.& Bonpl. - frailejón
Espejoa DC.
Erythrocephalum Benth.
Erythradenia (B.L.Rob.) R.M.King
Eryngiophyllum Greenm.
Erymophyllum Paul G.Wilson
Erodiophyllum F.Muell.
Erlangea Sch.Bip.

Eriotrix Cass.
Eriophyllum Lag. - woolly sunflower
Eriochlamys Sond. & F.Muell.
Eriocephalus L.
Erigeron L. - daisy, fleabane
Ericentrodea S.F.Blake & Sherff
Ericameria Nutt. - goldenbush, heath goldenrod
Eriachaenium Sch.Bip.
Eremothamnus O.Hoffm.
Eremosis (DC.) Gleason
Eremanthus Less
Erechtites Raf - burnweed
Erato DC.
Epitriche Turcz.
Episcothamnus H.Rob.
Epilasia (Bunge) Benth.
Epaltes Cass.
Enydra Lour - swampwort
Engleria O.Hoffm.
Engelmannia A.Gray ex Nutt. - Engelmann›s daisy
Endopappus Sch.Bip.
Endocellion Turcz. ex Herder
Enceliopsis (A.Gray) A.Nelson - sunray
Encelia Adans. - brittlebush
Emiliella S.Moore
Emilia (Cass.) Cass. - tasselflower
Elytropappus Cass.
Ellenbergia Cuatrec.
Eleutheranthera Poit. ex Bosc
Elephantopus L. - elephantsfoot
Elaphandra Strother
Elachanthus F.Muell.
Ekmania Gleason
Eitenia R.M.King
Eirmocephala
Egletes Cass. - tropic daisy
Edmondia Cass.
Eclipta L.
Echinops L. - globethistle
Echinacea Moench - coneflower
Eatonella A.Gray
Eastwoodia Brandegee

F

Fulcaldea Poir.
Freya V.M.Badillo
Foveolina Kallersjo
Formania W.W.Sm. & Small
Flyriella R.M.King - brickellbush
Flourensia DC - tarwort
Flosmutisia Cuatrec.
Floscaldasia Cuatrec.
Florestina Cass.
Fleischmanniopsis R.M.King
Fleischmannia Sch.Bip. - thoroughwort
Flaveria Juss. - yellowtops
Fitzwillia P.S.Short
Fitchia Hook.f.
Filifolium Kitam.
Filago L. - cottonrose
Ferreyrella S.F.Blake
Ferreyranthus H.Rob. & Brettell
Fenixia Merr.
Femeniasia Susanna
Felicia Cass.
Feldstonia P.S.Short
Feddea Urb.
Faxonia Brandegee
Faujasia Cass.
Farfugium Lindl.
Facelis Hemsl. - trampweed
Faberia Hemsl.

G

Gyrodoma Wild
Gyptis (Cass.) Cass.
Gyptidium R.M.King
Gypothamnium Phil.
Gynura Cass.
Gynoxys Cass.

Gymnostyles (synonym of *Soliva* Ruiz & Pav.) - burrweed
Gymnostephium Less.
Gymnosperma Benth.
Gymnopentzia
Gymnolaena (DC.) Rydb.
Gymnodiscus Less.
Gymnocoronis DC.
Gymnocondylus R.M.King
Gymnarrhena Desf.
Guynesomia
Gutierrezia Lag. - snakeweed
Gutenbergia Sch.Bip.
Gundlachia A.Gray
Gundelia L.
Guizotia Cass.
Guevaria R.M.King
Guayania R.M.King
Guardiola Cerv. ex Humb. & Bonpl.
Grosvenoria R.M.King
Grisebachianthus R.M.King
Grindelia Willd. - gumweed
Greenmaniella W.M.Sharp
Grazielia R.M.King
Grauanthus Fayed
Gratwickia F.Muell.
Graphistylis B.Nord.
Grangeopsis Humbert
Grangea Adans.
Goyazianthus R.M.King
Gossweilera S.Moore
Gorteria L.
Gorceixia Baker
Gonospermum Less.
Goniocaulon Cass.
Gongylolepis R.H.Schomb.
Gongrostylus R.M.King
Goldmanella Greenm.
Gochnatia Kunth
Gnephosis Cass.
Gnaphalium L. - cudweed
Gnaphaliothamnus Kirp.
Glyptopleura Eaton
Glossopappus Kunze
Glossocardia Cass.
Glossarion Maguire & Wurdack
Gladiopappus Humbert
Gilruthia Ewart
Gilberta Turcz.
Gibbaria Cass.
Geropogon L.
Gerbera L. - Gerbera or Transvaal daisy
Geraea Torr. & A.Gray - desert sunflower
Geissopappus Benth.
Geissolepis B.L.Rob.
Geigeria Griess.
Gazania Gaertn.
Garuleum Cass.
Gardnerina R.M.King
Garcilassa Poepp.
Garcibarrigoa Cuatrec.
Garberia A.Gray
Gamochaetopsis Anderb. & Freire
Gamochaeta Wedd. - everlasting
Galinsoga Ruiz & Pav. - gallant-soldier
Galeomma Rauschert
Galeana La Llave
Galatella
Galactites Moench
Gaillardia Foug. - blanketflower

H

Hystrichophora Mattf.
Hysterionica Willd.
Hypochaeris L. - catsear
Hypericophyllum Steetz
Hypelichrysum Kirp.
Hypacanthium Juz.
Hyoseris L.
Hymenoxys Cass. - rubberweed

Hymenothrix A.Gray - thimblehead
Hymenostephium Benth.
Hymenostemma Kunze ex Willk.
Hymenopappus L›Her
Hymenonema Cass.
Hymenolepis Cass.
Hymenoclea Torr. & A.Gray - burro-brush, burrobush
Hymenocephalus Jaub. & Spach
Hydropectis Rydb.
Hydroidea P.O.Karis
Hydrodyssodia B.L.Turner
Hybridella Cass.
Hyalosperma Steetz
Hyaloseris Griseb.
Hyalochlamys A.Gray
Hyalochaete Dittrich & Rech.f.
Hyalis D.Don ex Hook.& Arn.
Humeocline Anderb.
Hulsea Torr. & A.Gray - alpinegold
Hughesia R.M.King
Hubertia Bory
Huberopappus
Huarpea Cabrera
Hoplophyllum DC.
Homogyne Cass.
Holozonia Greene
Hololeion Kitam
Holocheilus Cass.
Holocarpha Greene - tarweed
Hofmeisteria Walp.
Hoffmanniella Schltr. ex Lawalree
Hoehnephytum Cabrera
Hispidella Barnadez ex Lam.
Hirpicium Cass.
Hippolytia Poljakov
Hippia L.
Hinterhubera Sch.Bip. ex Wedd.
Hilliardia B.Nord.
Hieracium L. - hawkweed
Hidalgoa La Llave
Heterotheca Cass. - goldenasters, camphorweed, telegraph weed
Heterothalamus Less.
Heterosperma Cav.
Heterorhachis Sch.Bip. ex Walp.
Heteroplexis C.C.Chang
Heteropappus Less.
Heteromma Benth.
Heteromera Pomel
Heterolepis Cass.
Heteroderis (Bunge) Boiss.
Heterocypsela H.Rob.
Heterocondylus R.M.King
Heterocoma DC.
Heteranthemis Schott - oxeye
Heteracia Fisch. & C.A.Mey.
Hesperomannia A.Gray - island-aster
Hesperodoria Greene - glowweed
Hesperevax(A.Gray) A.Gray - dwarf-cudweed
Hertia
Herrickia (synonym of *Aster* L.)
Herodotia Urb. & Ekman
Herderia Cass.
Heptanthus Griseb.
Henricksonia B.L.Turner
Hemizonia DC - tarweed
Hemisteptia Fisch. & C.A.Mey.
Helogyne Nutt.
Helminthotheca (*obsolete*)
Helminthia (synonym of *Picris* L.)
Heliopsis Pers.
Heliomeris Nutt. - false goldeneye
Heliocauta Humphries
Helichrysum Mill. - strawflower, everlasting
Helichrysopsis Kirp.
Helianthus L. - sunflowers
Helianthopsis H.Rob.
Helianthella Torr. & A.Gray
Helenium L. - sneezeweed

Hedypnois Mill.
Hecastocleis A.Gray
Hebeclinium DC - thoroughwort
Hazardia Greene - bristleweed
Hatschbachiella R.M.King
Hasteola Raf. - false Indian plantain
Hartwrightia A.Gray ex S.Watson
Harnackia Urb.
Harmonia
Harleya S.F.Blake
Haplostephium Mart. ex DC.
Haplopappus Cass.
Haploesthes A.Gray - false broomweed
Haplocarpha Less. - onefruit
Haplocalymma S.F.Blake
Handelia Heimerl
Haegiela P.S.Short
Haeckeria F.Muell.
Haastia Hook.f.

I
Ixodia R.Br.
Ixiolaena Benth.
Ixiochlamys F.Muell. & Sond.
Ixeris (Cass.) Cass.
Ixeridium (A.Gray) Tzvelev
Iva L. - marshelder, sumpweed
Isostigma Less.
Isopappus Torr. & A.Gray
Isoetopsis Turcz.
Isocoma Nutt. - goldenbush, jimmy weed
Isocarpha R.Br. - pearlhead
Ischnea F.Muell.
Irwinia Barroso
Iranecio B.Nord.
Iphionopsis Anderb.
Iphiona Cass.
Iostephane Benth.
Ionactis - Stiff-leaved Asters (?.)
Iogeton Strother
Iodocephalus Thorel ex Gagnep.
Iocenes B.Nord.
Inulanthera Kallersjo
Inula L. - yellowhead
Inezia E.Phillips
Imeria R.M.King
Iltisia S.F.Blake
Ighermia Wiklund
Ifloga Cass.
Idiothamnus R.M.King
Ichthyothere Mart.

J
Jurinea Cass.
Jungia L.f.
Joseanthus
Jensia
Jeffreya Wild
Jefea Strother
Jaumea Pers.
Jasonia (Cass.) Cass.
Jaramilloa R.M.King
Jamesianthus S.F.Blake & Sherff
Jaliscoa S.Watson
Jalcophila Dillon & Sagast.
Jaegeria Kunth
Jacmaia B.Nord.

K
Kyrsteniopsis R.M.King & H.Rob
Krigia Schreb - dwarf dandelion
Koyamacalia (L.) H.Rob. & Brettell
Koelpinia Pall.
Koehneola Urb.
Koanophyllon Arruda - thoroughwort
Kleinia Mill.
Kirkianella Allan
Kippistia F.Muell.
Kingianthus H.Rob.
Kinghamia C.Jeffrey
Keysseria Lauterb.

Kemulariella
Kaunia R.M.King
Kaschgaria Poljakov
Karvandarina Rech.f.
Karelinia Less.
Kalimeris (Cass.) Cass. - aster

L
Lygodesmia D.Don - skeleton weed
Lycoseris Cass.
Lychnophora Mart.
Lycapsus Phil.
Lundellianthus H.Rob.
Lulia Zardini
Luina Benth. - silverback
Lugoa DC.
Luciliocline Anderb. & Freire
Lucilia Cass.
Loxothysanus B.L.Rob.
Lourteigia R.M.King
Loricaria Wedd.
Lorentzianthus R.M.King
Lordhowea B.Nord.
Lorandersonia
Lophopappus Rusby
Lopholaena DC.
Lonas Adans.
Lomatozona Baker
Logfia Cass. - cottonrose
Llerasia Triana
Litrisa Small
Litothamnus R.M.King
Lipskyella Juz.
Lipochaeta DC - nehe
Lindheimera A.Gray & Engelm.
Limbarda Adans.
Ligulariopsis
Ligularia Cass.
Lifago Schweinf. & Muschl.
Lidbeckia Bergius
Libanothamnus Ernst
Liatris Liabum Adans. - blazing star, gay feather
Liabum Adans.
Leysera L.
Leuzea DC.
Leunisia Phil.
Leucoptera B.Nord
Leucopsis (DC.) Baker
Leucophyta
Leucomeris
Leucochrysum - sunray (?)
Leucheria Lag.
Leucanthemum Mill. - daisy, oxeye daisy
Leucanthemopsis (Giroux) Heywood
Leucanthemella Tzvelev
Leucactinia Rydb.
Lessingia Cham. - vinegarweed
Lescaillea Griseb.
Leptotriche Turcz.
Leptorhynchos Less. - scaly button
Leptoclinium (Nutt.) Benth.
Leptocarpha DC.
Leptinella Cass. - Brass Buttons, Creeping Cotula
Lepidostephium Oliv.
Lepidospartum (A.Gray) A.Gray - broomsage
Lepidophyllum Cass.
Lepidophorum Neck. ex DC.
Lepidonia S.F.Blake
Lepidolopsis Poljakov
Lepidolopha C.Winkl.
Lepidesmia Klatt
Lepidaploa
Leontopodium (Pers.) R.Br. ex Cass. - edelweiss
Leontodon - hawkbit
Lemooria P.S.Short
Lembertia Greene
Leiocarpa

Leiboldia Schltdl. ex Gleason
Leibnitzia Cass. - sunbonnets
Lecocarpus Decne.
Layia Hook. & Arn.ex DC - tidytips
Lawrencella Lindl.
Launaea Cass. - aulaga
*Lasthenia*Cass. - goldfield
Lasiospermum Lag. - cocoonhead
Lasiopogon Cass.
Lasiolaena R.M.King
Lasiocephalus Schltdl.
Lasianthaea DC
Lapsanastrum
Lapsana L. - nipplewort
Lantanopsis C.Wright
Langebergia Anderb.
Lamyropsis (Kharadze) Dittrich
Lamyropappus Knorring & Tamamsch.
Lamprocephalus B.Nord.
Lamprachaenium Benth.
Lagophylla Nutt. - hareleaf
Laggera Sch.Bip. ex Benth.
Lagenophora Cass. - island-daisy
Lagascea Cav.
Laestadia Kunth ex Less.
Laennecia Cass. - laennecia, laennicia
Lactucosonchus (Sch.Bip.) Svent.
Lactuca L. - lettuce
Lachnospermum Willd.
Lachnorhiza A.Rich.
Lachnophyllum Bunge
Lachanodes DC.

M

Myxopappus Kallersjo
Myripnois Bunge
Myriocephalus Benth.
Myriactis Less.
Myopordon Boiss.
Mycelis cass.
Mutisia L.f.
Muschleria S.Moore
Munzothamnus Raven
Munnozia Ruiz & Pav.
Mulgedium Cass.
Msuata O.Hoffm.
Moscharia Ruiz & Pav.
Morithamnus R.M.King, H.Rob. & G.M.Barroso
Moquinia DC.
Moonia Arn.
Montanoa Cerv.
Monoptilon Torr. & A. Gray - desertstar
Monopholis S.F.Blake
Monolopia DC
Monogereion G.M.Barroso & R.M. King
Monoculus
Monenteles Labill.
Monarrhenus Cass.
Monactis Kunth
Mniodes (A.Gray) Benth.
Miyamayomena
Misbrookia
Miricacalia Kitam.
Minuria DC.
Millotia Cass.
Milleria L.
Mikaniopsis Milne -Redh.
Mikania Willd. - hempvine
Microspermum Lag.
Microseris D.Don - silverpuffs, yam daisy
Micropus L. - cottonseed
Micropsis DC.
Microliabum Cabrera
Microgynella Grau
Microglossa DC.
Microcephala Pobed.
Micractis DC.

Mexianthus B.L.Rob.
Mexerion G.L.Nesom
Metastevia Grashoff
Metalasia R.Br.
Melanthera Rohr - squarestem
Melanodendron DC.
Melampodium L. - blackfoot
Megalodonta Greene - watermarigold
Mecomischus Coss. ex Benth.
Mausolea Poljakov
Mauranthemum Vogt & Oberpr.
Matudina R.M.King
Mattfeldia Urb.
Mattfeldanthus H.Rob. & R.M.King
Matricaria L. - mayweed
Marticorenia Crisci
Marshalljohnstonia Henr.
Marshallia Schreb. - Barbara›s buttons
Marasmodes DC.
Mantisalca Cass.
Malperia S.Watson
Malmeanthus R.M.King
Malacothrix DC - desert dandelion
Mairia Nees
Madia Molina - tarweed
Macvaughiella R.M.King
Macropodina R.M.King
Macronema Nutt. (synonym of *Ericameria* Nutt.)
Macroclinidium Maxim.
Macraea Hook.f.
Macrachaenium Hook.f.
Macowania Oliv.
Machaeranthera Nees - goldenweed, tansyaster
Macdougalia A.Heller

N
Novenia Freire
Nouelia Franch.
Notoseris C.Shih
Notoptera Urb.
Notobasis (Cass.) Cass. - Syrian thistle
Noticastrum DC.
Nothocalais Greene - prairie-dandelion
Nothobaccharis R.M.King
Nolletia Cass.
Nipponanthemum (Kitam.) Kitam.
Nikitinia Iljin
Nidorella Cass.
Nicolletia A.Gray - hole-in-the-sand
Nicolasia S.Moore
Neurolakis Mattf.
Neurolaena R.Br.
Nestotus
Nestlera
Nesomia
Neotysonia Dalla Torre & Harms
Neopallasia Poljakov
Neonesomia
Neomirandea R.M.King
Neojeffreya Cabrera
Neohintonia R.M.King
Neocuatrecasia R.M.King
Neocabreria R.M.King
Nemosenecio (Kitam.) B.Nord.
Nelsonianthus H.Rob. & Brettell
Neblinaea Maguire & Wurdack
Nauplius (Cass.) Cass.
Nassauvia Comm. ex Juss.
Narvalina Cass.
Nardophyllum (Hook. & Arn.)Hook. & Arn.
Nanothamnus Thomson
Nannoglottis Maxim.
Nananthea DC.
Nabalus Cass.

O
Ozothamnus R.Br.
Oyedaea DC.

Oxyphyllum Phil.
Oxypappus Benth.
Oxylobus (Moq. ex DC.) A.Gray
Oxylaena Benth. ex Anderb.
Oxycarpha S.F.Blake
Outreya Jaub. & Spach
Otospermum Willk.
Otopappus Benth.
Othonna L.
Oteiza La Llave
Otanthus Hoffmanns. & Link
Osteospermum L. - daisybush
Osmitopsis Cass.
Osmiopsis R.M.King
Osmadenia Nutt.
Osbertia Greene
Orochaenactis Coville
Oritrophium (Kunth) Cuatrec.
Oreostemma - aster
Oreoleysera Bremer
Oreochrysum Rydb. - goldenrod
Orbivestus H.Rob.
Opisthopappus C.Shih
Ophryosporus Meyen
Oparanthus Sherff
Oonopsis (Nutt.) Greene - false goldenweed
Oocephala (S.B.Jones) H.Rob.
Onoseris Willd.
Onopordum L. - cotton thistle
Ondetia Benth.
Oncosiphon Kallersjo
Omphalopappus O.Hoffm.
Omalotheca Cass. - arctic cudweed
Olivaea Sch.Bip. ex Benth.
Oligothrix DC.
Oligoneuron Small - goldenrod
Oligochaeta (DC.) K.Koch
Oligocarpus Less.
Oliganthes Cass.
Oligactis (Kunth) Cass.
Olgaea Iljin
Olearia Moench - daisy bush
Oldenburgia Less.
Oedera L.
Odontocline B.Nord.
Odixia Orchard
Oclemena - aster
Ochrocephala Dittrich
Oblivia Strother
Oaxacania B.L.Rob. & Greenm.

P

Pyrrocoma Hook - goldenweed
Pyrrhopappus DC - desert chicory
Pycnosorus - billy buttons
Pycnocephalum (Less.) DC.
Pulicaria Gaertn. - false fleabane
Ptilostemon Cass.
Pterygopappus Hook.f.
Pterothrix DC.
Pteronia L.
Pterocypsela C.Shih
Pterocaulon Elliott - blackroot
Pterocaulon Elliott
Pterachenia (Benth.) Lipsch.
Psychrophyton Beauverd
Psychrogeton Boiss.
Psilostrophe DC - paperflower
Psilocarphus Nutt. - woollyheads
Psilactis A.Gray - tansyaster
Psiadiella Humbert
Psiadia Jacq.
Pseudostifftia H.Rob.
Pseudonoseris H.Rob. & Brettell
Pseudokyrsteniopsis R.M.King
Pseudohandelia Tzvelev
Pseudogynoxys (Greenm.) Cabrera
Pseudognaphalium Kirp. - cudweed
Pseudoclappia Rydb. - false clapdaisy
Pseudocadiscus Lisowski
Pseudobrickellia R.M.King

Pseudoblepharispermum J. - P.Lebrun & Stork
Pseudobahia (A.Gray) Rydb. - sunburst
Pseudobaccharis Cabrera
Pseudelephantopus Rohr (synonym of *Elephantopus* L.) - dog›s-tongue
Psathyrotes A.Gray - turtleback, brittlestem
Psacalium Cass. - Indianbush
Psacaliopsis H.Rob. & Brettell
Proustia Lag.
Proteopsis Mart. & Zucc.ex Sch.Bip.
Prolongoa Boiss.
Prolobus R.M.King
Prionopsis Nutt.
Printzia Cass.
Prenanthes L. - rattlesnakeroot
Prenanthella Rydb.
Praxelis Cass.
Praxeliopsis G.M.Barroso
Porphyrostemma Benth. ex Oliv.
Porophyllum Adans. - poreleaf
Polymnia L.
Polychrysum (Tzvelev) Kovalevsk.
Polycalymma F.Muell. & Sond.
Polyarrhena Cass.
Polyanthina R.M.King
Polyachyrus Lag.
Pollalesta Kunth
Pojarkovia Askerova
Pogonolepis Steetz
Poecilolepis Grau
Podotheca Cass.
Podolepis Labill.
Podocoma Cass.
Podanthus Lag.
Podachaenium Benth. ex Oerst.
Plummera A.Gray
Pluchea Cass. - camphorweed, fleabane
Pleurophyllum Hook.f.
Pleuropappus F.Muell.
Pleurocoronis Benth.
Pleurocarpaea Benth.
Pleiotaxis Steetz
Plectocephalus D.Don
Plecostachys Hilliard & B.L.Burtt
Plazia Ruiz & Pav.
Platyschkuhria (A.Gray) Rydb. - basin daisy
Platypodanthera R.M.King
Platycarpha Less.
Plateilema (A.Gray) Cockerell
Planea P.O.Karis
Planaltoa Taub.
Plagius L›Her. ex DC.
Plagiolophus Greenm.
Plagiocheilus Arn. ex DC.
Plagiobasis Schrenk
Pladaroxylon (Endl.) Hook.f.
Pityopsis Nutt. - silkgrass
Pittocaulon H.Rob. & Brettell
Pithocarpa Lindl.
Pithecoseris Mart. ex DC.
Piqueriopsis R.M.King
Piqueriella R.M.King
Piqueria Cav.
Piptothrix A.Gray
Piptolepis Sch.Bip.
Piptocoma Cass. - velvetshrub
Piptocarpha R.Br. - ash daisy
Pippenalia McVaugh
Piora J.Kost.
Pinaropappus Less. - rock lettuce
Pilostemon Iljin
Pilosella Hill
Picrothamnus Nutt. - bud sagebrush
Picrosia D.Don
Picris L. - oxtongue
Picradeniopsis (synonym of *Bahia* Lag.) - bahia

Picnomon Adans.
Phymaspermum Less.
Phyllocephalum Blume
Phoebanthus S.F. Blake - false sunflower
Philyrophyllum O.Hoffm.
Philoglossa DC.
Philactis Schrad.
Phania DC.
Phanerostylis (A.Gray) R.M.King
Phaneroglossa B.Nord.
Phalacroseris A. Gray - mock dandelion
Phalacrocarpum (DC.) Willk.
Phalacraea DC.
Phalacrachena Iljin
Phagnalon Cass.
Phaeostigma Muldashev
Phaenocoma D.Don
Phacellothrix F.Muell.
Peyrousea DC.
Peucephyllum A.Gray
Petrobium R.Br.
Petradoria Greene - rock goldenrod
Peteravenia R.M.King
Petasites Mill. - butterbur
Petalacte D.Don
Perymenium Schrad.
Perymeniopsis H.Rob.
Pertya Sch.Bip.
Perralderia Coss.
Perityle Benth. - rock daisy
Peripleura Clifford & Ludlow
Pericome A.Gray
Pericallis D.Don - includes Florist›s Cineraria
Perezia Lag.
Perdicium L.
Pentzia Thunb.
Pentatrichia Klatt
Pentanema Cass.
Pentachaeta Nutt. - pygmy daisy
Pentacalia Cass.
Pelucha S.Watson
Pegolettia Cass.
Pectis L. - cinchweed, fetid marigold
Pechuel-loeschea O.Hoffm.
Pascalia Ortega (synonym of *Wedelia*)
Pasaccardoa Kuntze
Parthenium L. - feverfew, guayule
Parthenice A.Gray
Parastrephia Nutt.
Parasenecio W.W. Sm. & Small - Indian plantain
Paraprenanthes C.C. Chang ex C.Shih
Parapiqueria R.M.King
Parantennaria Beauverd
Paranephelius Poepp.
Paraixeris Nakai
Paragynoxys (Cuatrec.) Cuatrec.
Paracalia Cuatrec.
Papuacalia Veldkamp
Pappobolus S.F.Blake
Pamphalea Lag.
Pallenis Cass. (synonym of *Asteriscus*)
Paleaepappus Cabrera
Palafoxia Lag. - palafox
Palaeocyanus Dostal
Pacourina Aubl.
Packera A.Love & D.Love - includes some plants formerly *Senecio*
Pachythamnus (R.M.King) R.M. King
Pachystegia Cheeeseman
Pachylaena D.Don ex Hook.& Arn.

Q

Quinqueremulus Paul G.Wilson
Quinetia Cass.
Quelchia N.E.Br.

R

Rutidosis DC.
Russowia C.Winkl.
Rumfordia DC.
Ruilopezia Cuatrec.
Rugelia Shuttlew. ex Chapm. - Rugel›s Indian plantain
Rudbeckia L. - coneflower
Rothmaleria Font Quer
Rosenia Thunb.
Roldana La Llave - groundsel
Rolandra Rottb. - yerba de plata
Rojasianthe Standl. & Steyerm
Rochonia DC.
Robinsonia DC.
Rigiopappus A. Gray
Riencourtia Cass.
Richteria Kar. & Kir.
Rhysolepis S.F.Blake
Rhynchospermum Reinw.
Rhynchopsidium
Rhodogeron Griseb.
Rhodanthe Lindl. - sunray
Rhanterium Desf.
Rhanteriopsis Rauschert
Rhamphogyne S.Moore
Rhagadiolus Scop.
Revealia R.M.King
Rensonia S.F.Blake
Rennera Merxm.
Remya W.F.Hillebr. ex Benth.
Relhania L›Her.
Reichardia Roth. - brighteyes
Rayjacksonia - tansyaster
Raulinoreitzia R.M.King
Ratibida Raf. - prairie coneflower
Rastrophyllum Wild & G.V.Pope
Raouliopsis S.F.Blake
Raoulia Hook.f. ex Raoul - vegetable sheep
Rainiera Greene
Railliardia
Raillardiopsis Rydb.
Raillardella - silvermat
Rafinesquia Nutt. - California chicory
Radlkoferotoma Kuntze

S

Syreitschikovia Pavlov
Synurus Iljin
Syntrichopappus A.Gray - Fremont›s gold
Synotis (C.B.Clarke) C.Jeffrey & Y.L.Chen
Synosma (synonym of *Hasteola* Raf.)
Syneilesis Maxim.
Synedrellopsis Hieron. & Kuntze
Synedrella Gaertn.
Syncretocarpus S.F.Blake
Syncephalum DC.
Syncarpha DC.
Syncalathium Lipsch.
Symphyotrichum - aster (?)
Symphyopappus Turcz.
Symphyllocarpus Maxim.
Sventenia Font Quer
Stylotrichium Mattf.
Stylocline Nutt. - neststraw, woolly fishhooks
Stuessya B.L.Turner & F.G.Davies
Stuckertiella Beauverd
Stuartina Sond.
Struchium P.Browne
Strotheria B.L.Turner
Streptoglossa Steetz ex F.Muell.
Stramentopappus H.Rob. & V.A. Funk
Stomatochaeta (S.F.Blake) Maguire & Wurdack
Stomatanthes R.M.King
Stokesia L›Her.
Stoebe L.

Stilpnopappus Mart. ex DC.
Stilpnolepis Krasch.
Stilpnogyne DC.
Stifftia J.C.Mikan
Steyermarkina R.M.King
Steviopsis R.M.King
Stevia Cav. - candyleaf
Steptorhamphus Bunge
Stephanomeria Nutt. - wire lettuce
Stephanodoria Greene
Stephanochilus Coss. & Durieu ex Maire
Stenotus Nutt. - mock goldenweed
Stenops B.Nord.
Stenophalium Anderb.
Stenopadus S.F.Blake
Stenocline DC.
Stenocarpha S.F.Blake
Stenachaenium Benth.
Steirodiscus Less.
Steiractinia S.F.Blake
Stebbinsoseris K.L. Chambers - silverpuffs
Staurochlamys Baker
Stanfieldia Small
Standleyanthus R.M.King
Staehelina L.
Stachycephalum Sch.Bip. ex Benth.
Squamopappus Jansen, Harriman & Urbatsch
Spiroseris Rech.f.
Spiracantha Kunth - dogwoodleaf
Spilanthes Jacq. - toothache flower
Sphagneticola O.Hoffm. - «creeping-oxeye»
Sphaeromorphaea DC.
Sphaeromeria Nutt. - chickensage
Sphaeroclinium (DC.) Sch.Bip.
Sphaereupatorium (O.Hoffm.) Kuntze ex B.L.Rob.
Sphaeranthus L.
Spaniopappus B.L.Rob.
Soroseris Stebbins
Sondottia P.S.Short
Sonchus L. - sow thistle, sowthistle
Sonchella Sennikov
Sommerfeltia Less.
Soliva Ruiz & Pav. - burrweed
Solidago L. - goldenrod
Solenogyne Cass.
Solanecio (Sch.Bip.) Walp.
Soaresia Sch.Bip.
Smallanthus Mack.
Sinosenecio B.Nord.
Sinoleontopodium Y.L.Chen
Sinclairia Hook. & Arn.
Sinacalia H.Rob. & Brettell
Simsia Pers. - bush sunflower
Silybum Adans. - milk thistle
Silphium L. - rosinweed
Siloxerus Labill.
Sigesbeckia L. - St. Paul›s wort
Siebera J.Gay
Sidneya E.E.Schill. & Panero
Siapaea
Shinnersoseris Tomb - beaked skeleton weed
Shinnersia R.M.King
Sheareria S.Moore
Shangwua Yu J.Wang, Raab-Straube, Susanna & J.Quan Liu
Shafera Greenm.
Serratula L. - plumeless saw-wort
Seriphidium (Besser) Poljak.
Sericocarpus Nees - whitetop aster
Senecio L. - groundsel, ragwort
Semiria
Selloa Kunth
Selleophytum Urb.
Scyphocoronis A.Gray
Scrobicaria Cass.
Scorzoneroides Moench

Scorzonera L. - [salsify]
Scorzonella Nutt.
Scolymus L. - golden thistle
Sclerostephane Chiov.
Sclerorhachis (Rech.f.) Rech.f.
Sclerolepis Cass. - bogbutton
Sclerocarpus Jacq. - bonebract
Sciadocephala Mattf.
Schumeria Iljin
Schoenia Steetz
Schmalhausenia C.Winkl.
Schlechtendalia Less.
Schlagintweitia Griseb.
Schkuhria Roth - false threadleaf
Schizotrichia Benth.
Schizoptera Turcz.
Schizogyne Cass.
Schistostephium Less.
Schistocarpha Less.
Schischkinia Iljin
Scherya R.M.King
Scariola F.W.Schmidt
Scalesia Arn.
Scabrethia W.A.Weber
Saussurea DC - saw-wort
Sartwellia A.Gray - glowwort
Sartorina R.M.King
Sarcanthemum Cass.
Sanvitalia Lam. - creeping zinnia
Santosia R.M.King
Santolina L. - lavender cotton
Sampera V.A.Funk
Salmeopsis Benth.
Salmea DC - bejuco de miel
Salcedoa Jiménez Rodr. & Katinas
Sachsia Griseb.
Sabazia Cass.

T

Tzvelevopyrethrum Kamelin
Tyrimnus (Cass.) Cass.
Tyleropappus Greenm.
Tuxtla Villasenor & Strother
Tussilago L. - coltsfoot
Turczaninovia DC.
Turaniphytum Poljakov
Turanecio Hamzaoglu
Tugarinovia I/trljin
Tuberostylis Steetz
Tuberculocarpus Pruski
Troglophyton Hilliard & B.L.Burtt
Trixis P.Browne - American threefold, threefold, American trixis, California trixis
Triptilodiscus Turcz.
Triptilion Ruiz & Pav.
Tripteris
Tripolium - sea aster (?)
Triplocephalum O.Hoffm.
Tripleurospermum Sch.Bip. - mayweed
Trioncinia (F.Muell.) Veldkamp
Triniteurybia
Trimorpha - boreal daisy (?)
Trilisa (Cass.) Cass.
Trigonospermum Less.
Trigonopterum Steetz ex Andersson
Tridax L.
Tridactylina (DC.) Sch.Bip.
Trichospira Kunth
Trichoptilium A.Gray
Tricholepis DC.
Trichogyne Less.
Trichogoniopsis R.M.King
Trichogonia (DC.) Gardner
Trichocoryne S.F.Blake
Trichocoronis A.Gray - bugheal
Trichocline Cass.
Trichanthodium Sond. & F.Muell.
Trichanthemis Regel & Schmalh.
Trepadonia H.Rob.
Traversia Hook.f.

Tragopogon L. - goat›s beard, salsify
Tracyina S.F.Blake - Indian headdress
Toxanthes Turcz.
Townsendia Hook. - Townsend daisy
Tourneuxia Coss.
Tonestus A.Nelson - serpentweed
Tomentaurum G.L.Nesom
Tolpis Adans. - umbrella milkwort
Toiyabea R.P.Roberts, Urbatsch & Neubig
Tithonia Desf. ex Juss.
Tilesia G.Mey.
Tietkensia P.S.Short
Tiarocarpus Rech.f.
Thymopsis Benth.
Thymophylla Lag. - pricklyleaf
Thurovia (synonym of *Gutierrezia* Lag.)
Thiseltonia Hemsl
Thevenotia DC.
Thespis DC.
Thespidium F.Muell. ex Benth.
Thelesperma Less. - greenthread
Thamnoseris F.Phil.
Thaminophyllum Harv.
Tetraperone Urb.
Tetranthus Sw.
Tetraneuris Greene - four-nerve daisy
Tetramolopium Ness
Tetragonotheca L. - nerveray
Tetradymia DC - horsebrush
Tetrachyron Schltdl.
Tessaria Ruiz & Pav.
Tephroseris (Rchb.) Rchb. - fleawort
Tenrhynea Hilliard & B.L.Burtt
Telmatophila Mart. ex Baker
Telekia Baumg.
Telanthophora H.Rob. & Brettell
Teixeiranthus R.M.King
Tehuana Panero & Villaseñor
Tarlmounia H. Rob., S.C. Keeley, Skvarla & R.Chan
Tarchonanthus L.
Taraxacum Weber ex F.H.Wigg. - dandelion
Taplinia Lander
Tanacetum L. - tansy, feverfew
Tanacetopsis (Tzvelev) Kovalevsk.
Tamaulipa R.M.King - boneset
Tamania Cuatrec.
Tamananthus V.M.Badillo
Talamancaster Pruski
Talamancalia H.Rob. & Cuatrec.
Takhtajaniantha Nazarova
Taimingasa (Kitam.) C.Ren & Q.E. Yang
Tagetes L. - marigold

U

Ursinia Gaertn.
Urostemon B.Nord.
Urospermum Scop.
Uropappus Nutt. - silverpuffs
Urolepis (DC.) R.M.King
Urmenetea Phil.
Urbinella Greenm.
Urbananthus R.M.King
Unxia L.f.
Uniyala H.Rob. & Skvarla
Uleophytum Hieron.
Ugamia Pavlov
Uechtritzia Freyn

V

Volutaria Cass.
Vittetia R.M.King
Vittadinia A.Rich.
Virgulaster Semple = Aster L.
Vinicia Dematt.
Vilobia Strother
Villasenoria B.L.Clark

Villanova Lag.
Viguiera Kunth - goldeneye
Vigethia W.A.Weber
Viereckia R.M.King
Vieraea Sch.Bip.
Vicoa Cass.
Vernoniopsis Humbert
Vernoniastrum H.Rob.
Vernonia Schreb - ironweed
Vernonella Sond.
Vernonanthura H.Rob.
Verbesina L. - crownbeard
Venidium (synonym of *Arctotis* L.)
Venegasia DC
Vellereophyton Hilliard & B.L.Burtt
Varthemia DC.
Varilla A.Gray
Vanillosmopsis Sch.Bip.
Vanclevea Greene

W
Wyethia Nutt. - mule-ears
Wunderlichia Riedel ex Benth.
Wulffia Neck. ex Cass.
Wollastonia DC. ex Decne.
Willemetia
Wilkesia A.Gray - iliau
Whitneya A.Gray
Westoniella Cuatrec.
Werneria Kunth
Welwitschiella O.Hoffm.
Wedelia Jacq. - «creeping-oxeye»
Warionia Benth. & Coss.
Wamalchitamia Strother
Waitzia J.C.Wendl.
Wagenitzia Dostal

X
Xylothamia G.L.Nesom, Y.B.Suh, D.R.Morgan & B.B.Simpson
Xylorhiza - woody aster (?)
Xylanthemum Tzvelev
Xiphochaeta Poepp. & Endl.
Xerxes J.R. Grant
Xerolekia Anderb.
Xerochrysum Tzvelev
Xeranthemum L.
Xanthopappus C.Winkl.
Xanthocephalum Willd.
Xanthium L. - cocklebur
Xanthisma DC - sleepydaisy

Y
Youngia Cass.
Yermo Dorn - desert yellowhead
Yariguianthus S.Diáz & Rodr.-Cabeza

Z
Zyzyxia Strother
Zyzyura H.Rob. & Pruski
Zyrphelis Cass.
Zoegea L.
Zinnia L.
Zexmenia La Llave
Zemisia B.Nord.
Zandera D.L.Schulz
Zaluzania Pers.

3

Tribulus terrestris: Pharmacological and Nutraceutical Potential

Jyoti Singh*, Jaspreet Kaur, Mansehaj Kaur, Anvi Rana, Prasad Rasane and Sawinder Kaur

Lovely Professional University, Phagwara, Punjab, India

Abstract
Tribulus terrestris (TT) belongs to the Zygophyllaceae family, is a herbaceous noxious weeds grown in warm dry tropics all over the world. TT has been used in both Indian and Chinese systems for the treatment of diseases because of the presence of various phytochemical constituents including flavonoids, alkaloids, steroidal saponins and flavonol glycosides. The roots and fruits TT have been utilized for the preparation of medicine since ancient times. Recent studies have highlighted its application in its extract form for the treatment and prevention of various diseases due to hepatoprotective, anti-spasmodic, anti-bacterial, larvicidal, anti-cariogenic, anti-cancer, anti-inflammatory, and cardiotonic properties. Due to the presence of active components TT is also used for treating sexual disorders. The pharmacological and phytochemical activities of TT make it suitable for the development of functional and nutraceutical food products. The chapter focus on the medicinal and health properties of *Tribulus terrestris* and recommendations for its use.

Keywords: *Tribulus terrestris*, phytochemical, medicinal, toxicity, pharmacological

3.1 Introduction

Tribulus terrestris (TT) has been found in subtropical areas. It has shown many medicinal effects all around the world. It is a flowering plant, under the family of Zygophyyaceae. The plant is small, has a height of 10 to 60 cm. It is a

**Corresponding author*: jyotisingh9377@gmail.com

Prerna Gupta, Navnidhi Chhikara and Anil Panghal (eds.) Harvesting Food from Weeds, (79–112)

hairy shrub, having leaves at opposite sides, and is unequal. The arrangement of fruits is radial in nature, possesses diameter of 7 to 12 mm, and is hard in texture. The root is found cylindrical, fibrous, and branched. Roots and fruits of this plant have been regarded as folk medicine in China from thousands of years [1]. The herb is found in warm temperate and in tropical regions. It covers the area such as southern Europe, Vietnam, Middle East, South Africa, India, China, Pakistan, as well as Sri Lanka. In Sri Lanka, it is grown commonly as weed along the roadside areas. The plant is unable to grow in shade [2]. This actually comes under the category of fast-breeding species. It may have two or three generations in one season only. The soil preference for the plant may show variations from light sandy, medium loamy and heavy as clay soil. In Latin, the word "tribulus" is for "three- pointed a caltrop" in which caltrop means a military weapon, comprising an iron ball along with spike projections. The plant is capable of thriving in desert climate as well as in poor soils. The plant has given with title of "puncture vine" because of the potential of treating the various ailments [3].

Commonly, it is known as "Gokhru." Traditionally, Chinese medicine has been using this extract as treatment for eye impairment, abdominal distentions, edema, or sexual dysfunction. Its roots and fruits have been proved to be effective in the renal calculi, premature ejaculation, and rheumatism and also in piles. The constituents like saponin exhibit antimicrobial and cytotoxic effects [4]. It has the capability to protect against oxidative stress and has antitumor, antifungal, and anthelmintic properties [5]. Plant extractions have been commonly used now for various applications in the medicine. It ranges from skin to the regulation of hormones in human body as anti-bacterial, anti-inflammatory, anti-virus and anti-stimulant [6]. Luteinizing hormone (LH) is stimulated by *TT,* which signals the body to produce more of its testosterone, hence further imparts its essential function in reproductive system [7].

Botanical and ecological description of *Tribulus terrestris*

TT is found in warm temperatures, subtropical regions of southern Europe, west and south parts of Asia. It is widely found throughout Africa and Australia, China, Japan, and Korea. TT responds well in soils which are of light texture; however, it is capable of growing over a wide variety of soil types. *Tribulus terrestris* is found in cultivated crops, overgrazed pastures, roadside areas, lawns, and the areas which are of no use [5]. Preferences for the light soil as sandy, medium as loamy and heavy as clay soil. The plant demands well-drained soil and cannot show growth in shade [2].

Chhatre *et al.* [8] stated about the taxonomical classification as shown in Table 3.1. TT belongs to the family Zygophyllaceae, genus Tribulus and species terrestris Linn. TT is herb, having small silky, hairy and prostrate with stems up to 2 m long, as morphology is defined by Akram *et al.* [4], as shown in Table 3.2. The plant is also called by the English name as "Caltrop". It has been distibuted throughout India in an area of 5,400 m [8].

TT is herb, having small silky, leaves, and are pinnate, short, and approximately comprised of 1.25 cm in length, these are four to eight pairs of leaflets, opposite to each other and are spear shaped as shown in Figure 3.1. It is characterised as small yellow petals, having 8 to 15 mm diameter, comprising thorny fruits. Fruits consist of woody burr, having diameter of 1 cm, consisting of sharp spines of upto 6mm in length. Burr consists five segments, which are of wedge-shaped, whereas each segment

Table 3.1 Taxonomical classification of *Tribulus terrestris* [9].

Kingdom	**Plantae**
Division	Phanerogams
Class	Dicotyledonae
Subclass	Polypetalae
Series	Disciflorae
Order	Giraniales
Family	Zygophyllaceae
Genus	Tribulus
Species	terrestris Linn.

Table 3.2 Common classification of *Tribulus terrestris* [4].

Traditional (name)	**Gokhru**
English name	Caltrop
Botanical name	T. terrestris
Family	Zygophyllaceae
Part used	Seeds
Distribution	5,400m (throughout India)
Morphology	Perennial herb, slender branches, fine silky hair

Figure 3.1 Leaves and flowers of *Tribulus terrestris.*

is divided into two unequal pairs of spines. Seeds are encased in wood like star shaped structure called carpels, which are around 5 to 7 mm in length and 5 to 6 mm in width. Each carpel contains five seeds, and each seed is of 1.5 to 3 mm long and give yellow colour characteristics. Each plant has capacity of containing upto 2000 seeds [5]. Fresh root is slender, fibrous, and cylindrical shaped and give light brown colour [1]. Stomata of TT are anomocytic. Guard cells are not surrounded by any subsidiary cell. Fruits are comprised of dull greenish yellow shade, and possess spines. All the tips join together, forming pentagonal network around the fruit. Fruit possess aromatic properties and is slightly acrid in taste [9].

TT is considered as C4, summer annual by season, able to reproduce by seed, prostrate, and mat- forming in nature. It has been found in TT identified in California is of diploid nature, i.e., 2n=24 and it shows considerable seed dormancy, lasts over fall and winter months along with few seeds and stay dormant for longer duration of time period [10]. Emerging of seeds takes place in spring through summer and germination takes place on the arrival of monsoon season on any type of soil, which is barren in southern Arizona. In regions like Washington and Australia, germination appears to happen in the late spring to early summer when the condition of moist soil is satisfied. The roots of TT are capable of nitrogen fixing nodules. Studies stated that flowering appear within the 3 to 4 weeks of emergence, when there is rise in temperature above 68°F (20°C), regardless of planting date. Flowering begins primarily from July to August. Cross-pollination is done by the insects as coleoptra, dipteral, hymenoptera, Lepidoptera, thysanoptera. It has been found that TT is native to southern Europe. The origin of TT is the Saharan region, which spread across the Mediterranean region. It has been found that import of TT from Mediterranean region to United States on livestock was an accident. It was first spotted in California in early 1930s [11].

3.2 Chemical Composition and Active Constituents Possessed by *Tribulus terrestris*

TT is excellent source of various biological components including saponins, flavonoids, glycosides, alkaloids, and tannins, terpenoids, derivatives of amide, amino acids, and proteins. The potency of these active components is higher in concentrations above the ground area like in fruits and leaves [2]. Among all, the most important constituents of plan fruit are steroidal saponins names terrestrosins A, B, C, D, and E, F-gitonis, desgalacto-tigonis. It has been shown by the sources that geography of saponin content in TT varies. Steroidal saponins and flavonoids has been categorized as crucial metabolites that possess various bioactivities [12].

3.2.1 Saponins

There are 108 types of steroidal saponins, which have been isolated from TT. Of the total, 58 are spirostane saponins, another 50 are known as furostane saponins. Protodioscin and protogracillin category of saponins are well known for reflecting their unique biological activities [1]. The steroidal saponins of this plant have been found responsible for the conduction of biological responses. The response varies with the geographical region of the plant. Studies apply analytical tool as HPLC-ELSD, i.e., high performance liquid chromatography-evaporative light scattering detector and concluded that saponin composition and saponin content, both show variability according to the geographical area. Data shows that there is direct association of saponin composition with the bioactivity to be expressed by the herb [2].

3.2.2 Flavonoids

These are basically derivetives of quercitin, kaempferol, and isorhammnetin. Quercitin, isoquercitin, rutin, quercetin-3-O-gent, quercetin-3-O-gentr, and quercitin-3-O-rha-gent are some flavonoids which exhibit quercitin as basic parent structure. Likewise, for kaempferol, kaempferol-3-O-glu, kaempferol-3-O-gent, kaempferol-3-O-rutinoside share the same basic parent structure [14–16].

3.2.3 Alkaloids

There are some alkaloids which have been extracted from the stems, leaves, and fruits of TT named astribulusamide C, tribulusterine, tribulusin, harmine, Harman, harmmol, N-transcoumaroyltyramine, and N-trans-caffeoylyramine. The nuclear mainly belongs to beta-carboline alkaloids and amide alkaloids [17, 18].

3.2.4 Others

Some other components of TT are organic acids, amino acids, and other chemical substances found in very less amount. The isolation of organic acids from TT is benzoic acid, vanillic acid, 2-methyl benzoic acid, ferulic acids, succinic acid, palmitic acid monoglyceride, docosanoic acid. Alanine and threonine are considered as two main acids. Additional acids present in TT are like uracil nucleic acid, coumarin, emodin, physcion, and 4-ketopinoresinol [18–20].

3.3 Nutritional and Antinutritional Content of Leaves of *Tribulus terrestris*

The medicinal plant has both organic as well as inorganic constituents. The presence of some active chemical components validates vital physiological functions. Some of the essential chemical components are flavonoids, flavonol glycosides, steroidal saponins, and alkaloids, which exert medicinal effect. Also, the activity depend son the concentration, as well as the composition of active components, which again depends on region of origin of plant.

One of the studies conducted by Tkachencho *et al.*, [20] in which they collected four samples of the TT herb, from the different regions of Russia and China. After sampling, they subjected the samples to short term irradiations for 3 to 5 minutes to determine the short lived isotopes (Al, Ca, Cl, and Mg), as well as long-term irradiations for 4 to 5 days to detect long-term radionucleotidesunder the procedures of multielemental ENAA technique (epithermal neutral activation analysis). The studies found the value content for essential macronutrients and trace elements available in the TT leaves when compared to the "reference plant' datain which Potassium was found to be 41.5 mg per gram, calcium as 36.7, magnesium 6.1, and iron 0.3. The trace elements as the coefficient of accumulation of micronutrients are available as Zn, 2.49; Co, 0.90; Cr, 0.48; Al, 0.20; Se, 5.50; and Cd as 8.00; Ba, 11.83; Ni, 0.51 [21].

It was confirmed from the results that potassium and calcium accumulation is found in highest amount from the macronutrient category, obtained in the TT grass samples. 90% of total elemental content was found. Studies also stated that the distribution of elements also depends on geography, time, and conditions for growing plant. There has been observed that accumulation of elements varies with geographical region of the plant [20].

In consideration of nutritive content, anti- nutritional factors have crucial role. Phytate is present in higher amount. It was found in 1214 per 100 gm. High amount of phytate indicates the fact that consumption of leaves may decrease the bioavailability of minerals, especially calcium and zinc. A dosage of 4 to 9 mg/100 mg decrease Iron absorption by 4 to 5 folds in humans of phytic acid. Phytic acid and oxalic acid act as chelating agents for some essential minerals like calcium, magnesium, iron and zinc. Hence they both contributes to the non- bioavailability of these minerals.

The oxalic content of TT was found 3.15 mg/100g, to be lowered as compared to the content present in edible leafy vegetables. Oxalates may be one of the casual factor for the irritation in the mouth. It may interfere with the absorption of calcium by combining to form insoluble salts with them. Oxalates are also responsible for the occurrence of kidney disorders. According to Hassan and Umar [22] 63-159 kg of dried leaves is considered as toxic for the humans.

Hydrocyanic acid (HCN) is found low in quantity. Value, when compared to the raw leaves of celosia argenta and fruits of Nigerian cucurbitaceae species. Studies show evidences that high consumption of cyanide is linked with spastic paraparetis which is locally called "konzo." This results conclude that HCN level in TT leaves are safe and are edible.

Nitrate content was found in T.terrestrisabove the daily recommendations, i.e., 950/100 gm. An alternative of boiling is provided in order to reduce the nitrate content. When the vegetable like spinach, lettuce, onion, pumpkin, eggplant was compared with TT leaves, the content was found in TT was higher than these all. It is important to note that nitrate ion possess acute or low level of toxicity, whereas its conversion from nitrate to nitrite exhibits higher amount of toxicity.

It has been suggested that leaves of the plant should be given to the children with caution. Anti- nutritional factors may reduce to some extent, when leaves are subjected to boiling or may undergo fermentation [23].

3.4 Medicinal Benefits of TT Extracts

The compounds found in tribulus terrestrispossess a variety of biological properties due to their chemical framework and exerts medicinal effects. The components detected are flavonoids, glycosides, phytosterols, tannins, and steroidal saponins in which flavonoids and steroidal saponins are most critical metabolites and are biologically active components.

TT has been utilised as diuretic, tonic as well as aphrodisiac. The plant has shown its potency in the enhancement of sexual behavior in the animal studies. Since long, TT has been used as a constituent in ayurvedic medicine due to its great properties. It has shown inhibitory role in the formation of kidney stone. TT consists of three active phytochemicals; dioscin, protodioscin, diosgenin. All these active components are validated to increase sexual performance and also in the treatment of sexual disorders. The plant also shows hormone- balancing effects, which has shown its effective and promising results in premenstrual syndrome and also in menopausal syndrome [1].

TT being used as a medicinal benefit is whole plant, its roots as well as fruit. The properties shown by TT are diuretic, anti-urolithic, immunomodulatory, anti-diabetic, anti-inflammatory, antispasmodic, anticancer, antibacterial, anti-helmintic, and anticariogenic [9]. It has been proven the as best remedy for urogenital disease, helps in promoting the flow of urine, soothes mucosa, provides aid in the expelling stones and calculi. Studies stated that the plant as well as fruit is utilised in the infusion in the conditions like spermatorrhoea, infertility, dysuria, gonorrhoea, gleet, chronic cystitis, postpartum hemorrhage, renal calculi, gout, and in menstrual disorders. Studies indicate that TT has potential for the treatment of urinary stones, Parkinson's disease, [24] malignant melanoma, eye liver disorders, and also in benign prostatic hyperplasia. Research evidences have been shown that it improves male sexual functionality and athletic performance.

Chinese, have traditionally used this as treatment for liver, kidney, urinary tract disease, also in skin problems, and in the headache. Colic pain, hypertension, and hypercholesterolemia type disorders can be treated well with the regulatory use of this plant extract. Australian register of therapeutic goods has now listed it as reliever of pre-menstrual syndrome, and menopausal syndrome. Bulgarian Olympic athletes have claimed it as one of the reason for their success [2].

3.5 Ayurvedic Importance and Recommendations

Tribulus terrestris has been utilized as folk medicine for various therapeutic conditions. It includes aphrodisiac, analgesic, diuretic, anti-urolithiasis, and as an astringent. The leaves of this plant are believed to enhance menstrual flow, cure the problem gonorrhoea, and also have potential to reduce the gum inflammations [9]. A potent ayurvedic medicine, named as gokshuradi guggul, which is an extract of dried fruit of this herb, is actually responsible for the proper functioning of genitourinary tract, and also in removing urinary stones. Ayurveda has been using this for the treatment of sexual debility, impotence like problems. Ayurveda has contributed number of herbs, which has been proved effective in the most crucial stage of every woman. Gokshuraksheerpaka has been one of the neutraceutical recipes, extracted from the TT which has proven for its great nutritional content for the pregnant women along with the milk. The complication like preeclamptic toxaemia and other chronic complications can be prevented. It has been advised as best choice for the urogenital system disorders.

3.5.1 Gokshura Ksheerpaka

It is an antique recipe from the ayurveda, based on this herb which has health promoting properties, especially effective and proven nutritive for the pregnant women (shiddhinandanmishra). Milk plays a role of media for the formulation. Milk as emulsion is capable of dissolving water soluble, protein soluble as well as fat soluble ingredients, from herb being utilised in the preparation. Milk also showed the property of antacid due to its alkaline nature. Therefore milk possesses the properties of dietetic regimen as well as therapeutic. Prolonged effect of this drug is shown due to the presence of colloidal substance i.e. fat is present in the milk [25]. It is important note that it has beneficiary effects in the improvement of digestibility, improving palatability and it increase the efficacy.

3.6 Biological Activities of *Tribulus terrestris*

3.6.1 Effect on Female Reproductive System

TTis well accepted herb since the ancient periods, it has shown promising effect is both male and female sexual problems. Clinical and experimental studies found the active content saponin (furostanol) along with protodioscin (quantity less than 45%) extracts have been found effective on the endocrine functions of pre and postmenopausal women group. It act as a libido enhancer. Evidences have shown its effectiveness in improving libido, alleviating vasomotor as well as neuropsychic symptoms of menopause. Results have shown that regular consumption of TT significant in decreasing the occurrence and intensity of problems like hot flashes, insomnia, sweating, irritability, depression, and also in loss of sexual interest [2]. Studies have shown that the two-thirds of women when tested after the treatment with *Tribulus terrestris* detected with increased sex drive. Hence, TT has proven effective for the relief of menopausal and PMT (premenstrual tension) syndrome. This leads to regulation of hormones in both men and women, with improved balance. According to *in vivo* studies 95% of the females were reported after the treatment with zero PMS symptoms, 80% with libido and sexual performance [26].

3.6.2 Effect on Male Reproductive System

Tribulus terrestris has been proved as a sexual enhancer. Singh *et al.* [27] studied the dosage for the administration of lyophilized aqueous extract of the dries fruit of TT (LAET), as recommendation dose of 50 and 100 mg/kg of body weight. It, therefore, acts as sexual enhancer and responsible for the management of sexual dysfunction in male rats [27]. There was improvement by dose dependent treatment shown by the LAET, as it was proved as more effective than the chronic administration of LAET. It was recorded that serum testosterone levels also rise. All these evidences show the traditional usage of TT in the sexual male dysfunction. Ethanol extraction of TT has shown protective response against testicular damaging (cadmium- induced). This protective response is carried either by the stimulation of testosterone by leydig cells, or direct response by inhibiting testicular tissue peroxidation through metal chelation or by antioxidant activity. Brown *et al.* [26] stated that active component saponins, namely protodioscin and protogracilin show positive effect on the phenomenon of spermatogenesis, by increasing LH hormones (from pituitary gland), which stimulates the secretion of testosterones. The male hormones called testosterone is responsible for the quality as well as quantity of sperms [28].

Chhatre *et al.*, [9] revealed the pro- erectile effect of TT as mentioned in Table 3.3. The study was conducted on rabbit's corpus cavernosum smooth muscle (done in *ex vivo* conditions). Oral doses were administered of 2.5,5, 10 mg/kg body weight for the time period of 8 weeks. Relaxation effect of 24% and 10 % was found with nitroglycerine and acetycholine respectively

Table 3.3 Pharmacological activities of *Tribulus terrestris.*

Name of the activity	Description	References
Improvement in sexual function	TT activates aphrodisiacs which further improves fertility in men Diet induced with extracts of Mucunapruriens and Ashwagandha along with TT were fed to rats for 8 weeks. The result showed that extract of TT has increased the level of testosterone in male rats. Ethanolic extracts of TT has improved spermatogenesis by changing tubular compartment of testes, increase in tube length, tubular volume and height of seminiferous epithelium.	[9] [29] [30]
Treats urinary diseases	The aqueous of TT has shown the protective capacity against urolithiasis Protein extracted from TT extract decreases the cell injury imparted by oxalate during kidney stone problem.	[48] [47]
Anti- diabetic action	TT has potent action against defective secretion of insulin. The constituent like saponins has been proved for its inhibitory action against the α-glucosidase. It has great potential in improving the symptoms of insulin dependent diabetes (IDD).	[1]
Role in improving Cardiovascular health	It has shown its effect in the myocardium protection. TT is validated to protect against myocardial ischemia. It has shown improvement in heart function rate, coronary flow and further increase in the adenosine triphosphate (ATP) activity in conditions like myocardial ischaemia.	[2]
Anti- fungal potential	Capable of microorganism destruction, has potential to destruct he cell membrane.	[17]

in the corpus caverrnosum smooth muscle tissues. Reason behind this relaxation effect was the releasing nitric oxide from the endothelium and nitregic nerve endings, which further act as an aphrodisiac.

Sahin *et al.*, [29] studied the effectiveness of this herb on sexual functionality, serum biochemical parameters, on oxidative stress and also levels of NF-kB, Nrf2, and HO-1 in the reproductive tissues. Male rats were taken, weight approx 180 gm total thirty five in number. These rats were kept in household cages, in controlled conditions, humid environment (55%). The 12; 12 dark- light mode was enabled at 22°C. They divided this total number in five groups as 1. Negative control: rats fed with standard diet+ tap water) 2.Positive control: rats were fed with standard diet+-silenafilciterate treatment) 3 Rates were fed with standard diet + Mucuna pruriens treatment) 4. Rats were fed with standard diet+ *Tribulus terrestris* 5. Rats were treated with standard diet +Ashwagandha group. The extraction was done, and treated orally either with Mucuna, Tribulus, or with Ashwagandha according to the concentration of 300 mg/kg of body weight for the period of 8 weeks. Sperm count was determined by hemocytometer. Sperm morphology as well as viability was tested, on the basis of which scoring was done as 100 to normal sperm, and 0 to abnormal ones. ELISA test was performed for the measurement of biochemical parameters. Western blot analysis test was an aTTempted to know protein status (NF-Kb, Nrf2, and HO-1). Out of all these, *Tribulus terrestris* was proved with its significant activity. It has been found that serum testosterone levels were recorded increasing in the TT and ashawagandha group, when compared with control group. *Tribulus terrestris* has the potential to reduce the levels of NF-kB, to increase the Nrf2 and HO-1 levels, from all of the rest groups. They concluded from the study that *Tribulus terrestris* shows more bioavailability from Ashwagandha extract, when compared with Mucuna extract.

Oliveira *et al.* [30] performed an experimental study on 20 male rates, weighed 295 gm which were taken from the central animal facility of federal university of lavras (UFLA). They divided five rats in each group. Four groups as G1, G2, G3, and G4. G1 was treated with distilled water; rest other were treated with ethanol, haxanicand methanol respectively at the concentration of 42 mg/kg/day. Morphometric analysis was performed on right testicles using 70% ethanol, slides were performed and studied. The average from tube diameter, height of seminiferous epithelium was estimated. It was concluded that methanol fraction of TT was recorded higher in the estimation of same when compared to the other control groups.

3.6.3 The Potency of *Tribulus terrestris* in Female Sexual Dysfunction

In sexual desire or feeling the pain during a sexual intercourse, Humulus lupulus and *Tribulus terrestris* were suggested for the abovementioned disorders. By investigation on these plants, they suggest that they might be beneficial as a possible substitute or an equivalent beneficial procedure of species of plant that has been tested during this method and desire to treat the sexual dysfunction of female [31].

Tribulus terrestris as a medicinal remedy appears to be advantageous for aphrodisiac desire, a condition that increase a sexual desire in numerous ways by conducting an experiment on human and animal. Another study was conducted as a randomized double-blind placebo-controlled trial, it is a medical study combining a human as a participant in which participants do not know who is getting what kind of treatment and placebo, a treatment that have no therapeutic value are given to control the group, and this method is done to assess the effectiveness and safety of TT in a women who is suffering hypoactive sexual disorder or inhibited sexual kind of disorder which is a dysfunction of sexual and is stated as a lack of non-appearance of sexual desire were automatically allocate in a given extract of TT (7.5mg per day) or 4 weeks of placebo. Therefore, by conducting this kind of method, a study revealed that 30 women those were in placebo group and thirty women were in drug group which completed the study. At the end of the fourth week of placebo in a group of TT group, they experienced an important improvement in their total sexual function index of female that is desire (p) was less than 0.001(1 in a thousand), in arousal ($p=0.037$), in lubrication and satisfaction (p is less than 0.001), and in pain state ($P=0.041$) were domains of sexual function index of female. Usually, the side effects were alike to two groups who were in drug and placebo. However, TT may successfully and safely improve a desire in women with inhibited sexual desire disorder [32].

3.6.4 The Potency of *Tribulus terrestris* in Male Sexual Dysfunction

Male sexual dysfunction also known as erectile dysfunction (impotence) is a condition in which a man cannot get sufficient potential for sexual intercourse. Several studies of the distribution have been stated on sexual problems that are common in society, that is erectile dysfunction, sexual complaint especially occur for about 30% to 40% in men. In most of the cases, erectile dysfunction was originally stated as a psychological disorder but currently consider as a primary organic complication [33, 34].

The extracts of *Tribulus terrestris* importantly improve an erectile function by conducting a study for an evaluation on effectiveness of TT in male dysfunction in placebo-controlled group and an importantly improvement was seen in an overview state of sexual function among those who were treated with an extract of TT. The purpose of this study was to compare the effectiveness of the plant extract of TT in difference which was associated with placebo group, for the treatment of men who were suffering with erectile dysfunction and with the presence or the absence of hypoactive sexual desire disorder (lesser sex drive) and also was to monitor the security profile of the drugs. The second purpose of this study was to assess the phase of lipids in the blood during the treatment. This study was included with 180 males who were aged between 18 and 65 years of old with the presence of moderate or less erectile dysfunction and with the presence and absence of hypoactive sexual desire disorder. Moreover, 90 were automatically randomized to TT group and rest 90 was associated to placebo group. Patients were not only suffering from erectile dysfunction and hypertension but also from a diabetic mellitus and metabolic syndrome which was also included in this study. In the trial, a medicine (herbal) was used (Tribestan, Sopharma AD), each Tribestan tablets contained (250 mg) active component of *Tribulus terrestris*, a herbaextraciumsiccum in the ratio of (35 to 45:1) which was standardized to levulosan saponins which was not less than 112.5 mg. Each patient received orally 3 times with two coated tablets regularly after the meals only after the treatment of 12th week and later after conducting a questionary by International index of erectile function, 86 patients who were in each group was satisfied and there was an improvement in TT group as compared to placebo group, where P is less than 0.001. The only difference in TT and placebo was 2.70 for the intention to treat population (randomized patients). In favor of TT, a statistically important difference between TT and placebo was revealed for intercourse satisfaction (p=0.005), organism function with (p=0,0325), sexual desire with (p=0.0038) values were observed. Consequently, in the 12th week of treatment period, sexual function desire significantly improved in with the presence of TT as compared to placebo group who were with moderate or less erectile dysfunction. TT was generally well authorized for the treatment of erectile dysfunction [35].

Infertility in human is described as the failure to achieve a pregnancy after 365 days or more of daily unsafe sexual intercourse [36]. Moreover, infertility influences the society at high and ultimately connected with psychological and social issues for the patient. Worldwide, approximately around 40%, the male factor is involved in all the infertility cases. Many sources of male infertility are estimated as an unknown cause and diagnosis leads to be graphic, often resulting in noneffective medical perspective for the treatment. Semen is a multiplex fluid composed of cellular elements and plasma, which is derived from the prostrate and seminal vesicles [37]. The parameter of semen is separated into analysis of macroscopic as a quality of semen on the basis of color, viscosity, pH and volume, and into analysis of microscopic as a quality of spermatozoa on the basis of concentration, motility, and viability. These processes of parameter may be affected by many phases like certain kinds of medication, pollution, products of toxic, and smoking as well as the consumption of alcohol in excessive form [38]. There are a few data to support a positive effect on oral TT objection on male sexual dysfunction [39] The extract of *Tribulus terrestris* on the basis of oral objection increased the erectile function in rabbit tissue (tissue of penile) [40] and in older rats, extract of TT in oral form has increased the sexual activity, number of sperm, serum testosterone in a quantity dependent manner [25]. The objection of TT increased in dihydrotestosterone (hormone with powerful androgenic properties) and testosterone which was conducted in rabbits, rats, and rhesus monkey [41].

Hence, the main focus of this study was to evaluate to know whether the treatment with regular oral objection of TT extract in men infertility factor affects the quality of semen or not and also the androgen related physiological parameters of lean muscle mass and body fat index. 65 male patients were conducted in this clinical study for about 12 weeks of total duration of time, and the selection of patient under the inclusion criteria were around 18 to 50 years of age with no disorder of hormonal or diagnosed with infertility or altered with sperm parameters; according to the World Health Organization (2000). The phases were that the patients who used anabolic steroid along with the hepatic abnormalities were observed by the presence of liver enzymes on blood sample estimation and also those who were performing in another clinical study. All the men who were participating underwent an initial evaluation with a day 0, where they started the treatment by using the extract of TT in the form of a prepared ill, (*Androsten*) (Herbarium LaboratorioBotanico Ltda., Brazil). Each prepared pill contains the extract of 250mg including 37.5 mg that is 15% of steroidal saponins (also called as *protodioscin*). From the day 1, the men were ruled to take a pill orally after every 08 hour, which consequently provided 112.5 mg of steroidal saponins every day. Safety was provided from 0 to 84 days of treatment and following exams were also performed like complete blood count, test of liver function, test of urea, creatinine and uric acid but no differences were revealed. Moreover, participants were also assessed for weight, lean muscle mass, and body fat index.

The result of this study was proposed that the treatment with the extract of *Tribulus terrestris* can expand the level of serum of dihydrotestosterone concentration and improved the level of serum count and sperm movement motility, but not in men with a less quality of sperm in a reproductive technology program and also it was observed that there was an consequence increase in free testosterone level and positive impact was also observed after the 90 days of treatment with the same amount of extract of TT which were used in a clinical study.

3.6.5 Effect on Urinary System

Diuretic activity: Chhatre *et al.*, [9] obtained the comparative assessment of the diuretic performance in various extracts of TT fruits in the experimental animal studies. The oral dose of TT extract of 5g/kg has exerted positive dieresis in the male rates. It has also emphasised the contractile activity on guinea pig ileum [42]. Recent evidences *in vitro* studies human urine of have shown that diuretic properties of TT can be effective in the prevention of urinary stone formation [43].

Antiurolithiac activity: Anand *et al.* [44] observed an ethanolic extract of TT fruit, shown as dose- dependent protection against the uroliths, which got induced by glass bead implantation method in albino rats. The extract delivers extraordinary protection against the calculogenic material around glass bead, as well as protects leucocytosis and contributes in the elevation in serum urea levels [2].

Shirfule *et al.* [45], in their studies, stated that GOX i.e. glycolate oxidase is a principal enzyme in the pathway of oxalate synthesis. It is capable of converting glycolate to glyoxylate via oxidation and leads to the final product as oxalate. The antiurolithic activity of TT has been regarded as GOX inhibitor. The active components, like quercitin and kaempherol, are considered not only as non-competitive, but also competitive inhibitors of GOX.

The study investigations provided by Kavitha and Jagadeesan [46] stated the effectiveness of methanolic fraction of TT fruit extract. The study attempts were carried out on the kidney tissues of mice, with mercury intoxication. The results were carried out that oral administration of methanolic fraction has effective response against mercuric chloride toxicity.

Aggarwal *et al.* [47] investigated the potency of TT on oxalates (calcium oxalate). They examined the capability of TT on the injury, induced by oxalate in the renal tubular epithelial cells of rats as mentioned NRK 52E in their studies. The cells were subjected to incubation in 1 mm of sodium oxalate as sample using various concentrations for the time period of

72 hours. Cell injury was detected by estimating the cell viability using trypan blue (cytotoxicity) and monitoring the lactate dehydrogenase (LDH) leakage into medium. It was concluded that TT extract has potential to inhibit the calcium oxalate crystalisation *in vitro*. This extract has been validated to exhibit cytoprotective properties toward the renal tubular epithelial cells by lowering the LDH leakage, and by increasing the cell viability. This study thus contributes experimental evidence for TT as a therapeutic agent in a condition like urilithiasis.

Sharma *et al.* [48] has worked on exploring the capability of gokhru as anti-urolithiac. This study was done on quantitative estimation basis, via applying analytical methods. Dried and powered form of gokhru fruit extract (about 500 gm) was taken and filtered. Percentage of yields under different polarity fractions (using hexane, toluene, DCM, i.e., dichloromethane, n-butanol, and water) was estimated and stores at 4°C. The extract samples were subjected to *In vitro* and *In vivo* anti- urolithiac activity at 100 to 1000 µg/Ml via aggregation assay in rat plasma, as well as in synthetic urine. An attempt to cofocal microscopy, nucleation assay was performed for 30 minutes. Quantitative idea was taken by performing the HPLC method of constituents, like catechin, rutin, gallic acid, and quercitin in the fraction samples and in mother extract. After the analysis, it was found that n-butanol fraction exhibit as potent antiurolithiac phytopharmaceutical.

3.6.6 Effects on Cardiovascular System

A clinical study trial was conducted on 406 coronary heart disease patients, and treated with saponins, the active constituent of TT. It has been found in the conclusion that total effective rate of remission angina pectoris was found 82.3%. The effective rate of ECG performance 52.7% was found even greater than the control group which was 35.8% [5] Saponin component present in the herb is responsible for the improvement of coronary circulation, although the saponin content may vary in terms of geographical location of the plant. Thus due to saponin content present in this herb has-shown the positive results in ECG myocardial ischemia [2] as mentioned in Table 3.3 of pharmacological activities of TT.

3.6.7 Effect on Diabetes Mellitus

Lamba *et al.* [49] observed TT ethanolic extract at 2 g/kg body weight show protective effect in streptozotocin induced diabetic rats, by the action of inhibitory oxidative stress. The TT extract exhibit approx 70% of inhibition of alpha glucosidase at 500 micro ml, by substrate as maltose and

100% inhibitory effect of aldose reductase at 30 micro ml by diglyceraldehyde as substrate [49]. It has been stated by the studies that TT has significant potential inhibitory effect against gluconeogenesis, and plays an influencing role in glycometabolism. It has been engaged in the reduction of triglycerides and total cholesterol levels [50]. Zhang *et al.*, in their studies, conducted on rat intestine, stated that TT extract (majorly saponin) are proved for decreasing blood sugar after the meal intake. Experimental studies revealed that GSTT, i.e., gross saponins of TT has shown great potential in reducing the postprandial blood glucose levels in rats (normal and type 2 diabetic rats) by administering sucrose intragastrically. The water extract of TT has also shown antidiabetic property. It has been also concluded from one study that the parameters like postprandial glucose (2 hours), glycosylated hemoglobin, and lipid levels in diabetic women, when treated with TT extract for 3 months (on dosage of 1000 mg per day) were found lowered when compared with placebo group.

3.7 Pharmacological Profiling of *Tribulus terrestris*

3.7.1 Antioxidant Activity

Tribulus terrestris indicates effectual antioxidant activity in a concentration-dependent manner by water, superoxide scavenging activity, and also 2,2-di-(4-tert-octpylphenol), 1-picrylhydrazyl (2,2-diphenyl-1-picrylhydrazyl) as well as Ferric reducing antioxidant capacity [51]. The experiment demonstrates that the antioxidant level effect of glutathione S-transferase is exceptional and that could be capable to improve superoxide dismutase activity level and malondialdehyde content for chronically more intraocular pressure which is the pressure of the fluid inside the eye, in rabbits. However, compared with the ethanolic extraction (single stream process), butanol extract (1 mg mL^{-1}) was high in saponin (triterpene glycosides) and had a notable extinguishing of nitricoxide (NO), (90.30%), hydroxyl radicals (90.02%), and also hydrogen peroxide radical's agent (89%) [52]. Diosgenin, a crystalline compound, from the hardened part of the soft tissues of *Tribulus terrestris*, was found to have a prominent antioxidant activity [53].

3.7.2 Anticariogenic Activity

Tribulus terrestris was validating to have an anti-carious effect. The anticarious effect of TT was assessed for inhibiting *Streptococcus mutans* bacteria rise in an amount of dependent manner. Baik *et al.* [54] recognized that extract of ethanol of fruits of TT of (0.1-0.5 mg per ml) possess important anticariogenic activity against *Streptococcus mutans* (voluntary anaerobic, gram-positive coccus which is found in human oral cavity), the pathogens which is liable for dental carries. *Streptococcus mutans* is a main oral pathogen that roots dental carries. The rise, production of acid, sticking and water insoluble glucan combination of Streptococcus mutans were actually appreciably obstruct in the presence of ethanol extraction of TT. For the time being, the extract of *Tribulus terrestris* extinguishes the adherence of saliva-coated hydroxyapatite (SHA), which restore the teeth and reserved the formation of water-insoluble glucans [16].

Tribulus terrestris belonging to Zygophyllaceous family is broadly used because of its medicinal properties. Another study was undertaken to evaluate the anticancer system of TT on Michigan cancer foundation-7 breast cancer cells [55]. Toxic effect of the herb was evaluated by 3-(4,5diethyl-thizaol-2-yl)-2,5-diphenyl tetrazolium bromide colorimetric assay. The apoptotic prospective was assessed by through the fragmentation of DNA and caspase activity. The result in MTT phase revealed that methanolic and saponin extract of the seeds and roots of TT were toxic to MCF-7 cells (the cell of human breast cancer which line with estrogen, progesterone). The study on peripheral blood mononuclear cells demonstrates that the extracts of TT was not malignant cells and also were not toxic. The resulted in fragmentation of DNA and the induction of apoptosis was done by the treatment of human breast cells with the methanol of seed and root and saponin extract of TT. The extract of TT also results in causing the importance increase in caspase 3 activity (also called as executioner caspase). Hence, the result suggests that the extract of TT might exert the activity of anticancer by both the process of extrinsic and intrinsic apoptotic pathway [56].

3.7.3 Anti-Inflammatory Activity

The methanolic extract of *Tribulus terrestris* influences an anti-inflammatory and antibacterial activity, pathway [57]. However, recognized that ethanolic extract of TT inhibited the appearance of cyclooxygenase-2 (an enzyme that is present in humans which is encoded by the gene, PTGS2) and nitric acid synthase in lipopolysaccharide stimulated RAW 264.7 (macrophage-like, Abelson leukaemia virus transformed cell). Also, it suppressed the appearance of proinflammatory cytokines, like tumor necrosis factor alpha (a cytokine) and interleukin (IL-4) (a cytokine which generate differentiation of helper T cells to T-helper type 2 cells) in macrophage cell. Consequently, the ethanolic extract of TT inhibits the appearance of mediators linked to inflammation and appearance of inflammatory cytokine, which actually has an advantageous effect on various inflammatory conditions [58].

Rathi *et al.* [59] probed that anti-inflammatory activity of steroidal saponins is isolated from methanolic extract of fruits of *Tribulus terrestris* and tested that the fruit of TT contains an active principle like phytoesteroides, flavonoids, glycosides steroidal saponins of furostanol kind, which develop anti-inflammatory effect. The inhibitors of prostaglandin biosynthesis and the development of nitric acid have been termed as potential anti-inflammatory agents and inhibition of COX-2 activity [58].

Moreover, the ethanolic extract of *Tribulus terrestris* and N-trans-p-caffeoyl tyramine isolated from TT had marked anti-inflammatory [60]. Ethanolic extract of TT and N-trans-p-caffeoyl tyramine reserve the development of nitric acid (NO), tumor necrosis factor alpha (an inflammatory cytokine), interleukin-6 (acts as an anti-inflammatory and proinflammatory myokine) and Interleukin-10 (anti-inflammatory cytokine) in lipopolysaccharides stimulated RAW264.7 cells (macrophage-like, Abelson leukaemia virus transformed cell) in a quantity-dependent manner. Additionally, N-trans-p-caffeoyl tyramine distinctly suppressed the appearance of cyclooxygenase and the development of prostaglandin E2 or also known as dinoprostone, through reducing p-JNK (c-jun N-terminal kinase)expression.

As the test method was done on rat paw, the methanol extract of M. denticulate of (200-400mg kg^{-1}) indicate amount dependent of raw paw area in carrageenan – reduced rat paw edema model. The TT extract and diclofenac sodium (COX1 and 2 inhibitor) which was injected 30 minutes earlier to carrageenan. The outcome showed that both the drugs can diminish the volume of rat paw volume, 1 to 4 hours after injection of carrageenan by inhibiting the free of serotonin, histamine level in early stage. Moreover, the anti-inflammatory effect of 400 mg kg^{-1} of *Tribulus terrestris* is same with that of 20 mg kg^{-1} of diclofenac sodium [61].

3.7.4 CNS Activity

Deole *et al.* [62] established that Swiss Albino mice exhibited anxiolytic activity (medication that reduces anxiety), and antidepressant on management of 260 mg per kg dose of Rasayana Ghana tablet (rejuvenation therapy) consisting of three vigorous well-accepted rejuvenator herbs,-Tinosporacardifolia (heart-leaved moonseed) stem, Embilica officinalis (Indian gooseberry) fruit and TT (root and fruit), present in minor quantities in the tablet. It was demonstrated that harmine (hamala alkaloid and beta-carboline) present in TT, which is one of the important active constituents that give to the above-mentioned activities. Harmine is an inhibitor of monoamine oxidase (type of enzyme that helps neurons fire throughout your body), which help to enlarge the level of dopamine in the brain.

Another study revealed that the gross saponin of *Tribulus terrestris* improved gain memory disorder, incorporates memory disorder and recovered memory disorder [63]. The result of water extract of TT fruits on learning and ability of memory in rodents was estimated by recording the time of reaching the reward chamber in the Hebb William Maze (behavioral task used for studying the working of the memory, and it consists of square area with internal wall which is moveable) and transfer latency (part of the drug that lowers the inflammation in the body and treat allergic rhinitis and other conditions). The result revealed that the water extract of TT fruits importantly decreases the time of the arrival at the method of Hebb William Maze in a dose-dependent manner [64].

3.7.5 Larvicidal Activity

The petroleum ether extract (the petroleum fragment composed of aliphatic hydrocarbons which is boiled in the range 35–60°C) of the leaves of *Tribulus terrestris* demonstrate better larvicidal activity in opposition to the third instar larvae (Production phase of arthropods) and adults of the mosquito, *Ades aegypti* (yellow fever mosquito) which is the aim of fever, dengue with LC50 (the aggregation of the chemical in the water or air that will diminish 50% of the test animals with a single exposure) of 64.4 ppm as differ to the crude ethanol and acetone extracts [65].

Another study revealed that methanolic extract of *Tribulus terrestris* in wormicidal activity by slows down the spontaneous motility (paralysis) and cause the death with the lower doses. The result was similar with the Albendazole or also called as anti-worm medication [66]. Additionally, *Tribulus terrestris* resulted high larvicidal activity. *Anopheles stephensi, Aedes aegypti, Culex quinque-fasciatus* have been observed as the primary vectors of dengue fever, malaria, and lymphatic filariasis. The larvicidal potential of *Tribulus terrestris* was estimated by calculating the mortality percent of *Anopheles stephensi, Ades aegypti* and *Culex quinque fasciatus*, and the result revealed that fruits were more powerful than the leaves on the basis of larvicidal activity [67].

3.7.6 Hyperlipidemic Activity

The test method was done on healthy mice which were established that *Tribulus terrestris* importantly reduces the triglyceride (three fatty acids) and total level of cholesterol [57]. Moreover, extracts of the water of a fruit was in action on cat papillary muscle, rabbit hearts, and frog. The anti-hyper lipidemic prospects of fruits of TT [68]. Tuncer *et al.* [69] recognized the impact of fruits of TT extract on liquid state profile and endothelial formation in production of atherosclerosis lesions in the aorta of rabbits on a diet given in a high cholesterol level.

The extract of fruits in aqueous form of *Tribulus terrestris* was estimated that for their hypolipidemic level in Wistar albino rats, a quantity of 580mg per kg of extract was revealed to reduce the cholesterol induced Hyperlipidemia, with a decline in cholesterol, three fatty acids, very low-density lipoprotein, low-density lipoprotein, and atherogenic index and at last high density lipoprotein phases in the blood. The activity of hypolipidemia may be because of presence of phenolic compounds resulting in increased level of lipoprotein lipases in the muscles and decline activity in the adipose tissue, moreover presenting that the plasma triglycerides are utilized for production of energy by the muscle and not for the storage of energy by the adipose [68].

Another study was done to investigate the comparative effect of ethanol extract of Cissus quadrangular (plant of the grape family and grows in Africa and parts of Asia) and also the extract of TT on lipid profile of rats. The extract of plant was orally subjected to rats with high fats feds in perspective of ghee with normal feed (2:1) at a dose of 583 mg per kg body weight for 1 month, and at the final, the animal were scarified and analysis of biochemical of serum was performed at the level of cholesterol, phospholipid, high-density lipoprotein and triglyceride, but there was an increase of cholesterol level, phospholipid, high-density lipoprotein, and triglyceride, which were brought to somewhat at normal state by the oral administration of Cissus quadrangular and TT. The examination of heart and liver revealed positive. The result revealed that the oral administration of Cissus quadrangular and TT for 1 month has usually antihyperlipidemic effect against the rats with high fat feds [70].

3.8 Mechanisms of Action of *Tribulus terrestris*

The extract of *Tribulus terrestris* has been exhibit to stimulate the luteinizing hormone, which is released from the pituitary gland. Although it may also exhibit some peripheral effects as demonstrate by increasing a hair of pubic in some hypogonadal effect subjects (hypogonadism is a situation in which the body does not develop sufficient hormones and play an important role in growth and production of masculine during the stage of puberty (testosterone) or production of sufficient sperm or both). It is theorized that the metabolite of protodioscine (it is a compound of steroidal saponin component which is found in numerous of plant species, particular in *Tribulus terrestris*) that may have a modest androgenic hormone (male hormone). The exact system are still indeterminate and current explanations and speculative at best.

The cinnamic amides which play an important role in inhibition of high tyrosinase in fruits of TT inhibit papain like protease which is an important proteolytic enzyme, plays a role in virus-mediated replication of RNA [71]. *Tribulus* utilize use a protective effect in diabetic rats by inhibiting the oxidative stress and decreasing the levels of glycosylated hemoglobin and cholesterol, in animal model [72]. It modulates the oxidative stress and inflammatory cytokine which is released in a dose-dependent manner by increasing in superoxide catalase, glutathione, and decreases the levels of lipid and glutathione, in diabetic neuropathic pain models [73].

The effects of antihypertensive occur by arterial smooth muscle relaxation and muscle hyperpolarization [74]. An extract of aqueous also demonstrated angiotensin-converting enzyme (ACE) inhibiting activity [75] that help lower blood pressure. Phenolic acid, like chlorogenic acid and caffeic acid, may be responsible for reputed cardioprotective properties [76].

3.9 Effects of Herbal Supplements with Medication Effects

Tribulus terrestris is one of such plant that has been utilized for a significant period in both the Indian and Chinese frameworks of medication, it is a dicotyledonous (also known as dicots) herbal plant of the family of *Zygophyllaceous*. The plant is also said to possess various other pharmacological properties. The extract obtained from the air-dried aerial parts of TT plant contains mainly steroidal glycosides, the major saponin being protodioscin [40]. In investigation on TT on plant item, it was seen that protodioscin delivered a moderate expansion in testosterone, dihydrotestosterone, and dehydroepiandrosterone sulphate levels in primates following bolus intravenous organization of the TT which separate at dosages of 7.5, 15, and 30 mg per kg body weight. It additionally improved libido, sexual action, and intracavernous pressure (IP is an experimental process done in rat model for the observation of penile erection) in following TT extract through orally was given for about 2 months at dosages of 2.5, 5, and 10 mg per kg body weight and had a proerectile impact on the corpus cavernosum smooth muscle (tissue of the penis) of rabbits was given orally for about 2 months at portions of 2.5, 5, and 10 mg/kg body weight [41].

Tribulus terrestris developed numerous effects on cardiovascular system. The rate of heart displays no change in appreciable. However, lesser doses seemed to enlarge rate of heart in frog while higher doses seem to decrease the rate oh heartbeat. In primates, as there was no sufficient transient rate of the heart that were exceeding above the normal resting rate (an increase of 18-20 beats per min) that returned to the basic baseline through the bolus intravenous subjection of TT at doses of around 75., 15 and 30 mg per kg of body weight. The hypodynamic frog heart (A heart of frogs, when state of isolation and cause to spread, passes into a weaken state, in which the force of contraction, and the rate of the conduction of stimulation, are impaired) was stimulated at less doses of extract of TT while high doses of extract developed inhibition. The same dual effect was also observed on the force of contraction of the increased circulatory volume of cat papillary muscle).

An extract of alcoholic, which is produced from whole plant of TT in an air-dried crush form subjected orally developed a strong vasodepressor (an anti-hypotensive agent) in anaesthetized dogs (dogs were not feeling any kind of pain during this process).

Studies organized on animal models and in human trial in the effect of sexuality and reproduction on the basis of medication effects shows that TT containing protodioscin improves numerous aspects of sexuality. TT of around 500mg for 72 hours regularly for about 1 or 2 months) was found improve the level of libido and sexual arousal function in hypogonadal men (testes in men) and also improved the seminologicalindices in the fertility of men. Also, TT was revealed to be effective in opposition to idiopathic oligo-astheno-teratozoo-spermia (OAT, a condition that contains low number of sperm).

In other clinical study, improvement of infertility in male patients was observed which was treated with the *Tribulus terrestris* of 250mg for 72 hours regularly for 1 month and their partners afterward were able toachievepregnancy.Thus,this was assigned to TT, which is increasing the quality of semen (seminal fluid) in terms of movement of sperm motility as also the quantity [77]. Protodioscin is also termed to increase the level of luteinizing hormone and the level of testosterone. dehydroepiandrosterone (androsterone, a precursor hormone). In pigs, given TT of 70 mg per kg for about 10 days were improved the sexual activity and libido, and the effect was more effectiveness in those animals with the less sexual effects.

A few numbers of studies on animal show an important increase in levels of serum testosterone after the *Tribulus terrestris* objection, but this result was only seen in humans when TT was the part of a collaborate supplement management [78]. The traditional medicine of Chinese employs of *Tribulus terrestris* to improve the visual acuity level (ability to detect the shapes and details of things) as an anticonvulsant (drug which suppress the neurons during seizures), whereas the herbal medicine of India takes an advantage of its diuretic (a drug that develops the diuresis and helps in production in urine) and its aphrodisiac properties (that increase sexual desire) [41]. It mostly grew in Eastern Europe region in 1970 once its effects were developed by Bulgarian Olympic weight lifting team and continue to multiply as it is recommended as a legal, effective, and safe to anabolic steroids. Thus, it is usually ingested in the form of extract which is composed of aerial portions of plants and fruits that normally include steroidal glycosides of furastonal type which is found in TT [13, 79]. The system by which the *Tribulus terrestris* can increase the levels of serum testosterone by producing the androgen receptor through the brain and cause the posterior gland of pituitary to release more hormone of luteinizing and, moreover, the production of testes is to synthesize much more testosterone. However, the receptor of androgen stimulates the effects in the brain which will result to misconception of the sex steroids levels of the body thus accepting the body levels which are lower than they are.

3.10 Herb-Drug Interconnection

Herbs-drug interconnection has been a complete breakthrough in the objection of erectile dysfunction and traditionally has known worldwide as an "instant treatment." *Tribulus* also stated as a dietary supplement to improve the sexual desire function and also for body building because of the belief that it is use as testosterone in the body [80]. However, the modern state of the administration of erectile dysfunction to a single cause, a wide number of pharamacological properties are orally consumed or vasoactive agents (which is compose of broad categories state of drugs that have vasoactive effects) is inserted by within the urethra or by within penial to rescue good erection phase.

Experimentally and studies on animal revealed that *Tribulus* properties possess medicinal effects against the high blood pressure, also against for diabetic, infection and cancer [81]. The use of *Tribulus terrestris* has been connected to poor effect in humans and animals and because of this patient with prostate cancer, power hormonal activities should consult their physician before using the *Tribulus*. Normally, TT traditionally used in for chest pain, as a diuretic, tonic, disorder of eye and dizziness. TT as in a form of diuretics may increase the effect of other drugs of diuretic [42] but till now clinical study is not known yet.

In hyper anti-intensive drug, TT has antagonistic converting enzyme and inhibitory activity which is also term as (ACE) may not have an additional hypotensive effect [74, 75], and yet clinical study has not been done yet.

In antidiabetic, *Tribulus terrestris* may have an additive lower sugar glucose level effect, [59] and clinical relevance has not been doneyet. Clopidogrel, may increase the risk of blood clots, ST (stent thrombosis), formation of blood clot inside a blood vessel to the patient who has stent and they follow concurrent use of clopidogrel along with the herbal extract of *Tribulus terrestris* [82]. P-glycoprotein substrate drug, the compound of saponins and phenolic from TT has revealed to inhibit the P-glycoprotein activity which may increase the risk the adverse effect of substrate drug [83].

3.11 Toxicity and Dosage

Researchers have used a numerous of doses to evaluate the effects of *Tribulus terrestris*. Some of the studies have investigated its potential blood sugar lowering the effect where it has been observed to use 1,000 mg per day, whereas research has examined the enhancement of libido and doses from 250 to 1,500 mg per day is recommended [84].

Saponins are the chemical compounds which are present in *Tribulus terrestris,* and they are responsible for its own health benefits. It is often common for TT supplements to contain an estimation of around 45% to 60% of saponins. Significantly, the higher percentage of saponin value means that there should be the use of lower dose because the supplement is quite more in a concentrated form.

Other studies advised the use of dosage relative to body weight, like in some studies, they have used doses of 4.5 to 0 mg per pound that is 10 to 20 mg per kg of body weight. If the person's weight is around 70 kg, 155 lbs then a dose of 700 to 1,400 mg per day is suggested [84].

Tribulus terrestris has been broadly used in medicine for the treatment of urinary stones, circulatory problem and also for skin problems. In Sudan, it is mainly used to remove the urinary stones, so the study was conducted mainly to evaluate the toxicity of TT fruits on rats by using the three different doses in 2 months by allocating the rats into groups of four with each of six rats. Groups of two, three and four were fed with a diet which contained different concentration extracts of TT plants. The weight gain, mortality, and hematological parameter were recorded during the investigation to determine the changes of lipid and pathological changes. Additionally, the antimicrobial activity of methanolic extract of TT fruits was tested against the gram-positive bacteria (*Staphylococcus aureus, Bacillus subtitles*) and gram-negative bacteria (*Salmonella, Escherichia coli*). The result revealed that there were no changes in all the parameters including the serum lipid profile and biochemical parameter as they were not different from the control group. Methanolic extract of TT fruits showed a wide spectrum of anti-microbial activity against the pathogens which were tested [85]. In conclusion, the traits of TT fruits are not toxic and observed to be a safe alternative to be used in the treatment of various kinds of diseases.

There is no sufficient toxicological evidence present to assess the effectiveness use of plant parts of TT in food supplement. In some case, the daily quantity of TT in high amount in food supplements should not exceed the amount which is used for pharamacological activities. The part of the plant which is used whether it is extracted or prepared, the saponin level should be made clear in the supplement [86].

3.12 Conclusion

Tribulus terrestris is capable of treating many ailments. Traditionally, it is being used as medicine in the Chinese culture. Botanical and ecological nature of the plant suggest that it is an herb, which possess silky hairs and is found along the roadside areas, in gardens, or other neglected places. Chemically, it is composed of some potent constituents as saponins, flavonoids, glycosides, alkaloids and tannins, terpenoids, derivatives of amide, amino acids, and proteins. All the constituents, especially saponins are responsible for exerting excellent bioactive properties. *T. terrestris* has shown significant role in sexual dysfunction and act as sexual enhancer, has positive effects on male as well as female reproductive system. Plant has proven effective for the relief of menopausal and premenstrual tension (PMT) syndrome and balanced regulation of hormones. TT also reflects its potential in the prevention of urinary stone formation, effective role in cardiovascular system by enhancing the ECG performance and coronary circulation, as well as positive results recorded for the diabetes mellitus. The plant exhibits anti-fungal properties against the resistant fungi, due to the capability of eliminating the pathogen. It also possesses anti-bacterial action, anti-oxidant activity, anti-inflammatory effects and has controlled CNS activity effects. There are no toxic evidences found yet to question the effectiveness of plant as extract, to be utilised in the food supplements. All over, TT has served well as a natural effective agent in the serious ailments. The regular and proper dosage of TT has exerted positive results with research evidences.

References

1. Zhu, W., Du, Y., Meng, H., Dong, Y., Li, L., A review of traditional pharmacological uses, phytochemistry, and pharmacological activities of *Tribulus terrestris*. *Chem. Cent. J.*, 11, 1, 1–16, 2017.
2. Sivapalan, S.R., Biological and pharmacological studies of *Tribulus terrestris* Linn-A review. *Int. J. Multidiscip. Res. Dev.*, 3, 1, 257–265, 2016.
3. Kostova, I., Dinchev, D., Rentsch, G.H., Dimitrov, V., Ivanova, A., Two new sulphated furostanol saponins from *Tribulus terrestris*. *Z. Naturforschung C*, 57, 1-2, 33–38, 2002.
4. Akram, M., Asif, H.M., Akhtar, N., Shah, P.A., Uzair, M., Shaheen, G., Ahmad, K., *Tribulus terrestris* Linn. A review article. *J. Med. Plants. Res.*, 5, 16, 3601–3605, 2011.
5. Hashim, S., Bakht, T., Marwat, K.B., Jan, A., Medicinal properties, phytochemistry and pharmacology of *Tribulus terrestris* L. (Zygophyllaceous). *Pak. J. Bot.*, 46, 1, 399–404, 2014.
6. Sansebastiano, G.P.D., Benedictis, M.D., Carati, D., Lofrumento, D., Durante, M., Montefusco, A., Piro, G., Quality and efficacy of *Tribulus terrestris* as an ingredient for dermatological formulations. *Open Dermatol. J.*, 7, 1–7, 2013.
7. Neychev, V.K. and Mitev, V.I., The aphrodisiac herb *Tribulus terrestris* does not influence the androgen production in young men. *J. Ethnopharmacol.*, 101, 1-3, 319–323, 2015.
8. Shail, M.D., Kumar, N., Tripathi, A.D., Nutritive estimation of (Tribulus terrestris) GokshuraKsheerpaka: A health supplement for pregnant women. *Int. J. Home Sci.*, 2, 2, 349, 2016.
9. Chhatre, S., Nesari, T., Somani, G., Kanchan, D., Sathaye, S., Phytopharmacological overview of *Tribulus terrestris*. *Pharm. Rev.*, 8, 15, 45, 2014.
10. Washington State Noxious Weed Control Board, 2002, https://www.nwcb.wa.gov/weeds/puncturevine. Assessed on 13th March 2021.
11. Moradi Kor, N. and Bayati Zadeh, J., Physiological and pharmaceutical effects of *Tribulus terrestris* as a multipurpose and valuable medicinal plant. *Int. J. Adv. Biol. Biomed. Res.*, 5, 3, 149–153, 2013.
12. Mitra, N., Mohammad-Mehdi, D., Reza, Z.M., *Tribulus terrestris* L. (Zygophyllaceae) flavonoid compounds. *Int. J. Mod. Bot.*, 2, 35–39, 2012.
13. Liu, T., Chen, G., Yi, G.Q., Xu, J.K., Zhang, T.L., Hua, H.M., Pei, Y.H., New pregnane and steroidal glycosides from *Tribulus terrestris* L. *J. Asian Nat. Prod. Res.*, 12, 3, 209–214, 2010.
14. Qu, N.N. and Yang, S.S., Extraction and determination of chemical constituents of flavonides in *Tribulus terrestris* L. *J. Liaoning Univ. Tradit. Chin. Med.*, 3, 182–183, 2007.
15. Su, L., Chen, G., Feng, S.G., Wang, W., Li, Z.F., Chen, H., Pei, Y.H., Steroidal saponins from *Tribulus terrestris*. *Steroids*, 74, 4-5, 399–403, 2009.
16. Zhang, X., Wei, N., Huang, J., Tan, Y., Jin, D., A new feruloyl amide derivative from the fruits of *Tribulus terrestris*. *Nat. Prod. Res.*, 26, 20, 1922–1925, 2012.

17. Lv., A.L., Zhang, N., Sun, M.G., Huang, Y.F., Sun, Y., Ma, H.Y., Pei, Y.H., One new cinnamic imide derivative from the fruits of *Tribulus terrestris*. *Nat. Prod. Res.*, 22, 11, 1007–1010, 2008.
18. Wang, F.X., Sun, X.G., Wang, Z.F., Kang, L.P., Wu, K.L., Zhao, Y., Ma, B.P., Chemical constituents of *Tribulus terrestris* L. *Planta Med.*, 82, 05, PC45, 2016.
19. Li, M., Qu, W., Chu, S., Wang, H., Tian, C., Tu, M., Effect of the decoction of *Tribulus terrestris* on mice gluconeogenesis. *Zhong Yao Cai J. Chin. Med. Mater.*, 24, 8, 586–588, 2001.
20. Tkachenko, K., Frontasyeva, M., Vasilev, A., Avramov, L., Shi, L., Major and trace element content of *Tribulus terrestris* L. wildlife plants. *Plants*, 9, 12, 1764, 2020.
21. Selvaraju, R., Thiruppathi, G., Raman, R.G., Dhakshanamoorthy, D., Estimation of essential and trace elements in the medicinal plant *Tribulus terrestris* by icp-oes and flame photometric techniques. *Rom. J. Biol. Plant Biol.*, 56, 1, 65–75, 2011.
22. Hassan, L.G., Umar, K.J., Usman, A., Nutrient content of the leaves of *Tribulus terrestris* (Tsaida). *J. Trop. Biosci.*, 5, 2, 77–82, 2005.
23. Hassan, L.G., Umar, K.J., Umar, Z., Antinutritive factors in *Tribulus terrestris* (Linn) leaves and predicted calcium and zinc bioavailability. *J. Trop. Biosci.*, 7, 33–36, 2007.
24. Badmaev, V., The evolving approach to Parkinson's disease. *NutraCos*, 1, 4, 14–16, 2002.
25. Sharma, P.C., Yelne, M.B., Dennis, T.J., Joshi, A., Billore, K.V., *Database on medicinal plants used in Ayurveda*, Central Council for Research in Ayurveda & Siddha, Deptt. of ISM & H, Min. of Health & Family Welfare, Govt. of India, 2000.
26. Greene, J.W., Menstrual irregularities associated with athletics and exercise. *Compr. Ther.*, *25*, 4, 209–215, 1999.
27. Singh, S., Nair, V., Gupta, Y.K., Evaluation of the aphrodisiac activity of *Tribulus terrestris* Linn. in sexually sluggish male albino rats. *J. Pharm. Pharmacol.*, 3, 1, 43, 2012.
28. Rajendar, B., Bharavi, K., Rao, G.S., Kishore, P.V.S., Kumar, P.R., Kumar, C.S., Patel, T.P., Protective effect of an aphrodisiac herb *Tribulus terrestris* Linn on cadmium-induced testicular damage. *Ind. J. Pharmacol.*, 43, 5, 568, 2011.
29. Sahin, K., Orhan, C., Akdemir, F., Tuzcu, M., Gencoglu, H., Sahin, N., Juturu, V., Comparative evaluation of the sexual functions and NF-κB and Nrf2 pathways of some aphrodisiac herbal extracts in male rats. *BMC Complement. Altern. Med.*, 16, 1, 1–11, 2016.
30. Oliveira, N.N.P.M., Félix, M.A.R., Pereira, T.C.S., Rocha, L.G.P., Miranda, J.R., Zangeronimo, M.G., Sousa, R.V.D., Sperm quality and testicular histomorphometry of wistar rats supplemented with extract and fractions of fruit of *Tribulus terrestris* L. *Braz. Arch. Biol. Technol.*, 58, 6, 891–897, 2015.

31. Mazaro-Costa, R., Andersen, M.L., Hachul, H., Tufik, S., Medicinal plants as alternative treatments for female sexual dysfunction: Utopian vision or possible treatment in climacteric women. *J. Sex Med.*, 7, 11, 3695–3714, 2010.
32. Akhtari, E., Raisi, F., Keshavarz, M., Hosseini, H., Sohrabvand, F., Bioos, S., Ghobadi, A., *Tribulus terrestris* for treatment of sexual dysfunction in women: Randomized double-blind placebo-controlled study. *DARU J. Pharm. Sci.*, 22, 1, 1–7, 2014.
33. Aversa, A., Bruzziches, R., Pili, M., Spera, G., Phosphodiesterase 5 inhibitors in the treatment of erectile dysfunction. *Curr. Pharm. Des.*, 12, 27, 3467–3484, 2006.
34. Yeoh, S.H., Razali, R., Sidi, H., Razi, Z.R.M., Midin, M., Jaafar, N.R.N., Das, S., The relationship between sexual functioning among couples undergoing infertility treatment: A pair of perfect gloves. *Comp. Psychol.*, 55, S1–S6, 2014.
35. Kamenov, Z., Fileva, S., Kalinov, K., Jannini, E.A., Evaluation of the efficacy and safety of *Tribulus terrestris* in male sexual dysfunction—A prospective, randomized, double-blind, placebo-controlled clinical trial. *Maturitas*, 99, 20–26, 2017.
36. Zegers-Hochschild, F., Adamson, G.D., de Mouzon, J. *et al.*, International Committee for Monitoring Assisted Reproductive Technology (ICMART) and the World Health Organization (WHO) revised glossary of ART terminology. *Fertil. Steril.*, 92, 1520–1524, 2009.
37. Duncan, M.W. and Thompson, H.S., Proteomics of semen and its constituents. *Proteom.-Clin. Appl.*, 1, 8, 861–875, 2007.
38. Du Plessis, S.S., Agarwal, A., Sabanegh Jr, E.S. (Eds.), *Male infertility: A complete guide to lifestyle and environmental factors*, pp. 1–268, Springer-Verlag, New York, 2014.
39. M. Tomova, R. Gyulemetova, S. Zarkova, An agent for stimulation of sexual function. Patent (11). 27584, A61K35, 1978.
40. Adaikan, P.G., Gauthaman, K., Prasad, R.N.V., History of herbal medicines with an insight on the pharmacological properties of *Tribulus terrestris*. *Aging Male*, 4, 3, 163–169, 2001.
41. Gauthaman, K., Ganesan, A.P., Prasad, R.N.V., Sexual effects of puncturevine (*Tribulus terrestris*) extract (protodioscin): An evaluation using a rat model. *J. Altern. Complem. Med.*, 9, 2, 257–265, 2003.
42. Al-Ali, M., Wahbi, S., Twaij, H., Al-Badr, A., *Tribulus terrestris*: Preliminary study of its diuretic and contractile effects and comparison with Zea mays. *Int. J. Ethno*, 85, 2-3, 257–260, 2003.
43. Joshi, V.S., Parekh, B.B., Joshi, M.J., Vaidya, A.D., Inhibition of the growth of urinary calcium hydrogen phosphate dihydrate crystals with aqueous extracts of *Tribulus terrestris* and Bergenialigulata. *Urol. Res.*, 33, 2, 80–86, 2005.

44. Meena, P., Anand, A., Kumar, V., A comprehensive overview of Gokshura (*Tribulus terrestris* Linn.). *J. Ayurveda Integr. Med. Sci.* (ISSN 2456-3110), 4, 6, 205–211, 2020.
45. Shirfule, A.L., Sangamwar, A.T., Khobragade, C.N., Exploring glycolate oxidase (GOX) as an antiurolithic drug target: Molecular modeling and *in vitro* inhibitor study. *Int. J. Biol. Macromol.*, 49, 1, 62–70, 2011.
46. Kavitha, A.V. and Jagadeesan, G., Role of *Tribulus terrestris* (Linn.) (Zygophyllaceae) against mercuric chloride induced nephrotoxicity in mice, Mus musculus (Linn.). *J. Environ. Biol.*, 37, 2, 397–400, 2006.
47. Aggarwal, A., Tandon, S., Kumar Singla, S., Tandon, C., A novel antilithiatic protein from *Tribulus terrestris* having cytoprotective potency. *Protein Pept. Lett.*, 19, 8, 812–819, 2012.
48. Sharma, I., Khan, W., Ahmad, S., *In vitro* and *ex vivo* approach for anti-urolithiatic potential of bioactive fractions of gokhru with simultaneous HPLC analysis of six major metabolites and their exploration in rat plasma. *Pharm. Biol.*, 55, 1, 701–711, 2017.
49. Lamba, H.S., Bhargava, C.H., Thakur, M., Bhargava, S., α- glucosidase and aldose reductase inhibitory activity *in vitro* and antidiabetic activity *in vivo* of *Tribulus terrestris*. *Int. J. Pharm. Pharm. Sci.*, 3, 270–2, 2011.
50. Li, M., Qu, W., Wang, Y., Wan, H., Tian, C., Hypoglycemic effect of saponin from *Tribulus terrestris*. *Zhong Yao Cai J. Chin. Med. Mater.*, 25, 6, 420–422, 2002.
51. Bhuvad, S. U. S. H. A. M. A. and Nishteswar, K., Assessment of free radical scavenging activity of ten madhuraskandha drugs through UV spectroscopic and chromatographic technique. *J. Pharm. Sci.*, 8, 3, 92–96, 2016.
52. Hemalatha, S. and Hari, R., Fertility enhancing effect of saponin rich butanol extracts of *Tribulus terrestris* fruits in male albino rats. *Int. J. Pharmacol. Clin. Res.*, 7, 36–43, 2015.
53. Yogendra, K.G., Vimala, Y., Yogendra, K.G., Antioxidant activity and RP-HPLC analysis of diosgenin from the callus of *Tribulus terrestris* Linn. *Int. J. Res. Ayurveda Pharm.*, 5, 3, 343–346, 2014.
54. Oh, J.S., Baik, S.H., Ahn, E.K., Jeong, W., Hong, S.S., Anti-inflammatory activity of *Tribulus terrestris* in RAW264. 7. *Cells*, 54.2, 2012.
55. Patel, A., Soni, A., Siddiqi, N.J., Sharma, P., An insight into the anticancer mechanism of *Tribulus terrestris* extracts on human breast cancer cells. *3 Biotech.*, 9, 2, 58, 2019.
56. Sharma, D. and Kumar, A., Traditional medicinal plants of high value: *Tribulus Terrestris*: A review, 9, 12, 729–735, 2020.
57. Reggie, R. and Jaya, K.P., Anti-inflammatory activity *in vitro,* extractive process and HPLC-MS characterization of total saponins extract from *Tribulus terrestris* L. fruits. *Ind. Crops Prod.*, 150, 112343, 2008.
58. Hong, C.H., Hur, S.K., Oh, O.J., Kim, S.S., Nam, K.A., Lee, S.K., Evaluation of natural products on inhibition of inducible cyclooxygenase (COX-2) and nitric oxide synthase (iNOS) in cultured mouse macrophage cells. *Int. J. Ethno.*, 83, 1-2, 153–159, 2002.

59. Rathi, B., Subhash, B., M.V., Anti-inflammatory activity of steroidal saponins (IND 002) isolated from methanolic extract of fruits of *Tribulus terrestris*. *Inflamm. Res.*, 56, S281–S281, Basel, Switzerland: Birkhauser, Verlag, AG, 2007.
60. Baburao, B., Rajyalakshmi, G., Venkatesham, A., Kiran, G., Shyamsunder, A., Gangarao, B., Anti-inflammatory and antimicrobial activities of methanolic extract of *Tribulus terrestris* Linn plant. *Int. J. Chem. Sci.*, 7, 3, 1867–1872, 2009.
61. Deole, Y.S., Chavan, S.S., Ashok, B.K., Ravishankar, B., Thakar, A.B., Chandola, H.M., Evaluation of anti-depressant and anxiolytic activity of Rasayana Ghana Tablet (A compound Ayurvedic formulation) in albino mice. *Ayu*, 32, 3, 375, 2011.
62. Zhang, J., Zhang, D.S., Yan, C.L., Effects of GSTT on dysmnesia in mice. *Pharmacol. Clin. Chin. Mater. Clin. Med.*, 23, 3, 47–49, 2007.
63. Prabhu, N., Hadigal, S., Sheetal, U., Shenoy, A., Effect of *Tribulus terrestris* on learning and memory in wistar rats. *Pharmacogn. J.*, 6, 4, 68–71, 2014.
64. El-Sheikh, T.M., Bosly, H.A., Shalaby, N., Insecticidal and repellent activities of methanolic extract of *Tribulus terrestris* L. (Zygophyllaceae) against the malarial vector Anopheles arabiensis (Diptera: Culicidae). *Egypt. Acad. J. Biol. Sci. A, Entomol.*, 5, 2, 13–22, 2012.
65. Zhu, W., Du, Y., Meng, H., Dong, Y., Li, L., A review of traditional pharmacological uses, phytochemistry, and pharmacological activities of *Tribulus terrestris*. *Chem. Cent. J.*, 11, 1, 1–16, 2017.
66. Bansal, S.K., Singh, K.V., Sharma, S., Larvicidal potential of wild mustard (Cleome viscosa) and gokhru (*Tribulus terrestris*) against mosquito vectors in the semi-arid region of Western Rajasthan. *J. Environ. Biol.*, 35, 2, 327, 2014.
67. Khan, S., Kabir, H., Jalees, F., Asif, M., Naquvi, K.J., Antihyperlipidemic potential of fruits of *Tribulus terrestris*linn. *Int. J. Biomed. Res.*, 2, 98–101, 2011.
68. Tuncer, M.A., Yaymaci, B., Sati, L., Cayli, S., Acar, G., Altug, T., Demir, R., Influence of *Tribulus terrestris* extract on lipid profile and endothelial structure in developing atherosclerotic lesions in the aorta of rabbits on a high-cholesterol diet. *Acta Histochem.*, 111, 6, 488–500, 2009.
69. Jiji, M.J., Visalakshi, S., Meenakshi, P., Rathi, M.A., Thirumoorthi, L.D., ChinnaGuravaiah, D., Gopalakrishnan, V.K., Antilipidemic activity of Cissus quadrangularis and *Tribulus terrestris* on obesity in high fat fed rats. *Pharmacol.*, 2, 1250–1258, 2009.
70. Kim, H.J., Kim, J.C., Min, J.S., Kim, M.J., Kim, J.A., Kor, M.H., Ahn, J.K., Aqueous extract of *Tribulus terrestris* Linn induces cell growth arrest and apoptosis by down-regulating NF-κB signaling in liver cancer cells. *J. Ethnopharmacol.*, 136, 1, 197–203, 2011.
71. Tantawy, El-W.H. and Hassanin, L.A., Hypoglycemic and hypolipidemic effects of alcoholic extract of *Tribulus alatus* in streptozotocin-induced diabetic rats: A comparative study with T. terrestris. *Indian J. Exp. Biol.*, 45, 785–790, 20072007.

72. Ranjithkumar, R., Prathab Balaji, S., Balaji, B., Ramesh, R.V., Ramanathan, M., Standardized aqueous Tribulus terristris (Nerunjil) extract attenuates hyperalgesia in experimentally induced diabetic neuropathic pain model: Role of oxidative stress and inflammatory mediators. *Phytother. Res.*, 27, 11, 1646–1657, 2013.
73. Phillips, O.A., Mathew, K.T., Oriowo, M.A., Antihypertensive and vasodilator effects of methanolic and aqueous extracts of *Tribulus terrestris* in rats. *J. Ethnopharmacol.*, 104, 3, 351–355, 2006.
74. Sharifi, A.M., Darabi, R., Akbarloo, N., Study of antihypertensive mechanism of *Tribulus terrestris* in 2K1C hypertensive rats: Role of tissue ACE activity. *Life Sci.*, 73, 23, 2963–2971, 2003.
75. Reshma, P.L., Lekshmi, V.S., Sankar, V., Raghu, K.G., *Tribulus terrestris* (Linn.) attenuates cellular alterations induced by ischemia in H9c2 cells via antioxidant potential. *Phytother. Res.*, 29, 6, 933–943, 2015.
76. Stanislavov, R. and Rohdewald, P., Improvement of erectile function by a combination of French maritime pine bark and roburins with amino acids. *Minerva Urol. Nefrol. J. Urol. Nephrol.*, 67, 1, 27–32, 2015.
77. Qureshi, A., Naughton, D.P., Petroczi, A., A systematic review on the herbal extract *Tribulus terrestris* and the roots of its putative aphrodisiac and performance enhancing effect. *J. Diet. Suppl.*, 11, 1, 64–79, 2014.
78. Kostova, I. and Dinchev, D., Saponins in *Tribulus terrestris*–chemistry and bioactivity. *Phytochem. Rev.*, 4, 2-3, 111–137, 2005.
79. Pokrywka, A., Obmiński, Z., Malczewska-Lenczowska, J., Fijałek, Z., Turek-Lepa, E., Grucza, R., Insights into supplements with *Tribulus terrestris* used by athletes. *J. Hum. Kinet.*, 41, 99, 2014.
80. Joshi, V.S., Parekh, B.B., Joshi, M.J., Vaidya, A.B., Herbal extracts of *Tribulus terrestris* and Bergenia ligulata inhibit growth of calcium oxalate monohydrate crystals *in vitro*. *J. Cryst. Growth*, 275, 1-2, e1403–e1408, 2005.
81. Li, M., Qu, W., Wang, Y., Wan, H., Tian, C., Hypoglycemic effect of saponin from *Tribulus terrestris*. *Zhong Yao Cai J. Chin. Med. Mater.*, 25, 6, 420–422, 2002.
82. Vatankulu, M.A., Tasal, A., Erdoğan, E., Goatskin, Ö., Three case reports of the use of herbal combinations resulted in stent thrombosis: Herbal combinations; friend or foe. *Turk Kardiyol. Dern. Ars.: Turk Kardiyoloji Derneginin Yayin Organidir*, 40, 3, 265–268, 2012.
83. Ivanova, A., Serly, J., Dinchev, D., Ocsovszki, I., Kostova, I., Molnar, J., Screening of some saponins and phenolic components of *Tribulus terrestris* and Smilax excelsa as MDR modulators. *In Vivo*, 23, 4, 545–550, 2009.
84. Neychev, V. and Mitev, V., Pro-sexual and androgen enhancing effects of *Tribulus terrestris* L.: Fact or fiction. *J. Ethnopharmacol.*, 179, 345–355, 2016.

85. Shama, I.Y., Nesreen, A., Ahdaf, M.A., Ahmed Sahar, Y., Maiada, M., Hamed, S.M., Evaluation of toxicity and anti-microbial activity of *tribulus terrestris* fruits in Sudan. *Asian J. Biol. Sci.*, 12, 130–139, 2019.
86. Pascual, V.C., Cañavate, J.L.R., Gallego, Á. J., Report of the Scientific Committee of the Spanish Agency for Consumer Affairs, Food Safety and Nutrition (AECOSAN) on the risk of using Tribulus terrestris in food supplements, 2015.

4

Eleusine Indica

Piyush Kashyap[1*], Deep Shikha[2], Sunakshi Gautam[3] and Umexi Rani[4]

[1]*Department of Food Technology and Nutrition, School of Agriculture, Lovely Professional University, Phagwara-Punjab, India*
[2]*Department of Food Technology, Bhai Gurdas Institute of Engineering and Technology, Sangrur-Punjab, India*
[3]*Department of Food Science and Technology, Dr YS Parmar University of Horticulture and Forestry, Nauni, Solan, Himachal Pradesh, India*
[4]*Hamdard Vocational Knowledge Centre, Moradabad Uttar Pradesh, India*

Abstract

Eleusine indica L., (wiregrass) family Poaceae is a valuable medicinal plant widely spread in Asia, Africa, and South America tropical regions. Wiregrass is a rapidly growing diverse weed, which can produce almost 50,000 seeds per plant. These growing characteristics need a profounder understanding of the genetic deviation in different ecosystems. The plant is broadly used for deworming; mitigate cough and lung infections, dysentery, heart attacks, and high blood pressure, spleen and liver problems, blood and kidney stones as well as dislocation of bones and lumbago. Boiled plant decoctions are commonly used as anti-helminthic and febrifuge treatments. Phenolic compounds, essential oils, fatty acids, flavonoids, steroids, alkaloids, coumarins, and various chemical compounds are present in Eleusine indica. Schaftoside and vitexin are two major flavonoids isolated from their aerial part. The presence of these chemical compounds and their analysis by various *in vitro* and *in vivo* studies had proven its effect so effectively as compounds to other plants of its type. The natural fibers present in abundance in this plant are used as reinforcement material for various industrial applications and as a hybrid composite. This chapter provides insight into history, production, phytochemical composition, and antioxidative properties of different parts of *E. indica*. This chapter also summarizes the industrial use and health benefits with systematic scientific reports, which are highlighted.

Keywords: Eleusine indica, wire grass, antioxidant activity, health benefits, pharmacology

---

**Corresponding author*: p18.kashyap@gmail.com

Prerna Gupta, Navnidhi Chhikara and Anil Panghal (eds.) Harvesting Food from Weeds, (113–142)

4.1 Origin and History

Eleusine indica is a well-recognized medicinal plant that belongs to the family Gramineae or Poaceae. Around the world, it is known by various names, such as wiregrass, yard grass, Indian grass, and goosegrass. It is also known as Jongos grass in South Africa, which means that it will take a young animal, such as an ox, to root it out. *E. indica* have been continuously introduced in most of the nations where it is currently produced, which makes it hard to find its history from where its introduction starts into new habitat places. In the USA, the plant was introduced in 1800s and spread to different places *viz.* Cuba (1815), Martinique (1867), US Virgin Islands (1876), and Jamaica (1885). It is a well-known agricultural and environmental weed [1] and recognized among 42 nations as "serious weed" [2]. In a farming system, it is considered as a "dominant weed" and annual crop planted into rows which are broad enough to allow to be tilled where it can flourish well and grow explosively and provides seedlings in good quantity [2]. A single plant of goosegrass has the ability to produce 50,000 seeds, which are dispersed easily due to their light weight through wind and water, stick to skin or surface of animal or machine, and as a soil containment [3]. The suitable habitats for its growth are natural environment, grasslands, forests, coastal areas, stream banks, and marshes. This plant can easily grow along with roads, pavements, and poweline corridors [4].

4.2 Botanical Explanation

E. indica is an annual grass with prostrate and spreading nature. Depending on the nature of vegetation, it can develop up to 40 cm in height. The weed has well developed as well as strong roots, but there is commonly no rooting at the nodes. In earlier stages of its germination of first leaf (1 cm long), it tapers easily and can be pressed flat on the soil due to prostrate nature of spreading. On later stages, the leaves shape changes to V-shape with acute and boat shaped tip. It can grow up to 15 cm in height and 0.8 cm width, having smooth surface and bright grassy color. The ligule is a very tiny membranous brim about 1 mm long, casually bordered with little hairs. The stem and sheath bases are plated. Its inflorescence comprises of 3-8 racemes which are generally 5 to 10 cm long and 0.5 cm wide. These racemes are arranged in a digitate manner that means one racemes present 1 cm below other.

Figure 4.1 Growing stages of *E. indica* (goosegrass).

The slender rachis, around 1 mm wide, has two rows of smooth spikelet, 2.5 to 3 mm long each, 3 to 5 flowered and the lemmas very close in both texture and size to the upper glume. Rachis have keel, which are rough in texture and are acute but not waned. Its seeds are distinctly ridged and oval to round in shape, 0.1 cm long with reddish brown to black color. Figure 4.1 showed the different growing stages of Eleusine indica.

4.3 Production, Development, and Maturation

E. indica is a C4 plant and require proper sunlight for its growth. Second, shading has negative impact on its growth [5]. With 50% shading, its weight decreases up to 60%, whereas 90% reduction occurs at 80% of shady conditions [6]. Photoperiod is not serious in this plant and flowering can happen for all day between 6 and 16 hours [7]. However, 14 hours is the optimal time for its vegetative growth [8]. Cold temperature and dry condition can delay its flowering and in severe conditions can lead to death of young and emerging plants.

The cultivation of this plant is done by seeds as its growth is very rapid due to high prolificacy rate (one plant produce 140,000 seeds) [9]. Usually it prefers warm climate but can easily acclimatized to wider climatic situations attributed to its growth as a worldwide weed [10, 11]. Additionally, the broad emergence period of goosegrass makes its control very hard for cultivators. The seedlings of goosegrass start appearing from early April and lasts up to late September [12]. In earlier studies, researchers have also provided that its growth and appearance are directly influenced by season and time of sowing. The growth was observed between April and September in the cotton growing areas of China, more specifically June and mid-July for optimum growth condition [13]. However, less knowledge about the effects of appearance time on its growth pattern and development, which limits the construction of impactful weed control plans.

4.4 Nutritional Profile

Definition of traditional medicine given by the World Health Organization as: "the health practices, knowledge, approaches and mineral and animal-based medicines, beliefs incorporating plant, manual techniques, spiritual therapies, and exercises, practized singularly or in grouping to diagnose, treat and inhibit ailments or sustain well-being" [14]. The high nutritional profile (Table 4.1) of Eleusine indica makes it a dominant weed. Despite being a dominant weed it has many medicinal properties. It is being used as a local medicine in many regions due to its antioxidant, anti-cancerous and anti-inflammatory properties [15]. In the Philippines alone, its parts are being used in the cure of kidney problems and arthritis [16].

The phytochemicals such as flavonoids, tannins, alkaloids, saponins, cardiac glycosides, anthraquinones, and so on, are present in abundance in this plant [17]. The presence of these compounds was confirmed by Babu and Savithramma (2013) [18] in its extract. They have also confirmed the occurrence of coumarins, lignins, phenols and steroids. Due to the incidence of these bioactive compounds it is being used as source of antioxidants as well as phenolic compounds. In addition to these, its mineral profile is also suitable for the enrichment of various food products.

Table 4.1 Nutritional profile of *Eleusine indica.*

Nutrient compound		**Amount**	**Source**
Carbohydrates		80.19%	[19]
Fiber		27.56%	[19]
Lipid		7.14%	[19]
Ascorbic acid (aerial part)		13.1 ± 0.2 μg/Ml	[20]
Protein		15.55 ± 0.07% 12.40%	[20–22]
Minerals	Phosphorous	142.5 ± 3.53 mg/100 g	[20]
	Calcium	165 ± 14.41 mg/100 g	
	Iron	9.7 ± 0.14 mg/100 g	
	Manganese	0.020 ± 0.01 mg/100 g	
	Magnesium	3.5 ± 0.42 mg/100 g	
	Zinc	0.085 ± 0.01 mg/100 g	
Total Phenolic content		14.9 ± 0.002 mg GAE/g 0.410 ± 0.020 mgGAE/g (dw) 1.72 μgGAE/ml	[18, 23, 24]
Flavonoids		0.820 ± 0.020 mg/g (dw)	[18]
Tannins		0.316 ± 0.007 mg/g (dw)	[18]

4.5 Bioactives: Pharmacology and Bioactive

In the phytochemical analysis of goosegrass, the presence of sterol glycoside and C glycosylflavone was confirmed. In addition to these two, other bioactive components such as phenols, flavonoids, triglochinin, ocharatoin A, cyanogentic glucoside, α-amylase inhibitors, starch, albuminoids, and fatty acids were also present (Table 4.2) [25, 26].

The screening of phytochemicals in *E. indica* discovered the presence of about 1.80% alkaloids, 2% saponins, 6% flavonoids, and 21.76% tannins [26]. Among all, tannins were recorded as highest. Homburger (1989) [27] described that the plant extract with ethanol was proved to have insignificant acute toxicity and carries all the phytochemicals as described in their study. The antiviral and antifungal activity is due to the existence of alkaloids (in lesser amount) and saponins, respectively in this [28, 29].

The activity of polyphenolic secondary metabolites in this plant is due to the existence of flavonoids which act as antioxidants and modulatory agents. Its mechanism of action is to scavenge reactive oxygen species from the cells. The aqueous extract of goosegrass slow down the propagation of free radical oxygen species and prevent destruction of cell membrane, thus normalized the level of antioxidant defence [30]. Methanolic extract of *E. indica* with higher concentration of phenolic compounds have ability to inactivate Angiotensin Converting Enzyme (ACE). This is due to the presence of free hydroxyl group of phenolic compounds which act as the structural moieties that chelate zinc ions in their active site of ACE [31, 32]. Other secondary metabolites such as triterpenes, proanthocyanidins, and hydrolysable tannins also work in the same way [33]. The results of the phytochemical screening in a study was observed by Okokon *et al.* (2010) [34] exposed the existence of alkaloids, terpenes, flavonoids, tannins, anthraquinones, saponins, and cardiac glycosides in *E. Indica* extract. The similar ACE inhibiting ability was reported by Tutor *et al.*, (2018) [35] in methanolic and ethyl acetate fractions with percent inhibition of 68.84% and 51.51%, respectively. Other than these solvents, hexane can also be used but with decreased activity.

About 14.90 ± 0.002 mg/g of total phenolic content per GAE per gram of the extract was reported to be present in aqueous extract of *E. indica*. There was a decrease in DPPH free radical scavenging activity in a dose-dependent manner the half-maximum effective concentration (IC50) was observed to be 2350 μg/ml [23]. In another study, the results showed the FRAP total antioxidant as 18.96 ± 1.15 mg/g, FRAP free antioxidant activity as 1.94 ± 0.48 mg/g, folin total polyphenol content as .43 ± 1.69 mg/g, folin free polyphenol as 0.46 ± 0.10 mg/g, DPPH scavenging activity as 1.36 mg/g, nd DPPH free antioxidant activity as 1.64 ± 0.94 mg/g in the goosegrass extract [36].

The comparison between dichloromethane (DCM), hexane, methanol and ethyl acetate proved methanol as a solvent with maximum TPC (450 ± 210 GAE mg/g extract) among all. Also the methanolic extract showed free radical activity with 77.7% inhibition of DPPH absorption. In another study comparison between ethanol and acetone, extract was done, and results have shown ethanol as a better solvent than acetone. The ethanolic extract possessed an excellent prevention against DNA damage as flavonoid and phenol were present in higher amount in it. Maximum content of total phenols (201.115 ± 11.61 mg GAE/g) and flavonoids (32.621 ± 2.89 mg QE/g) was observed in ethanolic extract than acetone extract [17]. These secondary metabolites for example flavonoids and phenols also exert anti-viral and anti-bacterial and have anti-cancerous activity. Also, due to the existence of saponins in it cytotoxic potential was also observed in this plant [37, 38].

While comparing with all the cereal grains, *E. indica* which is a wild cereal reported with highest phenolic content (7.16 mg GA/100 mg sample) and β-carotene content (67.5%) [31]. The antioxidant activity of the two different extracts, i.e., 70% ethanol and watery extracts of *E. indica* (Sinngo-myet) also showed mild activity against oxidants with IC50 values of 100.61 and 86.14 μg/mL, respectively. The lower the IC50 values, the higher the antioxidant activities were found [39]. The methanolic extract of *E. indica* shown the most TPC value and DPPH scavenging activity was found to be 7.7%, which was followed by the ethyl acetate having 64.5% of activity. Hexane and dichloromethane extracts had 47.19% and 40.83% activity respectively [40]. The ethanol extract of *E. indica* leaves exhibited strong antioxidative attributes against both superoxide anion and hydrogen peroxide [41, 42].

4.6 Pharmacology

4.6.1 Antioxidant Activity

Antioxidants inhibit oxidative stress in cells by hindering the cellular rupture and providing molecular components to human body. They supress the activity of free radicals and stabilize them in the body by supplying extra free ions and thus help in lowering the risk of acute as well as chronic diseases such as cancer, ageing processes and various other health problems. Goosegrass as a medicinal herb is a potent source of natural antioxidants as oxidative compounds thus possess high antioxidant activity. These bioactive compounds extracted in pure methanol showed highest recovery (77.70%), ethyl acetate extract (64.50%), hexane extract (47.19%) and dichloromethane extract (40.83%). Highest phenolic content (450 mg GAE/g) was also observed in it which depicts that the amount of phenolic compounds directly influences the antimicrobial activity of goosegrass [40]. In another study, TP content as 14.90 ±0.002 mgGAE/g and DPPH free

Table 4.2 Bioactive compounds in *Eleusine Indica.*

Class	Compounds	Structure	Solvent used	Plant part	References
Flavonoids	Schaftoside	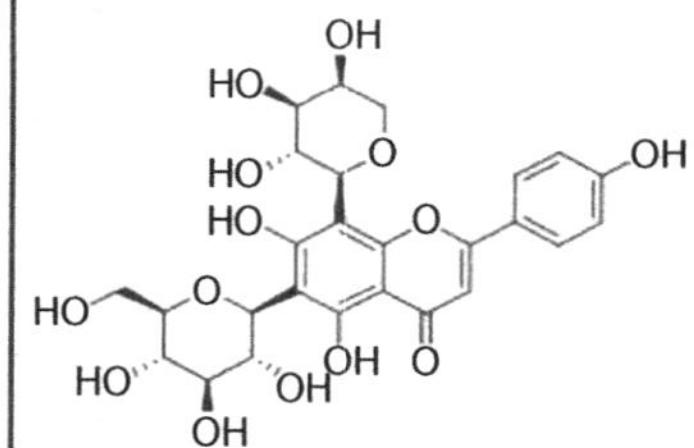	Ethanol, methanol, acetone, aqueous, n-hexane, ethyl acetate	Aerial parts	[43, 44]
	Vitexin	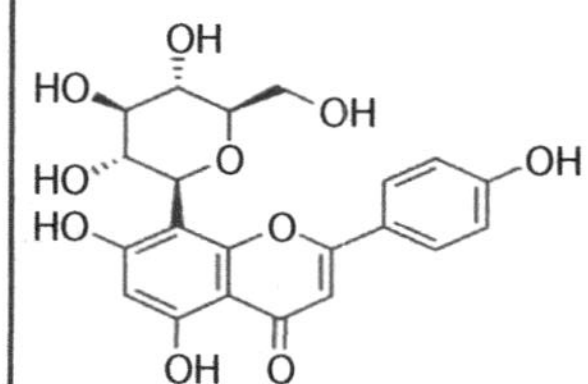	Ethanol, methanol, acetone, aqueous, n-hexane, ethyl acetate	Aerial parts	[43, 44]
	Orientin	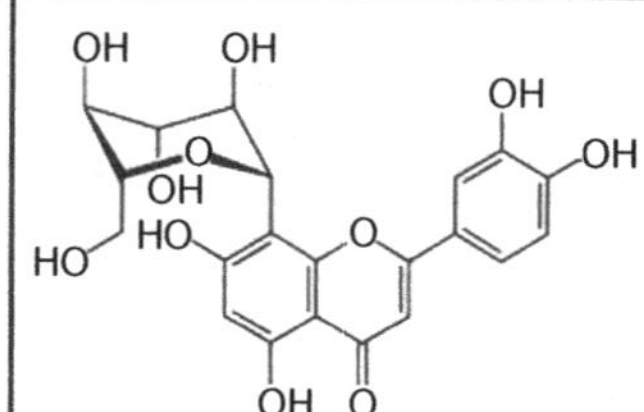	Ethanol, methanol, acetone, aqueous, n-hexane, ethyl acetate	Aerial parts	[17]

(*Continued*)

Table 4.2 Bioactive compounds in *Eleusine Indica.* (*Continued*)

Class	Compounds	Structure	Solvent used	Plant part	References
	Isoorientin	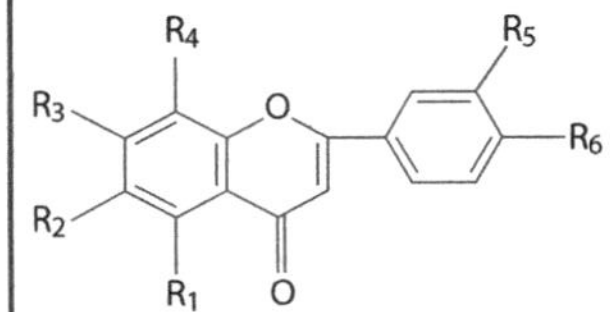	Ethanol, methanol, ethyl acetate	Aerial parts	[17]
Phenolic acids	p-coumaric acid	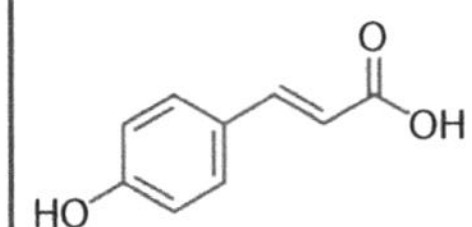	Hot water	Aerial parts	[17]
Steroid	3-0-beta-D-glucopyranosyl-Beta-sitosterol	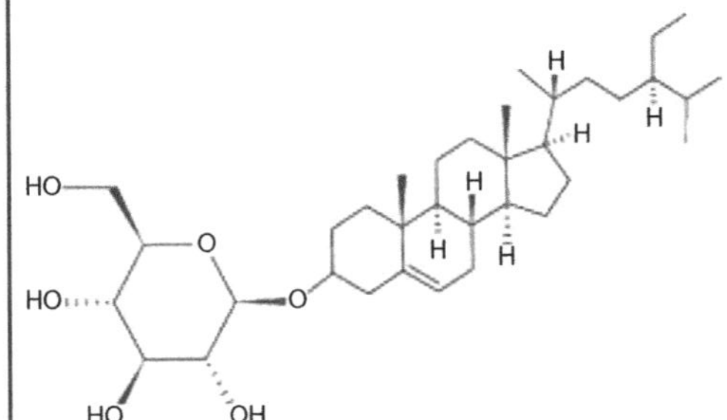	Ethanol, methanol, acetone, aqueous, n-hexane, ethyl acetate	Aerial parts	[45]

(*Continued*)

Table 4.2 Bioactive compounds in *Eleusine Indica*. (*Continued*)

Class	Compounds	Structure	Solvent used	Plant part	References
	1-[[[(2-aminoethoxy) hydroxyphosphinyl] oxy]methyl]-1,2-ethanediyl ester	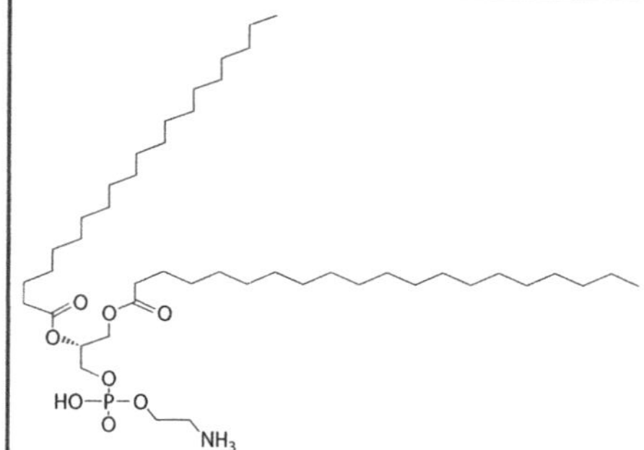	Chloroform	Aerial parts	[14, 46]
	Hexadecanoic acid	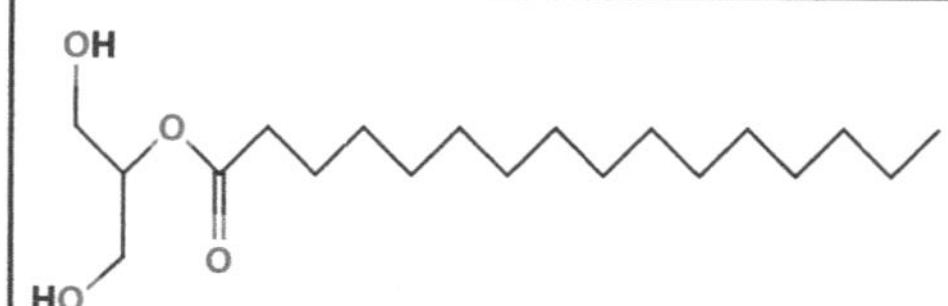	Methanol	Aerial parts	[14, 46]
	β-sitosterol	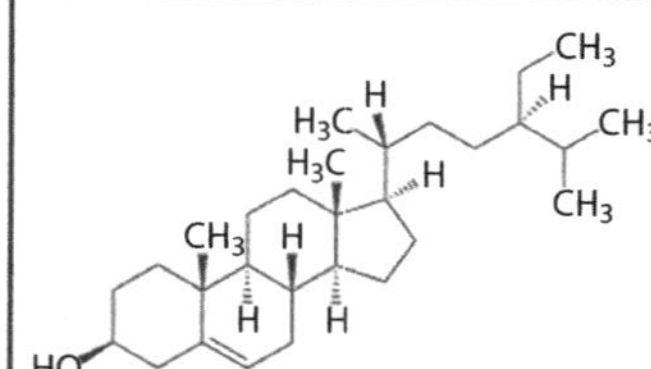	Hexane, Dichloromethane, ethyl acetate, butanol, water		[47]

(*Continued*)

Table 4.2 Bioactive compounds in *Eleusine Indica*. (*Continued*)

Class	Compounds	Structure	Solvent used	Plant part	References
	Stigmasterol	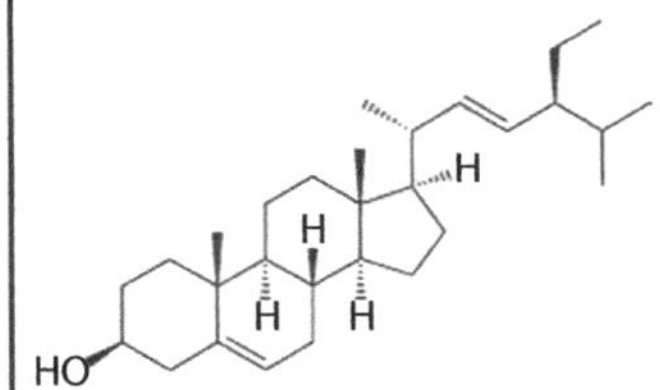	Hexane, Dichloromethane, ethyl acetate, butanol, water		[47]
	Lutein	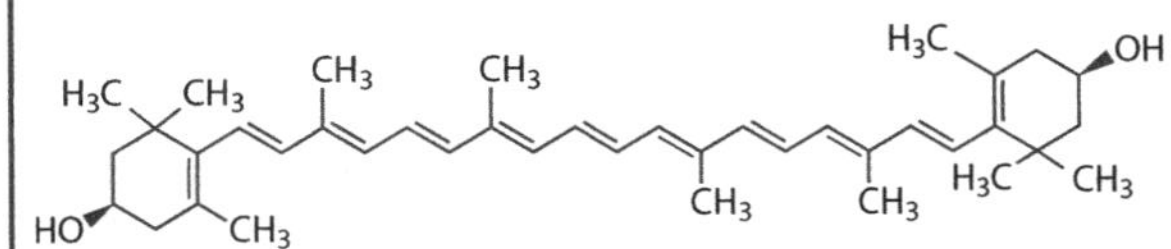	Hexane, Dichloromethane, ethyl acetate, butanol, water		[47]
	Saponins	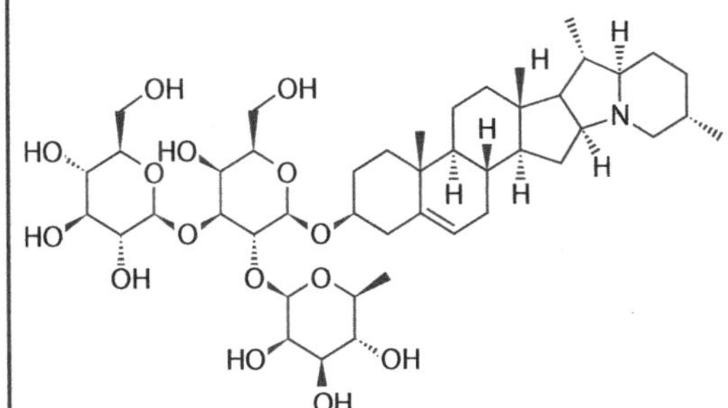			[29]

(*Continued*)

Table 4.2 Bioactive compounds in *Eleusine Indica.* (*Continued*)

Class	Compounds	Structure	Solvent used	Plant part	References
	C-glycosylflavone	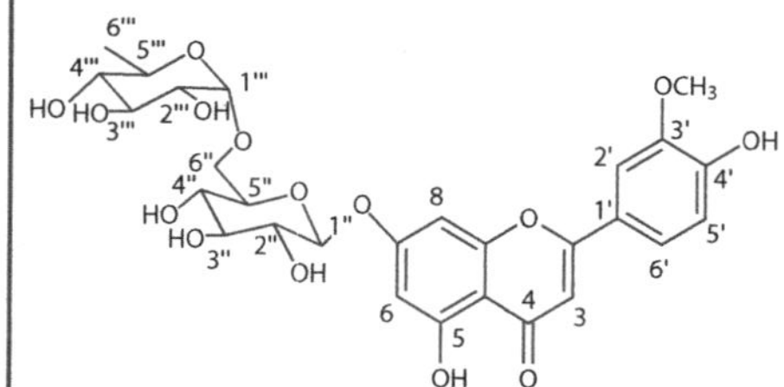	Aqueous extract		[25]
	Triglochinin	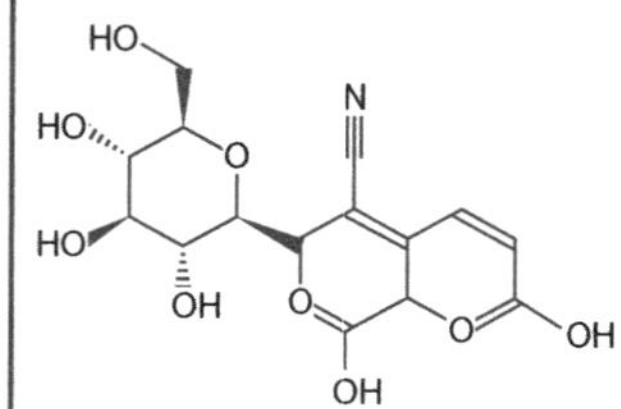	Aqueous extract		[25]
	Ochratoxin A	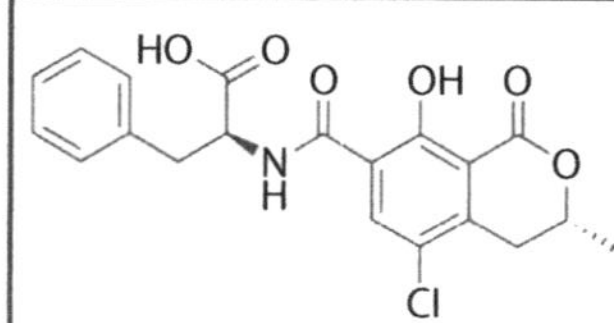	Aqueous extract		[25]

radical scavenging activity as 2350 μg/ml with a sequenced dose manner and IC50 was also observed [23]. Other than methanol, other solvent can also be used for extraction of bioactive compounds from *E. indica* like aqueous fractions, n-hexane, ethanol, butanol, ethyl acetate, chloroform, and so on. They have also assessed the influence of the extract in ethanol and its fractions on antioxidative enzymes and malondialdehyde (MDA) and met-hemoglobin, which are the final produce of lipid peroxidation. Among all the solvents used, the highest antioxidant activity as percent DPPH free radical scavenging activity was recorded in aqueous fraction as 79.00%, n-hexane (77.3%), crude (69%), butanol (63.1%), and chloroform (51.3%). Second, ethyl acetate as a solvent was least preferred as it showed the lowest antioxidant activity as 39.70%. A study was conducted by using crude extract and its fraction on albino rates and results exhibited that the higher amount of superoxide dismutase was found in serum sample and liver homogenate as compared with the control, whereas in similar studies, higher amount of reduced glutathione enzyme was recorded on animal treated by using the extract as compared to control. Likewise, higher level of catalase enzyme, met-hemoglobin and molondialdehyde (MDA) was observed in albino rats ingested with the same extract. This plant extract inhibits the activity of MDA, which directly represents the level of lipid peroxidation occur in the cells of membrane. Both MDA and MetHb are potential oxidants against antioxidant enzymes like glutathione, superoxide, and catalase. Due to which, the prooxidant effect of extract was recorded to be increased with an increase in the amount of MDA and MetHb, which proves their effect as an antioxidant [48].

A study conducted by Sagnia *et al.* [41] use chemiluminescence calculation against superoxide ion and hydrogen peroxide to check the antioxidant properties of leaves of goosegrass. To protect the DNA from damage in white blood cells, the comet assay was used. The results showed that by using 12.50 μg extract for single time, there was 45.50% reduction in the signals of luminol-amplified chemiluminescence. They have concluded 100 μg extract as best concentration to protect DNA damage from the negative effect of hydrogen peroxide. There was moderate reduction in tail intensity, tail moment, and tail length in the comet assay.

4.6.2 Antimicrobial Activity

The researchers show keen interest in medicinal herbs as they have antimicrobial properties due to the presence of microbial compounds. This antimicrobial resistance develops due to the nondiscriminated utilization of various chemically synthesized drugs or nondesirable negative effects of some antibiotics. The available standard drugs for curing diseases are available in markets at very high prices, which increases the need to utilize the plant as a replacement of these. The antimicrobial attributes of plant extracts as well as their physico-chemical characteristics have much significance in curing various diseases [49]. Goosegrass has higher antimicrobial activity against both gram-positive and gram-negative bacteria, such as *Bacillus subtilis* B29 and methicillin-resistant *Staphylococcus aureus* (MRSA), *Salmonella choleraesuis,* and *Pseudomonas aeruginosa* 60690. The presence of solvents in plant extract also boosts the antimicrobial activity like dichlomethane, methanol, ethyle acetate, and hexane. Difference in activity was found in hexane extracts and DCM extract as later one showed less activity on *P. aeruginosa* while most activity against most of the bacteria perceived was secured by the hexane extract on MRSA. Plant extract with ethyl acetate showed broad spectrum as compared to control (Streptomycin) specifically against *B subtilis,* which exhibited tolerance to plant extracts used. In this experiment, the methanolic extract does not show any activity towards MRSA, *S. choleraesuis, P. aeruginosa* 60690 and *B. subtilis* B29 [50]. In another study, the antimicrobial activity of chloroform and methanol extract against *S. aureus, E. coli, P. Vulgaris, Bacillus species, Aspergillus flavus, Enterobacter aerogens, Candida albicans, Klebsiella aerogenes, Pseudomonas aerogenes, Aspergillus niger,* and *Streptococcus species* by using agar diffusion method. To check minimum inhibitory concentration (MIC) of extract against these microorganisms, glucose is used as an indicator in the form of broth. The results showed a successfully exhibited activity against all the test microorganisms. Among all, most activity was inhibited by goosegrass extract against *Enterobacter aerogenes* with 33- to 38-mm zone of inhibition. Similar results were reported in case of fungal activity. An average width of zone of inhibition against *candida albican* was reported as 20 mm for chloroform and 18 mm for methanol extract. To check the minimum concentration required for inhibition of microorganism like bacteria and fungus (minimum fungal concentration [MIC/MFC]), a very low concentration of extract was sufficient. Methanolic and chloroform extracts were observed and found to have the compounds which are derivatives of phosphatidylethanolamine, ethane diol ester, and hexadecanoic acid. They are 1-[[[(2-aminoethoxy) hydroxyphosphinyl] oxy] methyl] 1-,2-ethandiylester for chloroform and hexadecanoic acid for methanol extracts, respectively. The anticancer, anticonvulsant, antifungal, and antiviral properties were confirmed in both compounds [46].

Ettebong *et al.* [51] also investigated the antimicrobial activity of extracts of *E. indica* plant against four bacteria *E. coli, P. aeruginosa, S. aureus, B. subtilis,* and one clinical strain of fungi, *C. albicans.* The control was taken as 0.4 mg/ml streptomycin and 50,000 IU/mL nystatin. The extract obtained using ethyl acetate resulted in major zone of inhibition (25.0 mm), after which, chloroform extract was tested for *Staphylococcus aureus.* Ethyl acetate extract showed broad-spectrum antimicrobial activity against *Pseudomonas aeruginosa, B. subtilis,* and *E. coli.* Inhibitory effect was not found against fungal strain of *Candida albicans.* Both extracts showed equivalent influences having MIC of 50 mg/ml against *E. coli and P. aeruginosa,* respectively. Recent studies of same plant tested the methanol leaf extract of *E. indica* against fungal and bacterial organisms. Fungal organisms include *Candida albicans, Cryptococcus neoformans,* and *Aspergillus fumigatus,* while bacterial organisms were *Staphylococcus aureus,* methicillin-resistant *S. aureus, E. coli, P. aeruginosa, Klebsiella pneumoniae,* and Vancomycin-resistant *Enterococci faecalis.* Modified version of the CLSI methods was used for *in vitro* antimicrobial assay in which drug controls (ciprofloxacin and amphotericin B) were added in the secondary assay. *In vitro* antimicrobial activity of methanolic extracts of *E. Indica* exhibited that the primary assay with *P. falciparum* D6Pinh resulted in a percentage inhibition of 37%. The extract exhibited antifungal activity with *Candida albicans* (0%), *Aspergillus fumigatus* (1%) and *Cryptococcus neoformans* (5%), while antibacterial activity exhibited MSRA (0%), *Escherichia coli* (7%) *Pseudomonas aeruginosa,* (1%) *Klebsiella pneumoniae* (0%), and vancomycin-resistant *Enterococci faecalis* (5%), respectively [52].

4.6.3 Anticancer Activity

Medicinal herbs are well accepted as an important source of more effective standard drugs which are well accepted for the therapy of various cancer types. These herbal plants are commonly utilized to produce the clinically functional antitumor into anticancer agents. One of several merits of using natural or dietary components as an ancillary for cancer is that lower down the toxicity and have negligible other negative effects. The cytotoxic activity toward human MCF-7 cells which are responsible for breast cancer, colonic carcinoma cells (HT-29) and human cancer lines (CEM-SS) was studied by using dichloromethane, hexane, methanol, and ethyl acetate extracts of *Eleusine Indica.* The cytotoxicity assay outcomes provided that IC_{50} value of more than 30 μg/ml was found in four extracts, so no cytotoxic effects were seen on all type of cancer cells after the incubation period of 72 hours in comparison to RPMI 1640 and DMSO. However, ethanolic *E. indica* extract showed growth inhibition and cytotoxic activity against HeLa cells through sulforhodamine-B assay. It has also been observed that plant extract showed less cytostatic activity against HeLa cells. The results also revealed that 54.03% cytostatic activity and 45.77% inhibition were observed but no cytocidal activity [50]. Six different concentrations of ethanolic and acetone extract of *E. Indica* were used to study cytotoxic brine shrimp assay and viable cell count were counted and their mortality (%) and LC50 values were determined after 24 hours. Vincristine sulfate with LC50 value as 0.839 ppm was considered as a control. *E. indica* acetone and ethanol extracts revealed significant cytotoxic activity with LC_{50} value of 16.857 ppm and 9.828 ppm for ethanolic and acetone extract, respectively, where lethality and concentration of extract were in proportional relationship with each other. The presence of cytotoxic components in the plant extract was observed by calculating the lethality of extract to shrimps kept in brine solution. It has been studied that the LC50 values of extract lower than 20 μg/ml have chances of producing cancer preventive components [53, 54].

Responte *et al.* [55] have evaluated the cytotoxic properties of *E. indica* extract by using Brine Shrimp Lethality Assay (BSLA). The most extensively known extraction methods, such as decoction with two solvents, one was absolute ethanol and other with equal proportion of water and ethanol (50:50) was used to find out the cytotoxic effects against the brine shrimp nauplii, and method of extraction was standardized. The LC50 value by using mixture of ethanol and water was observed as 153.99 ppm, whereas higher value (409.73 ppm) was observed with ethanolic extract. They have also reported enhanced rate of mortality of brine shrimps exposed for 24 hours in all the plant extract. The concentration of 100 to 1000 ppm was sufficient to enhance the rate of mortality up to 100% by using ethanolic extract of goosegrass. The existence of bioactive chemicals with chemical bonding was confirmed in this extract, indicating that *E. indica* possesses cytotoxic potential. In another study, screening of *E. indica* leaf methanolic extract was observed for the determination of cytotoxicity, which has been evaluated using fluorescence spectrophotometer. The result provided that the extract did not show any cytotoxicity THP1 cells [52].

4.6.4 Anti-Inflammatory Activity

Symptom of inflammation is usual for various chronic ailments which is the common reaction of the body to protect the tissue damage resulted from severe physical injury and entry of chemical or microbial agents in higher concentration to human body. The components that are able to prevent inflammatory reactions or conditions are accountable for minimizing the effect of pain and decreasing inflammation, hence assisting in the improving health condition of the body. Generally, for managing any type of inflammatory conditions, the drugs that are nonsteroidal in nature and have anti-inflammatory properties are used. But these chemically synthesized drugs are also responsible for their side effects, such as ulcer, gastric irritation, and various other. So, the plant-based medicines or medicinal herbs having anti-inflammatory effects can be a great substitute of chemical drugs in terms of safety and effectiveness [56]. The anti-inflammatory property of *E. indica* was inspected using TNF-α production by immature dendritic cells, using various plant extract concentrations. The result provided that *E. indica* extract dose-dependently inhibited the LPS-induced TNF-α production by imDC. High amounts of cytokines were produced by γδ T cells that can proliferate which were the result of phosphoantigen stimulation. *E. indica* extract was overserved for its effect on cytokine production and further proliferation.

Similarly, *E. indica* extract dose-dependently inhibited IPP-induced TNF-α production ($p < 0.05$). The above results show anti-inflammatory activity. However, when PBMCs (peripheral blood mononuclear cells) obtained from healthy donors were incubated for 10 days with IPP (isopentenyl pyrophosphate) and varied concentrations of extract in the presence of IL-2, the extract at 5 μg did not inhibit γδ T cells proliferation. Comparison of the effect of *E. indica* extract on acute inflammation in mice has been studied using three models: Cararrgeenin-induced hind paw edema, egg albumin-induced inflammation, and xylene-induced edema in ear of mice. Extract obtained from carrageenan (dose-dependent) was helpful against acute inflammation in mouse paw edema. The suppressing effect of 600 mg/kg of extract was maximum after 5 hours of ingestion of the phlogistic agent and causes the similar results as that of acetylsalicylic acid and showed statistically significant results in comparison with control.

Extract also found effective in egg albumin induced edema test where inhibition was noticed in mice for 5 hours. In relation with control sample results were found to be significant ($p < 0.01–0.001$) in terms of dose- and time-dependent effects. In xylene-induced ear edema in mice, edema was inhibited by dose-dependent manner. Result in xylene-induced ear edema in mice showed effect of extract was observed in comparison to standard drug dexamethasone and that was found significantly correlated ($p < 0.001$) to control. The extent of inhibition was completely similar to the standard drug dexamethasone [57]. A study of the effect of *E. indica* extract on lung inflammation has been revealed. Groups of male BALB/c mice six per cluster and weighing 250 to 300 mg were used in the study. Pre-treatment of mice were done with different doses of resuspended in saline flavonoid-enriched portion and with the flavonoids schaftoside and vitexin that is then resuspended in 1% DMSO, 1 hour before inhalation of 2.0 mL LPS (0.5 mg/mL) prepared in saline. Another way of treatment for next group was prepared in 10 mg/kg saline roligram with similar procedure and inhalation of LPS was given. Collection of bronchiolar lavage fluids was done at final volume of 1.5 mL after 3 hours of inhalation, and this was used to ascertain the number of neutrophils by differential counts under the microscope, and results are shown in form of percentage. The result provided that 1 hour before inhalation addition of 10 mg/kg of rolipram fully inhibited the LPS-induced recruitment of neutrophils. When *E. Indica* was administered to further fractionation, it provided two major flavonoids, namely, schaftoside (6-C-β-glucopyranosyl-8-C-α-arabino pyranosyl apigenin) and vitexin (8-C-β-glucopyranosyl apigenin) from spectral analysis. Administration of 400 μg/kg scaftoside or vitexin inhibited 62% and 80% neutrophil recruitment induced by LPS, respectively. Both flavonoids are potential anti-inflammatory components as shown by their high capacity to inhibit neutrophil migration in mice. The researchers deduced that c-clycosylflavones from aerial parts of *E. indica* inhibited LPS-induced mouse lung inflammation [43].

4.6.5 Antiplasmodial Activity

The antiplasmodial activity of *E. indica* was studied *in vivo* using the extract of whole-plant and its portions (which included n-hexane, chloroform, aqueous, butanol and ethyl acetate) of *E. indica* against *Plasmodium berghei* infected mice. Plant extract, i.e., .200, 400, and 600 mg/kg and) and fractions (40 g/kg) were analysed in suppressive (4-day), curative tests and prophylactic (repository). Pyrimethamine (1.2 mg/kg/d) and Chloroquine (.5 g/kg/d) were used as model drugs, while distilled water (10 mL/kg/d) assisted as control. From these models, the suppressive (4 days) test resulted in a dose-dependent reduction in the levels of parasitemia following use of the extract in comparison with control. This reduction was statistically significant ($p < 0.05$–0.001). However, the suppressive effect was less compared with the usual drug chloroquine. The portion having the most chemosuppressive attribute was the ethyl acetate portion (77.89%). The suppression in degree of parasitaemia was in this order: ethyl acetate > n-hexane > aqueous > chloroform > butanol. In the repository test, the reduction in parasitemia was dose dependent of extract compared to control. In the Curative Test (Rane Test), there was a progressive time and dose-dependent decrease in parasitemia when the extract was analysed on confirmed infection. The mean survival times of extract-treated groups of mice were dose-dependently and considerably lengthier. The extract elevated the mean persistence time from 12 to 14 days compared to the control. Though when compared with standard drug, chloroquine, the mean persistence time was shorter.

The study of the crude *E. indica* extract and its fractions provided that the plant has a abundant value as an antimalarial agent as found in its suppressive, repository, and curative activity against mice infected with *Plasmodium berghei berghei* [58]. The *in vitro* activity of *E. indica* was later screened against Chloroquine-resistant *P. falciparum* strain (Pf INDO) and chloroquine-sensitive *Plasmodium falciparum* strains (Pf 3D7). It resulted that due to the incorporation *E. Indica* had IC50, Pf INDO of > 100 µg/ml, and IC50, Pf3D7, of 85.60±43.23 (µg/ml) and its crude nature, it showed significant activity possibly due to the basic nature of the extract [59]. Further antiplasmodial assay of the *E. indica* extract using *P. falciparum* D6 (chloroquine-sensitive) and could not go further than the primary screening because the extract only showed 37% activity and did not show moderate inhibitory activity against the parasite [52]. Also, the presence of the active components like tannins, flavonoids and alkaloids lead to increase the antiplasmodial activity of extract [60].

4.6.6 Other Pharmacological Activities

E. indica also possesses many other pharmacological properties like anti-diabetic, hepatoprotective, analgesic, antipyretic, etc. The dose of leaf extract of *E. indica* was seen to directly influence the glucose level in blood in alloxan-induced diabetic rats. A comparison was done with commercial standard drug (glibenclaimide) which showed higher effectivity than plant extract [34]. The hepatoprotective effect was observed against rats pretreated with *E. Indica* which exhibited moderately elevated action of antioxidative enzymes compared to untreated CCl 4-intoxicated group. The dose of goosegrass extract moderately prevents the improved level of aspartate aminotransferase (AST) and alanine aminotransferase (ALT) serum. Also there was decreased degree of formation of malondialdehyde (MDA) and glutathione (GSH) was observed in comparison with CCl 4 – intoxicated group. In addition to this, reduced histopathological changes were also observed in liver by using *E. indica* extract [23]. The analgesic activity of *E. Indica* had been evaluated using three pharmacological models: hot plate test, acetic acid induced writhing and formalin induced paw licking in mice. The mice were pre-treated with 200 to 600 mg/kg of goosegrass extract intraperitoneally, whereas, 10 ml/kg distilled water and 100 mg/kg acetylsalicyclic acid were used as control and negative control, respectively.

The results showed that there was a decline in the count of abdominal constraints between pre-treated and control mice in case of acetic acid induced writhing. These stretching and constrictions of hind limbs of mice were decreased significantly with the increase in *E. indica* extract and this decrease was significant as compared to the control. In case of formalin test, the licking of paw was taken as indication of pain and the period occupied by each mice to lick their paws was recorded. The neurogenic response was noticed after 5 minutes of injection, whereas, inflammatory pain response was noticed in second phase after 15 to 30 minutes of formalin injection. These results showed moderate decrease in the frequency of licking of hind paw in a dose-dependent manner. In case of hot plate test which was used to quantify the response latencies and showed similar dose-dependent rise in response of analgesic effect in mice [61].

Pyrexia exhibits from infectious agents or injured tissues starting the expanded production of proinflammatory mediator cytokines such as α, β, interleukin 1β, and TNF-α which increase the development of prostaglandin E2 (PGE2) in the prostaglandin and the hypothalamus then acts on the hypothalamus to increase the body temperature. In Amphetamine – induced pyrexia in rats, the plant extract depicted a moderate and dose-dependent decrease in the increased body temperature. The maximum dose of the extract (60 g/kg) weighed up well with acetylsalicylic acid. In dinitrophenol (DNP)-prompted pyrexia in rats, the consequences also depicted a time-and dose-related reduction in temperature compared to control. It may be presumed that the plant extract may have action as a cyclooxygenase-2 (Cox-2) inhibitor through the inhibition of PGE2 creation in the hypothalamus or by increasing the production of endogenous antipyretic compounds in the body, such as vasodilatation of superficial blood vessels with resultant, or by vasopressin arginine, or by enhanced dissipation of heat after readjusting the hypothalamic thermostat [48].

E. indica extract was also established to possess anticonvulsant attributes by reducing the period of tonic convulsions as related to the control and elevation in the latency of colonic convulsions in a dose-dependent manner. The extract was found to be provide similar kind of protection against mortality as that of produce by the standard drug diazepam. The moderate elevation in the latency of clonic convulsions and reduction in the period of tonic convulsions resulted from the extract showed anticonvulsant property [51].

4.7 Health Benefits

Goosegrass has numerous health benefits like anti-inflammatory, diuretic as well as antipyretic effects due to which it is being used in Thai medicine [62–64]. Table 4.3 showed some of the health benefits of Eleusine indica.

Table 4.3 Health benefits of *Eleusine indica*.

Disease/Health condition	Technique/Methodology	Findings	Reference
Cancer	Phytochemical screening, Multivulva reduction assay	screen secondary metabolites including flavonoids, phenols, tannin and alkaloids Anticancerous property does not depend upon Wnt and Ras signaling pathways.	[16]
Cytotoxicity and/or Mutagenicity activity of *E. indica*	Allium test	Inhibition of meristems growth and stable mitotic index	[65]
Cytotoxic properties	Estimation of total phenol content	IC50 values of over 30 μg/ml against and CEM-SS cancer, HT-29 and MCF-7 cell lines	[40]
Cancer	HT-29 human colonic carcinoma cells, MCF-7 human breast cancer cells, and CEM-SS human cancer cell lines	IC50 values of >30 μg/ml 30 μg/ml and hence did not generate any cytotoxic effects on CEM-SS cancer cell lines and MCF-7, HT-29 cell lines	[50]
Cytotoxic and Antiproliferative effects of the butanolic and hexane extracts of *Eleusine indica*	ELISA-based apoptosis assay	Selective cytotoxic effect improve the apoptotic level in extract-treated A549 cells cytotoxic activity of the grass extracts against cervical and lung cancer cells is mediate through the initiation of apoptosis.	[63]

(*Continued*)

Table 4.3 Health benefits of *Eleusine indica.* (*Continued*)

Disease/Health condition	Technique/Methodology	Findings	Reference
Effect on pregnancy and child delivery	Phytochemical analysis	Have high iron content helps in prevention of anemia. High in Calcium and zinc – Stabilizes structure of protein and strong bones and teeth.	[21]
Diabetes	Rats as test organisms	Significant hepatoprotective effect against type II diabetes mellitus acts as a natural enzyme antioxidant protector	[70]
Antidiabetic	Ethanolic leaf extract (320 to 960 mg/kg) verified	(960 mg/kg, 58.85%) result of the maximum dose of the extract was unmatched to that of standard drug, glibenclamide, 10 mg/kg on day 15.	[34]
Arterial blood pressure regulator	capability to inhibit angiotensin-converting enzyme (ACE)-component of the Renin-Angiotensin Aldosterone structure	Antihypertensive activity with a percent inhibition of 51.51%	[35]
Anti-inflammatory activity	TNF-a production by imDC	Luminol amplified chemiluminescence signal reduced of 45.5% mainly inhibition was found with 1 mg of plants extract presence of C-glycosylflavones	[41]

(*Continued*)

Table 4.3 Health benefits of *Eleusine indica*. (*Continued*)

Disease/Health condition	Technique/Methodology	Findings	Reference
Genotoxicity	Micronuclei assay on Albino rats	Does not confer any genotoxicity in mammals can be safely use in food for human and animals	[26]
ACE Inhibitory Activity	Angiotensin converting enzyme inhibitory activity	500 ug/ml to 125 ug/ml	[67]
Anti-hypertensive activity	Phytochemical screening tail-cuff method with adrenalin induced hypertensive rats model was used	Presence of alkaloids, glycosides, phenol, flavonoids significantly inhibit hypertension	[14]
Hepatoprotective effect		Showed hepatoprotective effect against CCl_4	[23]
Anti-herpes	Virucidal assay	Antiviral activity towards HSV-1 with prophylactic effect, inhibits attachment and penetration of virus into cell and virucidal activity.	[68]
Lipid lowering effect	Toxicity and efficacy analyses	Hexane extract has significantly reduced body weight great potential in the inhibition of porcine pancreatic lipase (27.01 ± 5.68%).	[71]
Nephrotoxicity	Oral administration in rats as 100 mg/kg or 200 mg/kg, *per os*)	i) Defensive property in L-NAME-induced renal failure ii) Antioxidant capacity against kidney damages.	[72]

It is very helpful in curing respiratory diseases due to which it is extensively used in the Brazilian medicine system [65]. The plant has multibiological antioxidant action and consisting of various secondary metabolites [66]. Each fragment of this plant specially its roots is being used as anti-urolithiasis, diuretic, febrifuge, laxative, and depurative, and these properties are being used in curing various diseases such as treating hypertension, urine retention, influenza, and oliguria. It is the "basic remedy" used in curing various diseases in Vietnamese medicinal system [14, 40]. The plant is a natural alternative for anti-hypertensive agents [67] and helps in the regulation of the menstrual cycle [46]. There are various reports published in which the use of roots of goosegrass has been mentioned for curing illnesses related to kidney and urine [68]. In West Bengal, the tribal people use its fresh roots in the cure of gonorrhoea [69]. In addition to this, tribe of Nigeria use its decoction make from each portion of this plant is used as anti-inflammatory agent [57]. It is great source to fulfil the nutritional and energy requirement during pregnancy, thus preventing problems like malnutrition and supplements the important phytochemicals in therapeutic activities.

4.8 Future Prospectus and Conclusion

E. indica used as a medicinal plant showed the presence of various phytochemical compounds that are accountable for antioxidant activities that neutralize free radicals in the human body and also other beneficial effects, such as anti-microbial activity, anti-cancerous activities, anti-inflammatory activities, etc. the plant is also having a good nutritional profile. Moreover, a few revisions have been conducted for this plant, so future investigations could be applied for analysis to explore the characteristics of goosegrass for its maximum beneficial effects in different areas, such as medicinal, as well as in the cure of various chronic diseases.

References

1. Randall, R.P., *A global compendium of weeds*, p. 1124, Department of Agriculture and Food Western Australia, Perth, Australia, 2012.
2. Holm, L., Pancho, J.V., Herberger, J.P., Plucknett, D.L., *A geographical atlas of world weeds*, John Wiley and Sons, United Kingdom, 1979.
3. Waterhouse, D.F., *The major arthropod pests and weeds of agriculture in Southeast Asia: Distribution, importance and origin (No. 435-2016-33732)*, 1993.

4. Queensland Department of Primary Industries and Fisheries, Special edition of Environmental Weeds of Australia for Biosecurity Queensland: The University of Queensland and Department of Primary Industries and Fisheries, Australia, 2011.
5. Ampong-Nyarko, K. and De Datta, S.K., *A handbook for weed control in rice, Int. Rice Res. Inst.*, Philippines, 1991.
6. Bantilan, R.T., Palada, M.C., Harwood, R.R., Integrated weed management: 1. Ley factors affecting crop-weed balance. *Philipp. Weed Sci. Bull.*, *1*, 2, 14–36, 1974.
7. Nakatani, K. and Kusanagi, T., Effect of photoperiod and temperature on growth characteristics, especially heading or flower bud appearance of upland weeds. *Weed Res. (Tokyo)*, 36, 1, 74–81, 1991.
8. Holm, L.G., Plucknett, D.L., Pancho, J.V., Herberger, J.P., *The world's worst weeds. Distribution and biology*, University Press of Hawaii, USA, 1977.
9. Chin, H.F., Weed seed-a potential source of danger, in: *Proceedings of the Plant Protection Seminar*, L.T. Kwee (Ed.), pp. 115–119, Malaysian Plant Protection Society, Kuala Lumpur, Malaysia, 1979.
10. Ismail, B.S., Chuah, T.S., Salmijah, S., Teng, Y.T., Schumacher., R.W., Germination and seedling emergence of the glyphosate-resistant and susceptible biotype of goosegrass (Eleusine indica [L.] Gaertn.). *Weed Biol. Manage.*, 2, 177–185, 2002.
11. Ismail, B.S., Chuah, T.S., Salmijah, S., Teng, Y.T., Effects of shade and watering frequency on growth and development of glyphosate-resistant and susceptible biotypes of goosegrass (Eleusine indica [L.] Gaertn.). *Plant Prot. Q.*, 18, 30–34, 2003.
12. Chuah, T.S., Salmijah, S., Teng, Y.T., Ismail, B.S., Changes in seed bank size anddormancy characteristics of the glyphosate-resistant biotype of goosegrass (Eleusine indica [L.] Gaertn.). *Weed Biol. Manage.*, 4, 114–121, 2004.
13. Ma, X.Y., Ma, Y., Xi, J.P., Jiang, W.L., Ma, Y.J., Li, X.F., Mixed weeds and competition with directly seeded cotton, north Henan province China. *Cotton Sci.*, 24, 91–96, 2012.
14. Desai, A.V., Patil, V.M., Patil, S.S., Kangralkar, V.A., Phytochemical Investigation of Eleusine indica for *in-vivo* anti-hypertensive activity. *Int. J. Innov. Sci. Res. Technol.*, *2*, 6, 405–416, 2017.
15. Luchian, V., Georgescu, M.I., Savulescu, E., Popa, V., Some aspects of morpho-anatomical features of the invasive species *Eleusine indica* (L.) Gaertn. *Scientific Papers. Series Agronomy*, pp. 529–537, 2019.
16. Nas, J.S., Dangeros, S.E., Chen, P.D., Dimapilis, R.C., Gonzales, D.J., Hamja, F.J., Evaluation of anticancer potential of *Eleusine indica* methanolic leaf extract through Ras- and Wnt-related pathways using transgenic *Caenorhabditis elegans* strains. *J. Pharm. Negat. Results*, 11, 42–46, 2020.
17. Ettebong, E.O., Ubulom, P.M., Obot, D., A systematic review on Eleucine indica (L.) Gaertn.): From ethnomedicinal uses to pharmacological activities. *J. Med. Plants*, 8, 4, 262–274, 2020.

18. Babu, R.H. and Savithramma, N., Phytochemical screening of underutilized species of Poaceae. *BioMedRx: An Int. J.*, 10, 947–951, 2013.
19. Ettebong, E., Ubulom, P., Etuk, A., Antiplasmodial activity of methanol leaf extract of Citrus aurantifolia (Christm) Swingle. *J. Herbmed Pharmacol.*, 8, 4, 274–280, 2019.
20. Chiribagula, V.B., Bakari, A.S., Ndjolo, P.O., Kahumba, B.J., Simbi., J.B.L., Phytochemical screening and antioxidant activity of methanolic extracts of 53 antimalarial plants from Bagira in Eastern DR Congo. *GSC Biol. Pharm. Sci.*, 12, 2, 099–118, 2020.
21. Gbadamosi, I.T. and Otobo, E.R., Assessment of the nutritional qualities of ten botanicals used in pregnancy and child delivery in Ibadan, Nigeria. *Int. J. Phytomedicine*, 6, 1, 16, 2014.
22. Regmi, P.R., Devkota, N.R., Timsina, J., Re-growth and nutritional potentials of Eleusine indica (L.) Gaertn. (Goose Grass). *J. Inst. Agric. Anim. Sci.*, 25, 55–63, 2004.
23. Iqbal, M. and Gnanaraj, C., *Eleusine indica* L possesses antioxidant activity and precludes carbon tetrachloride (CCl4)-mediated oxidative hepatic damage in rats. *Environ. Health Prev. Med.*, 17, 307–315, 2012.
24. Aliyu, M.A., Abdullahi, A.A., Ugya, A.Y., Antioxidant properties of selected poaceae species in Kano, Northern Nigeria. *Eur. J. Biomed. Pharm. Sci.*, 4, 5, 577–585, 2017.
25. Ghani, A., *Medicinal Plants of Bangladesh*, 2nd Edition, pp. 1–16, 138, Asiatic Society of Bangladesh, Bangladesh, 2003.
26. Etta, H.E., Ikpeme, E.V., Offiong, E., Investigating genotoxicity of Eleusine indica by micronuclei assay in albino rats. *Afr. J. Biol. Sci.*, *1*, 1, 33–40, 2019.
27. Homburger, F., *In vivo* testing in the study of toxicity and safety evaluation, in: *A Guide to General Toxicology*, 2nd E, J.K. Marquis, (Ed.), EdnKarger, New York, 1989.
28. Abul farah, M., Ateeq, B., Niamat Ali, M., Ahmad, W., Introduction of Micronuclei and erythrocytealterations in the catfish clarias acid and butacheior. *Mutat. Res.*, 518, 135–144, 1996.
29. Cowan, M.M., Plant products as antimicrobial agents. *Clin. Microbiol. Rev.*, 12, 4, 564–582, 1999.
30. El-Sawi, S.A. and Sleem, A.A., Flavonoids and hepatoprotective activity of leaves of *Senna surattensis* (burm.f.) in CCl4 induced hepatotoxicity in rats. *Aust. J. Basic Appl. Sci.*, 4, 1326–1334, 2010.
31. Chitindingu, K., Chitindingu, J.J.F., Benhura, M.A.N., Marume, A., Mutingwende, I., Bhebhe, M., Antioxidant capacity of bioactive compounds extracted from selected wild and domesticated cereals of Zimbabwe. *Afr. J. Biochem. Res.*, 6, 5, 62–68, 2012.
32. Loizzo, M.R., Saab, A.M., Tundis, R., Statti, G.A., Menichini, F., Lampronti, I., Doerr, H.W., Phytochemical analysis and *in vitro* antiviral activities of the essential oils of seven Lebanon species. *Chem. Biodivers.*, 5, 3, 461–470, 2008.

33. Braga, F.C., Wagner, H., Lombardi, J.A., Braga, D.O.A., Screening of Brazilian flora for anti-hypertensive plant species for *in vitro* angiotensin-I-converting enzyme inhibiting activity. *Phytomedicine*, 7, 3, 245–250, 2000.
34. Okokon, J.E., Odomena, C.S., Imabong, E., Obot, J., Udobang, J.A., Antiplasmoidal and antidiabetic activities of *Eleusine indica*. *Int. J. Drug Dev. Res.*, 2, 3, 493–500, 2010.
35. Tutor, J.T. and Chichioco-Hernandez, C.L., Angiotensin-converting enzyme inhibition of fractions from Eleusine indica leaf extracts. *Pharmacogn. J.*, 10, 1, 25–28, 2018.
36. Agbor, G.A., Kuate, D., Oben, J.E., Medicinal plants can be good source of antioxidants: Case study in Cameroon. *Pak. J. Biol. Sci.*, 10, 4, 537–544, 2007.
37. Montoro, P., Braca, A., Pizza, C., De Tommasi, N., Structure-antioxidant activity relationships of flavonoids isolated from different plant species. *J. Food Chem.*, 92, 349–55, 2005.
38. Negi, J.S., Negi, P.S., Pant GJ, G.J., Rawat, M.S.M., Negi, S.K., Naturally occurring saponins: Chemistry and biology. *J. Pois. Med. Plant Res.*, 1, 1, 001–6, 2013.
39. Aye, M.T., Win, P.P., Than, N.N., Ngwe, D.H., Bioactivity study of cleome burmanni L. Merr. (Taw-Hingala) and Eleusine Indica L.Gaertn. (Sinngo-Myet). *J. Myanmar Acad. Arts Sci.*, XVI, 1B, 179–191, 2018.
40. Al-Zubairi, A.S., Abdul, A.B., Abdelwahab, S.I., Chew, Y.P., Mohan, S., Elhassan, M.M., Eleusine indica possesses antioxidant, antibacterial and cytotoxic properties. *Evid-Based Complement. Alternat. Med.*, 2011, 1–6, 2011.
41. Sagnia, B., Fedeli, D., Casetti, R., Montesano, C., Falcioni, G., Colizzi., V., Antioxidant and anti-inflammatory activities of extracts from Cassia alata, Eleusine indica, Eremomastax speciosa, Carica papaya and Polyscias fulva medicinal plants collected in Cameroon. *PLoS One*, 9, 8, e103999, 2014.
42. Lim, T.K., Eleusine indica, in: *Edible Medicinal and Non-Medicinal Plants*, pp. 228–236, 2016.
43. De Melo, G.O., Muzitano, M.F., Legora-Machado, A., Almeida, T.A., De Oliveira, D.B., Kaiser, C.R., Costa, S.S., C-glycosylflavones from the aerial parts of Eleusine indica inhibit LPS-induced mouse lung inflammation. *Planta Med.*, *71*, 04, 362–363, 2005.
44. Agrawal, P.K., Carbon-13 NMR of flavonoids. *Stud. Org. Chem.*, 39, XVI–564, 283–355, 1989.
45. Phuong, N.M., Van Sung, T., Ripperger, H., Adam, G., Sterol glucosides from *Eleucine indica*. *Planta Med.*, *60*, 5, 1994.
46. Alaekwe, I.O., Ajiwe, V.I.E., Ajiwe, A.C., Aningo, G.N., Phytochemical and anti-microbial screening of the aerial parts of Eleusine indica. *Int. J. Pure Appl. Biosci.*, *3*, 1, 257–264, 2015.
47. Ong, S.W., HuiMah, S., Lai, H.Y., Porcine pancreatic lipase inhibitory agent isolated from medicinal herb and inhibition kinetics of extracts from *Eleusine indica* (L.) Gaertner. *J. Pharm.*, 2016, 1–9, 2016.

48. Ettebong, E.O. and Nwafor, P.A., Antipyretic and antioxidant activities of Eleucine indica. *J. Phytopharmacol.*, 4, 235–242, 2015.
49. Moyo, B., Masika, P.J., Muchenje, V., Antimicrobial activities of *Moringa oleifera* Lam leaf extracts. *Afr. J. Biotechnol.*, *11*, 11, 2797–2802, 2012.
50. Adel, A.M., El-Wahab, Z.H.A., Ibrahim, A.A., Al-Shemy, M., Characterization of microcrystalline cellulose prepared form lignocellulisic material. Part II: Physico chemical properties. *Carbohydr. Polym.*, 83, 676–687, 2011.
51. Ettebong, E.O. and Bassey, A.I., *In vitro* antimicrobial evaluation of whole-plant extracts of Eleucine indica. *J. Med. Plants*, *5*, 4, 99–102, 2017.
52. Ogbole, O.O., Segun, P.A., Fasinu, P.S., Antimicrobial and antiprotozoal activities of twenty-four Nigerian medicinal plant extracts. *S. Afr. J. Bot.*, *117*, 240–246, 2018.
53. Wong, F.C., Tsun-Thai, C., Yee-Wei, H., Antioxidation and cytotoxic activities of selected medicinal herbs used in Malaysia. *J. Med. Plant Res.*, 6, 16, 3169–3175, 2012.
54. Fatima, I., Kanwal, S., Mahmood, T., Evaluation of biological potential of selected species of family Poaceae from Bahawalpur, Pakistan. *BMC Complement. Altern. Med.*, *18*, 1, 1–13, 2018.
55. Responte, M.A., Maria, R., Dacar, B., Olga, M.N., Mylene, M.U., Brine shrimp lethality assay of whole plant extracts of Eleusine indica. *Adv. Agric. Bot. Int. J. Bioflux Soc.*, 7, 2, 90–95, 2015.
56. Alhakmani, F., Kumar, S., Khan, S.A., Estimation of total phenolic content, *in-vitro* antioxidant and anti-inflammatory activity of flowers of Moringa oleifera. *Asian Pac. J. Trop. Biomed.*, *3*, 8, 623–627, 2013.
57. Obute, G.C., Ethnomedicinal plant resources of south eastern Nigeria. *Afr. J. Interdiscip. Stud.*, 3, 1, 90–94, 2007.
58. Ettebong, E.O., Nwafor, P.A., Okokon, J.E., *In vivo* antiplasmodial activities of ethanolic extract and fractions of Eleusine indica. *Asian Pac. J. Trop. Med.*, 5, 673–676, 2015.
59. Okokon, J.E., Okokon, P.J., Sahal, D., *In vitro* antiplasmodial activity of some medicinal plants from Nigeria. *Int. J. Herb. Med.*, 5, 5, 102–109, 2017.
60. Builders, M.I., Isichie, C.O., Aguiyi, J.C., Toxicity studies of the extracts of Parkia biglobosa stem bark in rats. *Modern Advances in Pharmaceuticals Research*, 1, 95–109, 2019.
61. Ettebong, E.O. and Nwafor, P.A., Anti-inflammatory and analgesic potentials of Eleucine indica. *J. Phytopharm.*, *3*, 2, 130–138, 2014.
62. Ettebong, E.O. and Nwafor, P.A., Antipyretic and antioxidant activities of *Eleucine indica. J. Phytopharm.*, *4*, 235–242, 2015.
63. Hansakul, P., Ngamkitidechakul, C., Ingkaninan, K., Sireeratawong, S., Panunto, W., Apoptotic induction activity of Dactyloctenium aegyptium (L.) P.B. and Eleusine indica (L.) Gaerth. Extracts on human lung and cervical cancer cell lines. *Songklanakarin J. Sci. Technol.*, 31, 273–279, 2009.

64. Pattanayak, S. and Maity, D., Use of *Eleusine indica* (L.) Gaertn. *(Kechila ghas)* as an antipyretic medicine of herbivores. *Explor. Anim. Med. Res.*, 7, 1, 94–96, 2017.
65. Oliveira, A.A. and Romao, N.F., Growth inhibition and pro-apoptotic action of *Eleusine indica* (L) Gaertn extracts in allium test. *Eur. J. Med. Plants*, 8, 3, 121–127, 2015.
66. Damasceno, D.C., Volpato, G.T., Calderon, I.D.M.P., Aguilar, R., Rudge, M.C., Effect of Bauhinia forficata extract in diabetic pregnant rats: Maternal repercussions. *Phytomedicine*, 11, 196–201, 2004.
67. Cometa, J.C., Mancol, L.F.R., Yen, A.N., Talens Manlusoc, J.K., *ACE* inhibitory activity and functional group analysis of solvent- partitioned fractions of Eleusine indica. *Pharm. Sci. Res.*, *7*, 1, 57–65, 2020.
68. Iberahim, R., Bahtiar, A.A., Ibrahim, N., Anti-herpes simplex virus type-1 activity of Eleusine indica methanol extract. *Malays. J. Microbiol.*, *12*, 471, 2016.
69. Chowdhury, T., De Sarker, D., Roy, S.C., Local folk use of plants in Dakshin Dinajpur district of West Bengal, India. *Int. Res. J. Biol. Sci.*, 3, 5, 67–79, 2014.
70. Souza, I.D., Melo, E.S., Nascimento, V.A., Pereira, H.S., Silva, K., Espindola, P. R., Tschinkel, P.F.S., Ramos, E.M., Reis, F.J.M., Ramos, I.B., Paula, F.G., Oliveria, K.R.W., Lima, C.D., Nunes, A.A., do Nascimento, V.A., Potential health risks of macro-and microelements in commercial medicinal plants used to treatment of diabetes. *BioMed Res. Int.*, 1–11, 2021.
71. Ong, S.L., Nalamolu, K.R., Lai, H.Y., Potential lipid-lowering effects of *Eleusine indica* (L) Gaertn. Extract on high-fat-diet- induced hyperlipidemic rats. *Pharmacognosy*, 13, 49, 1–9, 2021.
72. Huguette, T.T., Tom, N.L.E., Florence, N.T., Farouck, A.B., Joseph, N., Théophile, D., Preventive effects of aqueous extract of the whole plant of *Eleusine indica* (Linn) Gaertn. (Poaceae) against L-NAME induced nephrotoxicity in rat. *J. Phytopharm.*, 8, 1, 28–32, 2019.

5

Hemp (*Cannabis sativa* L.) Agronomic Practices, Engineering Properties, Bioactive Compounds and Utilization in Food Processing Industry

Vipul Mittal, Anil Panghal* and Ravi Gupta

Department of Processing and Food Engineering, AICRP-PHET, Chaudhary Charan Singh Haryana Agricultural University, Hisar, India

Abstract

Different plant parts accounting to their nutritive, economic, and social importance are always being one of the favorite topics for food scientists and engineers, and Hemp is one of those. The hemp, which was considered as a weed in the past, and was removed by the farmers from the field as an unwanted plant is now grown commercially, where it is legalized and the rest of the world is demanding to get it legalized. However, the commercial hemp known as industrial hemp must have tetrahydrocannabinol (THC) content of less than 0.3%. The study of agronomic principles recognizes it as a sustainable crop. Every part of the hemp plant gives alternative solutions to many products already available in the market but at very high cost. Further, there are many compounds found, which are only associated with hemp and have shown potential health benefits in human and animal trials. Hemp seed protein and oil in hempseed, hemp fiber and hurds from the stem, hemp cannabinoid oil, and other extracts from leaves and flowers, triterpenoids, and phytosterols from hemp roots are some common associations. This review begins with a comprehensive summary on agronomic practices for hemp cultivation, Phytomorphology of different parts of hemp plant including engineering properties of hemp seed, bioactive compounds and further investigates its pharmacological properties, i.e., the use of hemp in treating different ailments. Finally, this study focuses on the many processing technologies that have been successfully used to hemp plant parts and demonstrated positive outcomes. The main objective

**Corresponding author*: anilpanghal@gmail.com

Prerna Gupta, Navnidhi Chhikara and Anil Panghal (eds.) Harvesting Food from Weeds, (143–182)

of this review is to thoroughly analyze the present-day information available on the abovementioned topics to help the researcher's research and explore hemp as a functional food in foodstuffs and fiber in textiles on a broader aspect. This will help the world to get aware of the truth behind the weeds like cannabis.

The primary goal of this study is to thoroughly analyze current knowledge on the aforementioned subjects to aid the researcher's research and to explore hemp as a functional food in foodstuffs and fiber in textiles in a larger sense.

Keywords: Hemp, processing, cannabinoids, engineering properties, therapeutic, weed food

5.1 Introduction

Hemp (*Cannabis sativa* L.) belongs to Cannabaceae family is an annual herbaceous weed crop, globally distributed, and suitable for any type of habitat. *Cannabis sativa* L. can be conceived as one of the majority of studied plants with more than 25,000 papers published and a lot of great research work done on it. Because of its favorable influence on the environment, production of feed, high-nutrition food, and high-strength fiber, commercial hemp growing has recently gained a lot of traction [11]. The importance of hemp can be estimated from its nutritional composition (Table 5.1) associated with different parts (Figure 5.1). The use of hemp stem as hempcrete as a natural building material has also gained importance in recent years in European countries. According to a survey report by Kolodinsky and Lacasse [68], there is a surge increase seen in production and consumption of hemp and hemp-based products from 2019 to 2020 after the first year of federal legalization in the U.S. Theoretical constraints like as price volatility, oversupply, and risk management are becoming real-world issues. The cannabis plant was firstly observed in the Persian region, where it is subjected to great alternations of heat and cold; and that it has passed on the one hand into Europe, and on the other into India [31]. Hemp can be regarded as a crucial plant in human history because it is one of the earliest cultivated plants [72]. Pollen grains from a site in northern Italy dated 3450 BC; samples from central and northern Germany, Scandinavia, England, and France were dated between 2900 and 1700 BC; and pollen from central Germany showed a continuous presence of hemp from 2000 to 530 BC. The arrow's point progressed from stone to bronze to iron, but hemp remained the binding fiber [6]. Bhanga, or hemp, is one of the five sacred plants listed in the Atharvaveda (c. 1500–1000 BCE) as a way to cure worry. In the Sayana, it is also called a chemical makeup [60].

Table 5.1 Nutritional composition of different parts of hemp plant.

Parameters (%)	Whole plant	Leaves	Stalk	Hemp flower	Seed heads
Dry matter	69.9	87.6	63.8	90.5	88.6
Crude protein	6.7	12.9	5.1	21.0	22.9
Fat	2.5	8.8	1.1	12.3	13.0
Ash	8.6	21.0	6.1	13.9	16.5
Sugar	2.5	5.6	2.0	4.9	2.6
Starch	0.2	0.9	0.1	0.6	0.7
ADF	60.5	20.7	63.5	25.9	28.9
NDF	81.3	43.9	84.1	52.3	53.0
Non-fiber carbohydrate	2.54	15.1	5.2	6.1	-

ADF - Acid detergent fiber; NDF - Neutral detergent fiber; Kleinhenz *et al.* [67].

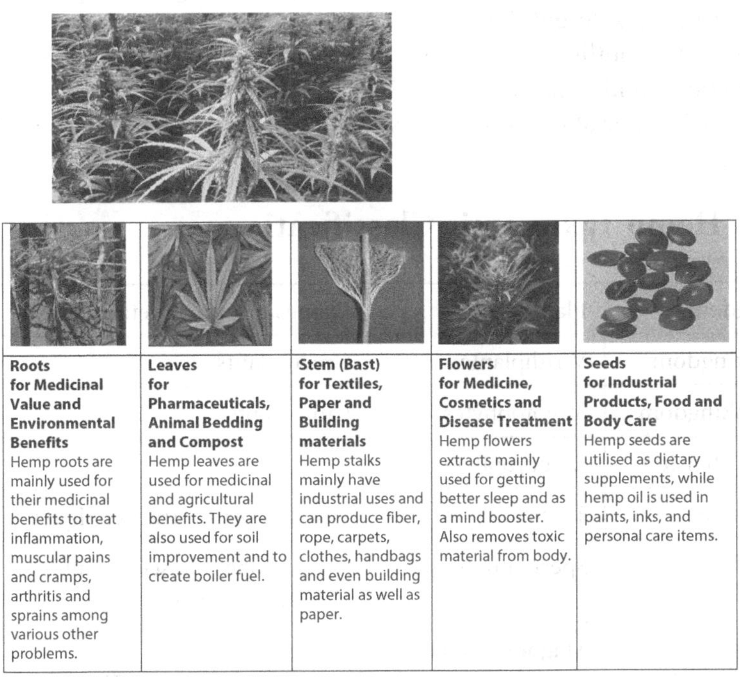

Figure 5.1 Different parts and their uses of *Cannabis sativa.*

The hemp issue is no exception when it comes to scientific evidence being misconstrued, over-interpreted, and misrepresented. Marijuana and hemp both come from the same plant species, Cannabis sativa, but have distinct properties. The former is commonly used for therapeutic and recreational purposes, while the latter is used for both, whereas the latter The food and fiber industries rely on it. Hemp, on the other hand, is genetically distinct and identified by its use. Hemp can be used to extract over 100 distinct phytochemicals. Tetrahydrocannabinol (THC) and cannabidiol (CBD) are the two most important cannabinoids (CBD). Marijuana plants have been used for herbal medicine and intoxication since antiquity [61]. THC levels in hemp are as low as 0.3%, but THC levels in marijuana can reach up to 20%. Certain hemp types have higher CBD levels, and the non-psychoactive component of the plant has medical qualities. Hemp would be more useful as a medical treatment for diseases if it had a high CBD to THC ratio, however ideas on how CBD levels affect psychoactive effects are conflicting. Traditionally, the wild cultivar of the hemp plant is commonly known as Bhang in many parts of India and shows great historical significance and religious values. In India, at the rural level, hemp cultivation is done at a small scale and is used to make different products like apparel, handicrafts, cloth, etc. by doing the manual processing of hemp. Further, it is used to cure different illnesses and injuries.

This paper outlines the agronomic principles, phytomorphology, bioactive compounds pharmacological properties, and their potential health benefits with special focus on processing technologies and its industrial use.

5.2 Hemp Taxonomic Classification

Kingdom	Plantae	Plantes, planta, vegetal, plants
Subkingdom	Viridiplantae	Green plants
Infrakingdom	Streptophytes	Land plants
Superdivision	Embryophyta	
Division	Tracheophyta	Vascular plants, tracheophytes
Subdivision	Spermatophytina	Spermatophytes, seed plants, phanérogames
Class	Magnoliopsida	

(*Continued*)

(*Continued*)

Superorder	Rosanae	
Order	Rosales	
Family	Cannabaceae	Hemp
Genus	Cannabis L.	Hemp
Species	Cannabis sativa L.	Hemp, grass, hashish, Mary Jane, pot, marijuana

(https://www.itis.gov/servlet/SingleRpt/SingleRpt?search_topic=TSN&search_value=19109#null)

5.3 Agronomic Practices/Growing Condition for Hemp Cultivation

Most hemp-growing countries grew one or more types, which were mostly indigenous cultivars. Hemp types grown for CBD (cannabidiol), seed, and fiber are distinct from one another [2]. The FINOLA industrial hemp cultivar is a dioecious, auto-flowering oilseed crop that was developed in Finland for the production of hempseed grain. Finola samples collected in the field were erect, annual, dioecious, and grew to a height of 1.8 m. Green, hollow, cylindrical, and longitudinally ridged stems [40]. Crop rotation can help to prevent erosion, enhance soil health, regulate soil and fertility, boost nutrients available for crops, and function as a cure for Orobanche patches known as "spots of death" [108]. Hemp can be suitably rotated with vegetable and fodder crops. Hemp itself is considered an early weed so it is minimally affected by other weeds. Crop's vigorous growth suppresses the growth of disease and pests, so no pesticides are used, hence decreasing the energy, as well as cost requirement of the crop. In addition, it can be used on soil banks to suppress the growth of the weed or can be rotated with pasture [82]. Hemp naturally acts in accordance with the demand of organic farming as there is no requirement for pesticides, herbicides, and fungicide [88].

Soil Condition

A fine and homogeneous seedbed for cultivation is a necessary condition for achieving the optimal and uniform plant density [105]. Nonuniform crops leave more plant nutrients in the soil than necessary, which might be leached, and they compete poorly with weeds, requiring more pesticides [48]. It is important to break all the soil clods to allow free draining of water as hemp is particularly sensitive to water logging, and it requires well-drained soil to maximize production [109]. The soil should be deeply tilted, rich in capillary action to allow free water flow, aeration and have plentiful nutrients. A drought condition should not arise to avoid the wilting of crops so it requires a minimum level of soil moisture. For best results, the pH of the soil should be between 6 and 8. If the pH of the soil is too low, it will need to be limed, as a neutral pH is ideal [33]. A pH of 6 to 7.5 may be more appropriate [8]. Roots gathered more heavy metals than above-ground biomass in the majority of the types. In acid soil, the removal of Cd, Ni, Pb, Hg, Co, Mo, and As was higher. Alkaline soil had the ability to phytostabilize in the sequence Cu>Cr>Cd>Mo>Hg>Zn>Ni>Co>As>Pb, whereas acid soil had the ability to phytostabilize in the order Zn>Cd>Cr>Ni>Hg>Cu>Mo>As>Co>Pb [44]. The optimum soil for hemp development is sandy loam, followed by clay loam, whereas heavy clay soil and sandy soil are not well suited as former results in water logging problem and later have very low water holding capacity, which can affect the emergence of seedling [8].

Sowing Condition

The best sowing date and time from sowing to flowering is a key driver of hemp fiber yield potential, as well as a source of input savings. To guarantee the accurate germination and a rapid crop establishment, the date of sowing is an important factor that is mainly dependent on the temperature and moisture content of soil [76]. Hemp is typically planted from mid-March (early) until the last week of May (late). Further, their types of early sowing trends for high stem and seed yield have been reported for monoecious hemp [42]. Maximum stem yield was attained in a Mediterranean environment by sowing in February or March; however, earlier seeding led in significant irrigation water savings, but dry stalk yield reduced dramatically [35]. The soil temperature must be high enough (8–10°C) to guarantee that vegetative development begins quickly and that weed competition is minimal [33]. Hemp row cropping has been employed on a small scale. Hemp has traditionally been produced in Western Canada, where row cropping is not common. Weed control would be the principal benefit of row cropping. The row spacing can be changed to meet the tillage equipment. Because the plant population will be denser within the row and the plants will self-thin to a sustainable population, the seeding rate can be reduced [24]. In the comparison of broadcasting and row cropping (with row width of 8 to 30 cm) of hemp with the same amount of seed, there are fewer plant densities found with a former method and an increase in plant densities with the decrease in-row spacing. Further row spacing shows more homogeneous results [109].

5.4 Hemp Phytomorphology

Hemp's sex expression

Although hemp is a dioecious plant, monoecious cultivars have been developed by selective breeding. Hemp's inflorescence can be either feminine (leafy, stocky, with no branches) or masculine (few leaves, heavily branched). Solely female flowers, only male flowers, or flowers of both sexes in varying proportions may be found in each of these two categories. Shortly after anthesis, male plants die. Until the seed is ready, female plants live 2 to 4 weeks longer than male plants. In the pre-flowering stage, female plants form a pair of white hairs in a V shape coming from the green calyx, whilst male plants form simple tiny balls (Figure 5.2). The chromosome study of the dioecious hemp plant was done and genome size was found to be 1636 and 1683 Mbp of male and female plants, respectively. Furthermore, karyotype analysis revealed that the X and Y chromosomes were submetacentric and subtelocentric, respectively, in the same study [100].

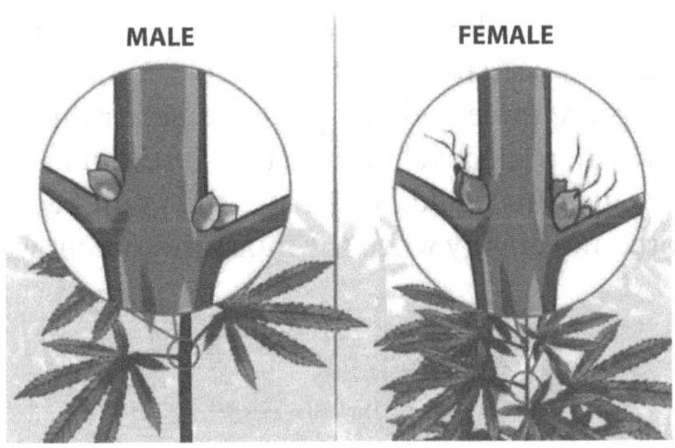

Figure 5.2 Sex determination in hemp.

5.5 Hemp Plant Parts

Seed

The textural properties of hemp seed are smooth from the outer surface seed coat but hard in nature, which requires high force to rupture it, while the kernel part is extremely soft due to high protein and oil content. There is a small space called cavity found between the seed coat and kernel in which moisture gets trapped, and this cavity gets increased or decreased according to the available moisture content (Figure 5.3). The ten major grown varieties were studied and shown the max and mean value of 306 g/kg and 292 g/kg of oil and 280 g/kg and 256 g/kg of crude protein respectively [110]. Hemp seed oil has similar oil qualities to drying oils like linseed and tung and is primarily composed of three fatty acids: linoleic (54–60%), linolenic (15–20%), and oleic (15–20%) (11–13%) [55].

The physical and mechanical properties of hemp seed have been studied extensively. Taheri-Garavand *et al.* [106] tested the seed at moisture content (MC) levels ranging from 5.39% to 27.12% MC (db) and revealed that when MC rises, the linear dimensions of length, width, and thickness increased. Moisture uptake in the intracellular spaces within the seeds leads the seeds to expand in size. With the rise in MC, the sphericity and surface area increased from 79.10% to 80.37% and 41.64 to 49.81 mm^2 respectively, while the bulk density and true density reduced from 563.67 to 556.23 kg/m^3 and 1034.63 to 902.35 kg/m^3, respectively. This demonstrates that the volume expansion is greater than the mass gain. Emptying and filling angles of repose increased from 21.75 to 23° and 34.30 to 37.53°, respectively, in the hemp seeds tested. In the moisture range of 5.39% to 27.12% db on three structural surfaces, the static coefficient of friction increased: glass (0.347 to 0.459), galvanized iron sheet (0.286 to 0.438), and plywood (0.267 to 0.410). The rupture force and energy decreased from 36.65 to 18.67 N and 10.25 to 5.41 mJ, respectively, as the moisture content increased from 5.39% to 27.12% db. This could be because the grain becomes softer and required less force and energy when the moisture content was higher. With

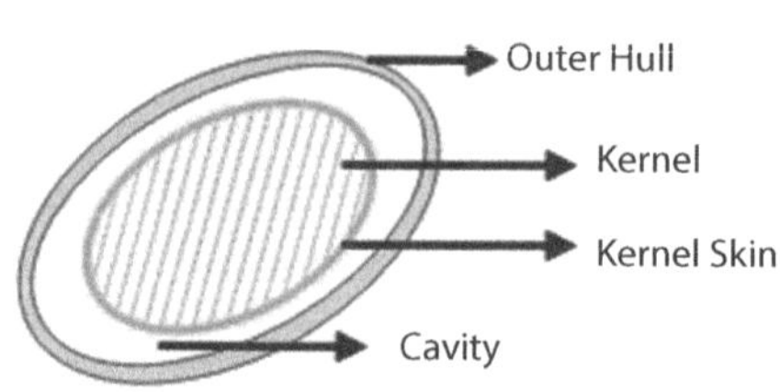

Figure 5.3 Hemp seed.

an increase in loading rate, the rupture force and loading rate both dropped. Furthermore, the maximum rupture force (37.20 N) was measured at the lowest loading rate (1 mm/min) and moisture content (5.39% db). At a loading rate of 3 mm/min, lowest rupture energy (4.39 mJ) was found. In the main axis, the medium axis and the small axis were increased to 8.44%, 6.51%, and 14.17% in the range 8.62% to 20.88% db, respectively [99]. In comparison to its two other main axes, hemp seeds expanded more along its minor axis. The thousand-grain mass increases from 15.3 g at 8.62% db to 16.9 g, at 20.84% db on a linear basis with moisture content. The mass of thousands of seeds rises linearly from 15.3 g at 8.62% db to 16.9 g at 20.84% db. The size of the seed area is increasing linearly and the volume and the actual densities have reduced correspondingly from 557.5 to 512.3 kg/m^3 and 1043.0 to 894.8 kg/m^3. The predicted porosity was reduced from 46.5% to 42.7% with the increased MC in the relevant experimental data. The terminal speed is increased linearly from 5.5 m/sec at a humidity content of 8.62% db to 6.4 m/sec at a moisture value of 20.88% db. The static friction coefficient for rubber, plywood, and galvanized metal was between 0.42 and 0.49, 0.30 and 0.43, and 0.34 to 0.38 respectively. The dynamic friction coefficient values of 0.34 to 0.41, 0.31 to 0.37, and 0.27 to 0.32 have been found to be equivalent. Eight different types of oil seeds utilized in the food processing sector were studied for their physical impacts [69]. The maximum mass of 1,000 sunflower seeds was discovered to be 43.3 g, significantly less for hemp seeds (21.2 g), and less than 0.5 g for cobble seeds compared to 6.5 g for flax seeds. The obtained sliding angle (fluidity) values for watercress seeds were 30.30 at 500 for hemp seeds and 50.30 for rape seeds. The size, construction, form, humidity, and infection of seeds all influence the extent of this angle. The perspective coefficient of external friction for hemp seeds (0.2742) and the coefficient for sesame seeds with the highest coefficient were reported (0.6080). The greatest tap density of waterspout and poppy seeds was 750.1 and 726.8 kg/m^3 and 589.5 kg/m^3 were reported for sunflower seeds as the lowest value of this attribute. The remaining seeds were 615.8 kg/m^3 for flax seed, with the results of hemp seeds being 702.6 kg/m^3. For watercress (707.8 kg/m^3), hemp was found to have a high density of 540 kg/m^3. To break sunflower seeds, the greatest pressure was required (124.1 N). The compression values for the remaining seeds were significantly lower, ranging from 2.6 N for poppy seeds to 37.1 N for hemp seeds.

Hemp Stem

The basic cross-sectional analysis of hemp stem gives different layers, which have different representations. The center of the stem is mainly

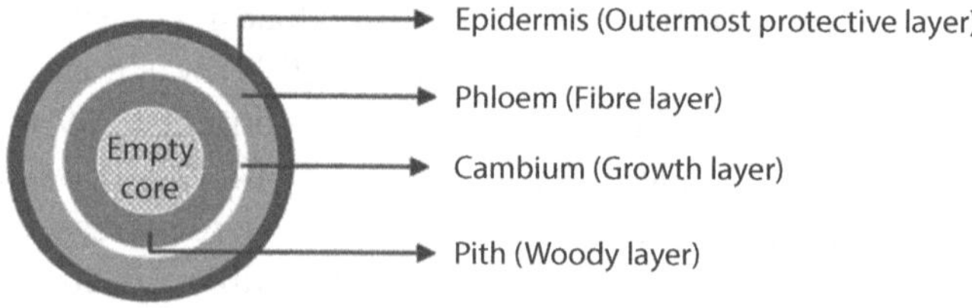

Figure 5.4 Cross-section of hemp stem.

found hollow. The outermost layer is called epidermis from the direct effect of anything on phloem, i.e., the bast fiber 30% to 40% depending on the variety of plant. The epidermis and fiber layer are in direct contact with each other with no space b/w them. The third layer in this series is the cambium named as growth layer 10 to 50 mm, which signifies its work, i.e., growing the woody layer inside of it and fiber layer outside of it. The woody layer is called pith and is used as fuel wood after removing fiber from it. The central part is called the core or cortex is 100 to 300 mm, which is hollow in the cannabis plant's stem. The hemp stem are mainly 1.5 to 2.5 m tall and 5 to 15 mm in diameter [101] the best fibers are further divided into primary and secondary fibers and a variation in the bottom and top dimensions of the fiber was observed [78]. The cross-sectional view of hemp stem can be seen in Figure 5.4.

5.6 Bioactive Compounds

Essential and nonessential substances (e.g., vitamins or polyphenols), which occur naturally, are part of the food chain and can be proven to have an influence on human health and are known as bioactive compounds [17]. Cannabis sativa L. has a number of bioactive chemicals, many of which have previously been identified, and study continues in many parts of the world to uncover the rest of the plant's secrets. Hundreds of different compounds with potential biological activity have been identified as a result of the extensive research that has made Cannabis the most studied plant in human history (Table 5.2), including more than 120 terpenoids, 100 cannabinoids, 50 hydrocarbons, 34 glycosidic compounds, 27 nitrogenous compounds, 25 noncannabinoid phenols, 22 fatty acids, eleven steroids, nine trace elements, seven simple alcohols, two colors, and vitamin K [27].

Table 5.2 Bioactive compounds in various parts of hemp plant.

Plant parts	Bioactive compounds	Classes	Bioactive properties	Reference
Inflorescence	Cannabinoids	Cannabidiol (CBD)	Antioxidant, anti-inflammatory, anticonvulsant, anxiolytic, neuroprotective, and antibiotic properties	[3, 4, 32, 50]
		Cannabigerol (CBG)	Anti-inflammatory, antimicrobial , analgesic activities	[4, 20, 50, 56]
		Cannabidiolic acid (CBDA)	Antimicrobial and anti-nausea activities	[4, 50, 56]
		Cannabigerolic acid (CBGA)	-	[87]
Inflorescence	Flavanoids	Cannflavin A and Cannflavin B	Anti-inflammatory property, anti-inflammatory, antimicrobial, neuroprotective, anti-proliferative properties	[9]

(*Continued*)

Table 5.2 Bioactive compounds in various parts of hemp plant. (*Continued*)

Plant parts	Bioactive compounds	Classes	Bioactive properties	Reference
Inflorescence	Polyphenols	Dihydrostilbenoid (canniprene)	Anti-inflammatory activity Inhibit the production of pro-inflammatory eicosanoids	[7, 91]
Inflorescence	Terpenes	Monoterpenes (myrcene)	Anti-inflammatory, analgesic, and anxiolytic activities	[9, 16]
		Sesquiterpene (β-caryophyllene)	Anti-inflammatory agent, gastric cytoprotector activity	[97, 98]
Seeds	Phenolic compounds	N-trans-cumaroyltyramine	Anti-neuroinflammatory property	[118]
Seeds	Phenolic compounds	3,3'-demethyl-grossamide	DPPH radical scavenging property Anti-neuroinflammatory property	[117, 118]
Seeds	Phenolic compounds	Grossamide	Anti-neuroinflammatory activity	[119]

(*Continued*)

Table 5.2 Bioactive compounds in various parts of hemp plant. (*Continued*)

Plant parts	Bioactive compounds	Classes	Bioactive properties	Reference
Seeds	Phenolic compounds	N-trans-caffeoyltyramine	Arginase inhibitory activity, antioxidant activity, LDL protection against oxidation	[19, 28, 117]
Seeds	Phenolic compounds	Cumaroylaminobutanol glucopyranoside	Inhibition of TNF α release Inhibition of the NFkB inflammatory pathway	[114, 119]
Seeds	Flavonoids	Quercetin, kaempferol and isorhamnetin	Antioxidant, anticonvulsant, anti-proliferative, antihypertensive and anti-inflammatory properties	[41]
Leaves	Cannabinoids	Cannabispirketal, α-cannabispiranol 4'-O-β-D-glucopyranose	Broad-spectrum antitumor property Inhibiting cell proliferation and promoting apoptosis	[47]
Roots	Triterpenoids	Friedelin and Epifriedelinol	Antioxidant activity	[59]
	Phytosterols	Campesterol, Stigmasterol, β-Sitosterol, Stigmastanol, Fucosterol	Antioxidant activity	[59]
Stem	Triterpenoids	Friedelin and epifriedelinol	Antioxidant activity	[59]

Cannabinoids

The main bioactive compound of the hemp plant, which makes it different from other weed crops, is the presence of cannabinoids. Cannabinoids are the main active compounds found naturally in the cannabis plant also called phytocannabinoids differentiating themselves from other cannabinoids consisting of medicinal properties in them. They are found in different structures and they are compiled under 10 major structural groups. It shows great pharmacological importance as they get bind with cannabinoid receptors in the human body. However, cannabis belongs to terpephenols, which generally come under alcohol with the chemical formula $C_{21}H_{26}O_2$. The cannabinoids are further classified into 11 major groups as: cannabichromene (CBC), cannabidiol (CBD), cannabielsoin (CBE), cannabigerol (CBG), cannabicyclol (CBL), cannabicyclol (CBL), cannabinol (CBN), cannabidiol (CBND), cannabitriol (CBT) Δ8-trans-tetrahydrocannabinol (Δ8-THC), Δ9-trans tetrahydrocannabinol (Δ9-THC), and miscellaneous-type cannabinoids (Figure 5.5). Δ9-tetrahydrocannabinol (Δ9-THC) and cannabidiol (CBD) are the two main cannabinoids, which are the most researched out of all the cannabinoids and also found most prominent of all. When compared to plants cultivated indoors, cannabis grown outdoors has lower levels of cannabinoids. Cannabinoids could also provide protection against UV radiation, pathogens (antibiotic effect) and predators. The cannabinoid concentrations (in the percentage of dry weight plant material) in different plant parts in descending order Flowers > Leaves > Stems > Roots (Table 5.3). The concentrations in seeds are found in traces and majorly on the outer side of the seed coat that is mainly from contamination with Cannabis resin from the flowers or leaves. Also, the concentration in drug type cannabis is found more than the fiber type [97]. The hemp seed oil doesn't contain CBD itself but it gets contaminated with it at the time of extraction of oil by pressing method [73].

Alkaloids

Alkaloids are the class of nitrogen-containing organic compounds that are not only found in plants but also found in microorganisms like fungi, which shows different types of physiological effects on humans and other animals. The nitrogen atoms, be single or more, can also contain oxygen, sulfur, chlorine, bromine, or phosphorus atom attached to compound. Spermidines seem to be widely distributed in the plant kingdom majorly substituted with cinnamic acid derivatives, caffeic acid, ferulic acid, and sinapic acid-forming aromatic amide substituents [46]. The spermidines isolated from *Cannabis sativa* include cannabisativine

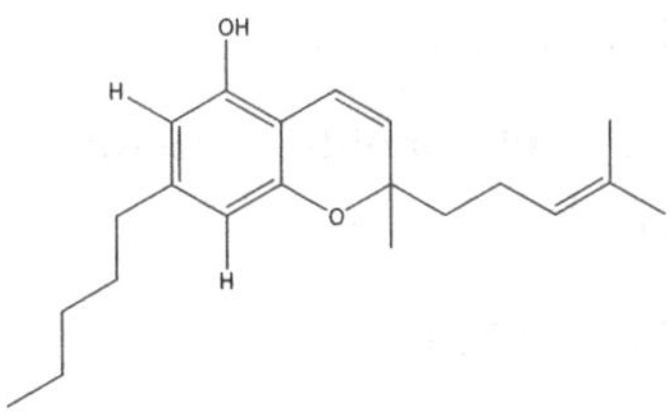

cannabichromene (CBC)

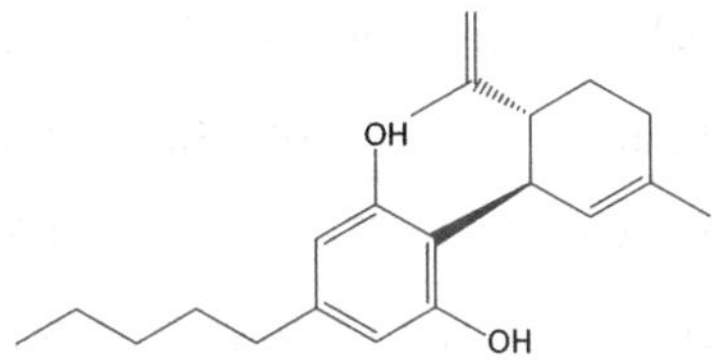

cannabidiol (CBD)

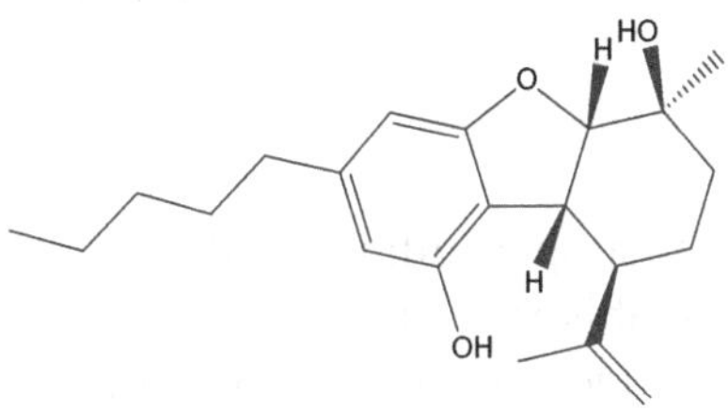

cannabielsoin (CBE)

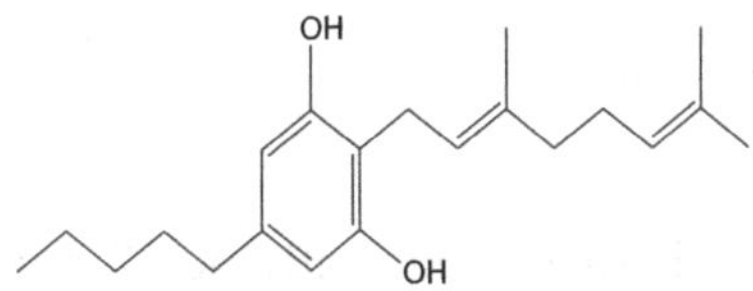

cannabigerol (CBG)

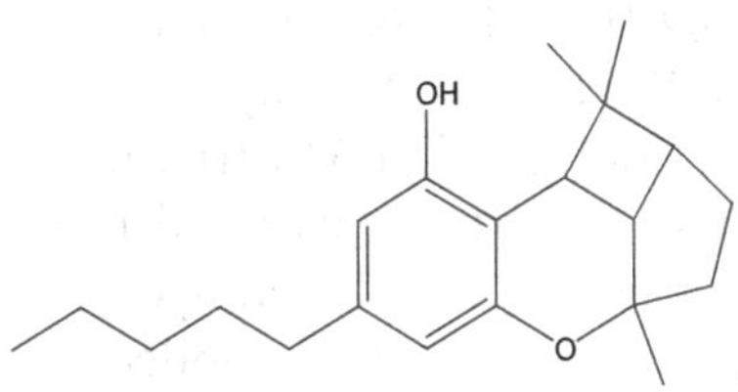

c

annabicyclol (CBL)

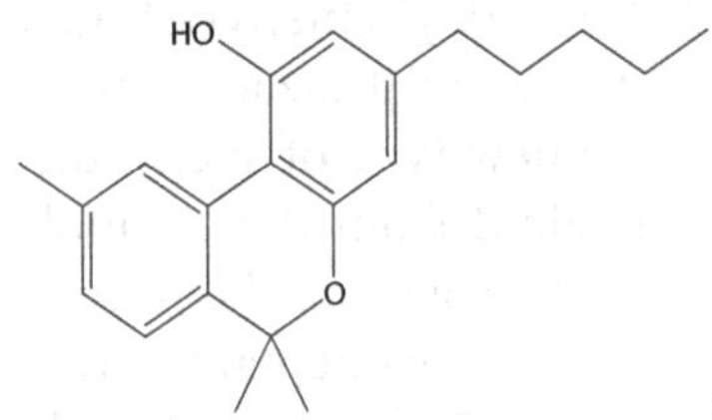

c

annabinol (CBN)

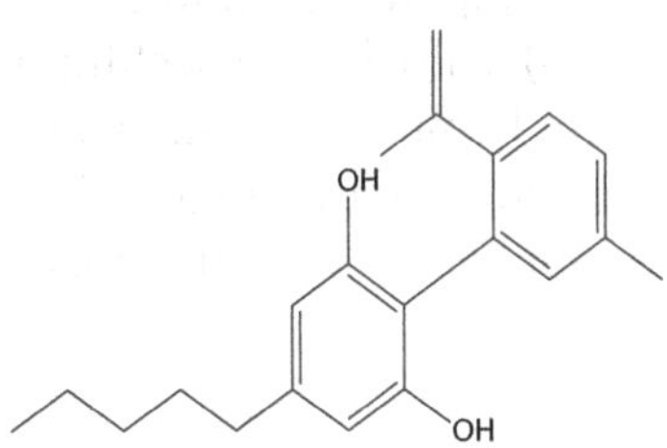

cannabinodiol (CBND)

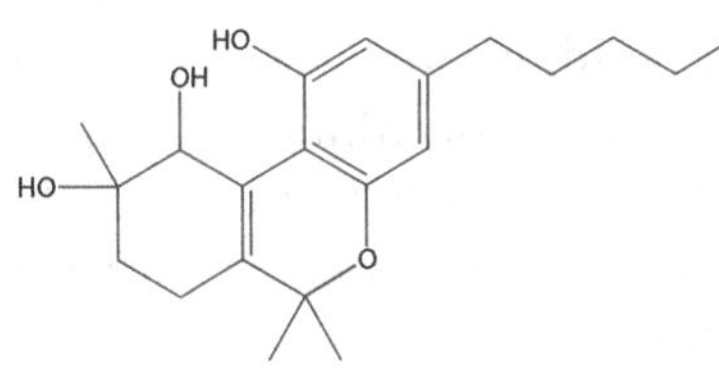

cannabitriol (CBT)

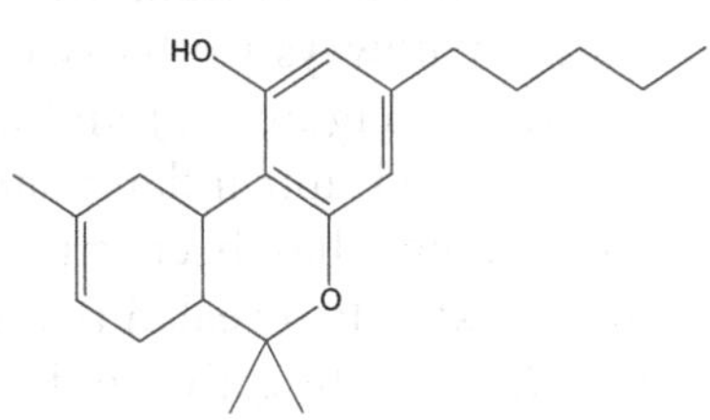

Δ8-trans-tetrahydrocannabinol (Δ8-THC)

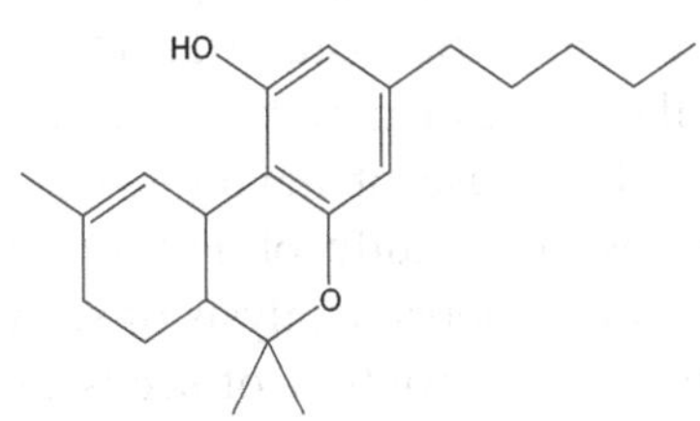

Δ9-trans tetrahydrocannabinol (Δ9-THC)

Figure 5.5 Structure of cannabinoid compounds in cannabis.

Table 5.3 Bioactive compounds in different plant parts of hemp.

Plant parts	Cannabinoids	Terpenoids	Flavonoids	Sterols
Inflorescence	15.77-20.37%	1.28- 2.14%	0.07-0.14%	-
Leaves	1.10-2.10%	0.13-0.28%	0.34-0.44%	-
Stem bark	-	0.05-0.15%	-	0.07-0.08%
Root	-	0.13-0.24%	-	0.06-0.09%

Jin *et al.* [59].

and anhydrocannabisativine [71], the first spermidine alkaloid was isolated from a methanolic extract of cannabis roots from a Mexican variant grown in Mississippi by using repeated chromatography and X-ray crystallography and was identified as cannabisativine with chemical name 13-(1,2-dihydroxyheptsnyl)-1,4,5,6,7,8,9,10,11,13,16,16a-dodecahydropyrldo[2,1-~][1,5,9]trlazacyclotrldec~n-2(3H)-one. Elsohly *et al.* [37] isolated the new type of spermidine (pyrido[2,1-d] [1,5,9] triazacyclotridecine) alkaloid from the ethanol extract of dried leaves and roots of Mexican variety of *Cannabis sativa* L. (marijuana) called as anhydrocannabisativine [13-(2-oxoheptanyl)-1,4,5,6,7,8,9,10,11,13,16,16a dodecahydropyridoI2,l-d] [1,5,9]triazacyclotridecine-2(3H)-one and also compared the structure of compound with the former. Yan *et al.* [116] have reported the isolation of two pairs of stereoisomers of diketopiperazine indole alkaloid (12S, 22R)-dihydroxyisoechinulin A, (12S, 22S)-Dihydroxyisoechinulin A and (12R/S)-Neoechinulin A (Figure 5.6) from hemp seeds which are also extracted previously from *Aspergillus* and *Eurotium* type fungi. Alkaloids Occurring from *Cannabis sativa* are commonly called cannabinaceous alkaloids.

Flavonoids

Radwan *et al.* [94] have reported 34 flavanoids in their review. The latest research by Bautista *et al.* [14] reported the pathways for biosynthesis of flavonoids cannaflavin A and cannaflavin B (Figure 5.7) and also discussed the use of flavonoids in antioxidative and anti-inflammatory properties. The study of Futura-75 variety of hemp has been studied. Spectrophotometric analysis of seed and sprout extracts revealed that the total flavonoids content of seeds extract was 2.8 mg QE/g FW and 2.93 QE/g DW [43]. Hemp flavonoids show the property of natural antioxidants. The extraction of flavonoids from hemp leaves was done by using

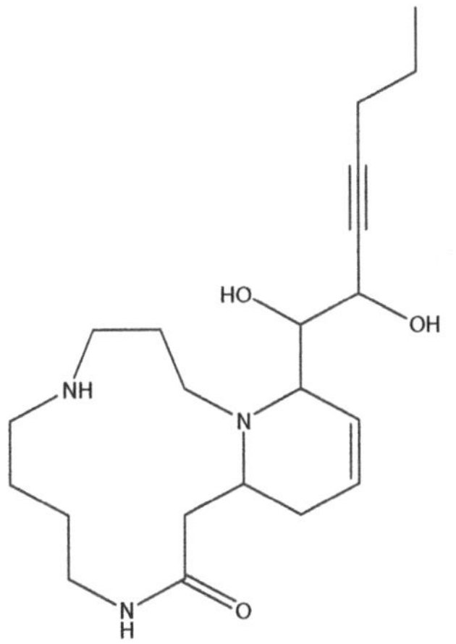

[13-(1,2-dihydroxyheptsnyl)-1,4,5,6,7,8,9,10,11,13,16,16a-dodecahydropyrido[2,1-d][1,5,9]triazacyclotridecine-2(3H)-one]

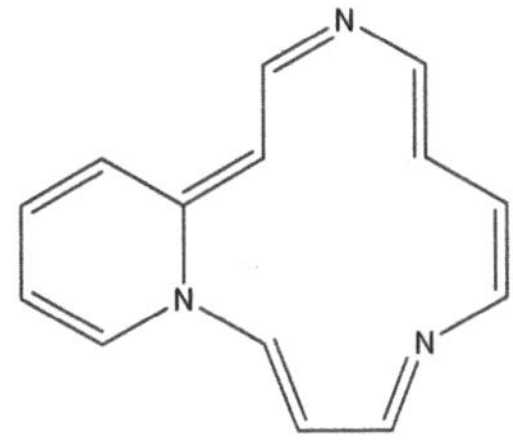

(pyrido[2,1-d] [1,5,9]triazacyclotridecine)

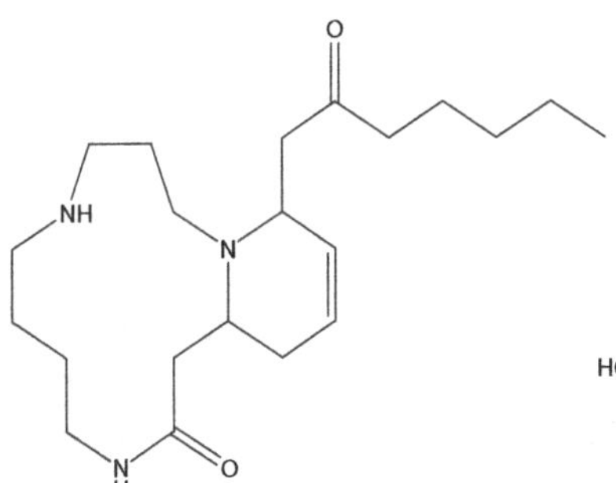

anhydrocannabisativine[13-(2-oxoheptanyl)-1,4,5,6,7,8,9,10,11,13,16,16a-dodecahydropyrido[2,1-d][1,5,9]triazacyclotridecine-2(3H)-one]

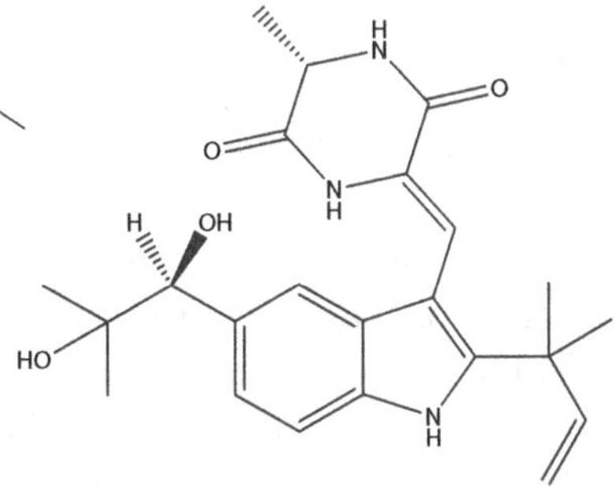

(12S, 22R)-Dihydroxyisoechinulin A

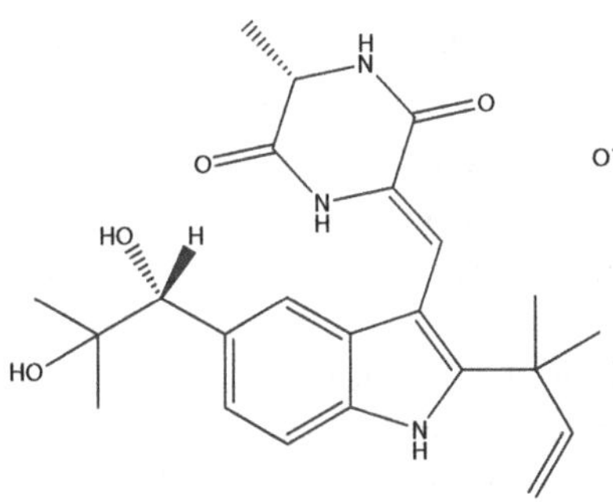

(12S, 22S)-Dihydroxyisoechinulin A

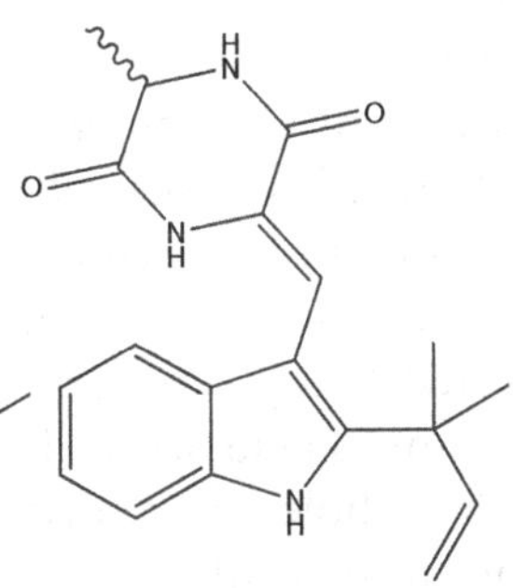

(12R/S)-Neoechinulin

Figure 5.6 Structures of different alkaloid compounds in hemp.

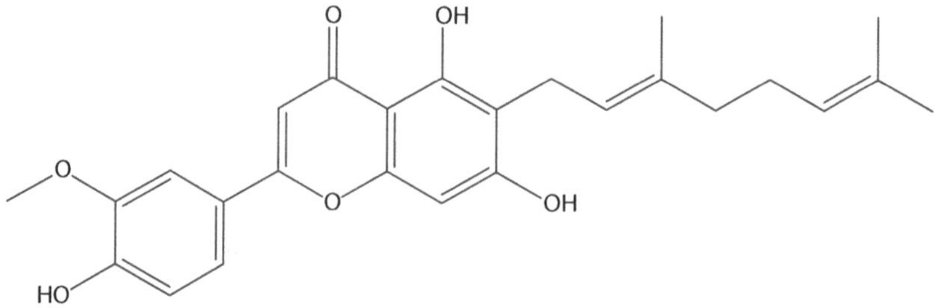

cannaflavin A

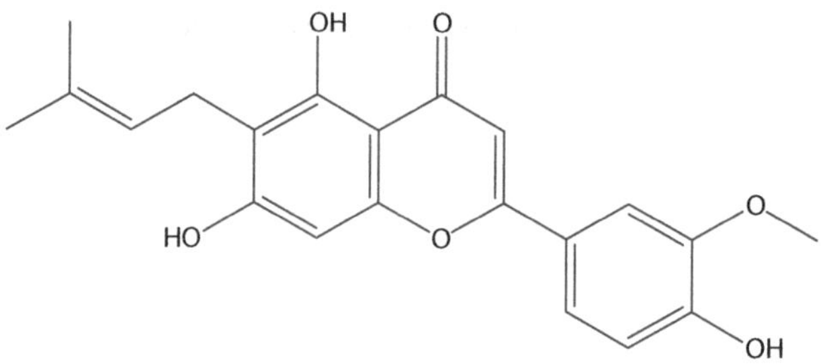

cannaflavin B

Figure 5.7 Structure of different flavonoid compounds in hemp.

microwave-assisted extraction technique and maximum yield was at 69.15% ethanol, 31.69 mL/g solvent-to-solid ratio, extraction duration of 25.14 minutes, and extraction temperature of 69.96°C. The experimental yield of 3.04% was consistent with the projected extraction yield of 3.06% [25]. The percentage flavonoid concentration in different plant parts of hemp is given in Table 5.3.

Proteins, glycoproteins, and enzymes

The amino acids are mainly present in leaves, stem and seeds of hemp plant [10]. The seed of hemp generally contains 25% protein content by weight, which mainly consists of high-quality storage proteins and is easily digestible in nature named as edestin (commonly known as globulin) and albumin [22]. House *et al.* [52] investigated the limits for percentage. For decorticated hemp seed, protein digestibility and protein digestibility corrected amino acid score (PDCAAS) values of up to 92% and 66%, respectively, for whole hemp seed, respectively. Lysine and leucine are the main limiting amino acids with scores of less than 1.0, indicating that they cannot complete the amino acids requirement of an individual. The highest amount of glutamine and arginine is found in dehulled hemp seed with an amino acid score ranging from 5.17 to 7.21 and 4.04 to 5.31, respectively

[52]. Because of the increased needs during pathological situations, both arginine and glutamine, which are nonessential amino acids, have now become "conditionally essential," and metabolism may not be able to maintain their concentration at appropriate levels to fulfil metabolic requirements [36].

Terpenoids

Terpenoids are organic aromatic compounds that give cannabis smell and this 5 carbon isoprene compound is called terpenes and shows vast usage in herbal remedies [29, 86]. Percentage terpinoids in different plant parts of hemp mentioned in Table 5.3. Terpenes should be extracted from hemp plants parts initially as it get usually get lost during the thermal processing of cannabis foods and are largely used to add flavors after processing. The maximal extraction temperature was reported to be 40°C without any loss of monoterpenes (β-myrcene) and sesquiterpenes (β-caryophyllene) (Figure 5.8) in hemp plant inflorescences using the HPLC method. Semiquantitative profile monoterpenes were the most represented class of compounds among volatiles using the GC method of volatile terpenes extraction, accounting for 47% to 89% of the total peak area. However, in one case, the total area of sesquiterpenes (52% of the total area) was found to be higher than that of monoterpenes [87]. Bakro *et al.* [12] discovered that enhanced rapid gas chromatography with flame ionization detection made it possible to analyze 29 terpenes and CBD in one run.

Phenolic compounds and ketones

There are 42 phenols are identified in hemp so far out of which 16 are spiro-indans, 12 are dihydrostilbenes, 7 are dihydrophenanthrenes, 7 are simple phenols. The new forms of phenol of cannabis are also called cannabispirans [94]. Caffeoyltyramine, cannabisin A, B, and C, as well as amino acids and saccharides, are abundant in the seeds and sprouts of Cannabis

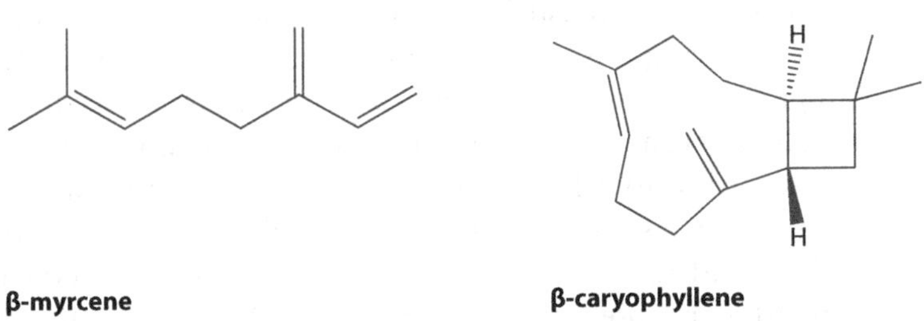

Figure 5.8 Structure of different terpene compound in hemp.

sativa. Ketones are organic molecules that include the carbonyl group, which is a covalent link between a carbon atom and an oxygen atom.

5.7 Pharmacological Properties

Clinical trials for hemp by using phytochemicals present in the different parts of the hemp crop are been done from a long time ago and have shown great results for the treatment of different illnesses (Table 5.4). Beyond basic nutrition, phytochemicals are defined as the release of bioactive plant compounds in fruits, vegetables, grains, and other plant foods, as well as in the structural portions of the plant, that may provide desirable benefits such as illness cure and infection defense [77]. These compounds are secreted by plants to defend themselves against pollution, stress, dehydration, UV radiation, and other pathogenic attacks. The flavour, texture, smell, and color of the plant are also influenced by these bioactive compounds [30]. *Cannabis sativa* L. not only contains a number of phytochemicals also each one of them is found to be successful to cure different chronic diseases in clinical trials done so far. The main bioactive compounds of the hemp plant, i.e., the cannabinoids are found to have a lot of therapeutic benefits. Different cannabinoids show up remedial effects on particular diseases. Further, the usage of hemp phytochemicals in the drug industry should also be checked and inclusions of them in remedial drugs should be done after proper clinical trials. The studies till now are not found to be sufficient for the pharmacological approach of hemp in many chronic diseases as it shows many adverse drug effects (ADE) in many trials and also be screened for potential Drug-drug interactions (DDI) [21].

Migraine

A severe case of migraine can completely affect a person's daily life as it creates mild to extreme pain usually on the one side of the head which can be one time or recurring. Visual disturbances such as flashes of light, blind spots, and tingling on one side of the face, arm, or leg, as well as difficulties speaking, are all signs of an aura. Migraine is also referred to as a severe headache disease. According to WHO, the global prevalence of current headache disorder among adults is estimated to be over 50%. Half to three-quarters of adults in the world aged 18 to 65 years had had a headache in the previous year, with 30% or more reporting migraine. 1.7% to 4% of the world's adult population suffers from headache on 15 or more days per month.

Table 5.4 Potential health benefits of hemp and its plant parts.

Plant part/ products	Health effect	Experimental model	Outcome of the study	Reference
Hempseed	Hypocholestermic effect	Rats	Significantly reduced the average fasting serum low-density lipoprotein, Increased the fasting serum high-density lipoprotein improved total protein content	[63]
Hempseed and hempseed oil	Improvement in nutritional profile	Hen	Increased omega-3 polyunsaturated fatty acid content Improved color intensity of egg yolks No negative impact on the sensory profiles of the cooked eggs	[45]
Whole dried hemp flowers	Treatment for headache & migraine	Human studies (699 participants, both males and females)	94 participants experienced symptom relief within a two-hour observation Males experienced greater relief than females Younger participants (< 35 years) experienced greater relief than older participants	[104]
Ethanolic extract of leaves and stems	Antimicrobial property	Bacteria	Very good antibacterial activity against *Staphylococcus auerus*	[18]

(*Continued*)

Table 5.4 Potential health benefits of hemp and its plant parts. (*Continued*)

Plant part/ products	Health effect	Experimental model	Outcome of the study	Reference
Hemp seed oil	Antimicrobial activity	Bacteria	Highest antibacterial activity against *Bacillus subtilis, Staphylococcus aureus* Moderate activity against *Escherichia Coli*	[5]
Hemp plant (Resin)	Multiple sclerosis	Human studies	Administration of THC and CBD in a 1:1 ratio has proven to be a well-tolerated antispasticity therapeutic solution	[93]
Hemp hurd powder	Antibacerial effect	Bacteria	Inhibition the growth of *Escherichia coli*	[65]
Hemp protein hydrolysate	Antioxidant effect	Human studies	Significant reduction in radical scavenging activity,Fe^{+2} chelating activity at 2 hour of hydrolysis	[96]
CBD solution derived from hemp	Prevention of seizures	Human studies (Age group: 30 years) Dose: 2–5 mg/kg/day	Decreased motor seizures by 36.5%	[34]
CBD gel derived from hemp	Prevention of pain during osteoarthritis	Human studies Dose: 250 mg/day	Qualitative reductions in knee pain scores	[53]
CBD solution	Anti-diabetic effect	Human studies Dose: 100 mg, twice daily for 16 weeks		[57]

The endocannabinoid system, which is involved in inflammatory and neuropathic pain processes, is made up of the central and peripheral nerve systems. Cannabinoids of cannabis plants act on the nervous system from where the pain springs and are proved for treatment of classified pain disorders of headache, including chronic headaches, medication overuse headache, cluster headache, and migraine headaches, especially chronic migraine headaches Trigeminal neuralgias associated with multiple sclerosis (MS) and idiopathic intracranial hypertension [13]. The study of the effect of consumption of cannabis flowers on headache and migraine on the different age groups of people and among males and females. The best relief in symptoms is found in younger ones as compared to older ones and at low THC levels < 10%. According to the "dopamine pathway hypothesis," cannabis can also be used as a dopamine antagonist because dopaminergic activation can generate headache symptoms [104].

Antidiabetic

Diabetes mellitus is a chronic and progressive disease that happens when blood glucose levels are too high. Blood glucose is our body's main source of energy, which is obtained from the food we eat via insulin. Diabetes can affect any region of the body and even cause death if left untreated. According to a WHO research [26], the yearly cost of diabetes worldwide is more than US$ 827 billion, and the International Diabetes Federation (IDF) predicts that total global health care spending on diabetes more than quadrupled between 2003 and 2013. Till now, no direct studies found to exist for the cure of diabetes with hemp but indirect approaches have been made by many scientists by taking metabolism as a cause of diabetes [70, 102]. The problem of obesity was found to be a major cause of diabetes and cannabinoid CB1 antagonists represented as a potential approach for the reduction of body weight and treatment of obesity and metabolic disturbances [54]. Similarly, ENV-2 is a molecular hybrid of rimonabant, a well-known selective CB1 receptor antagonist that could lead to the development of new CB1 receptor antagonists that could help treat diabetes, obesity, and hyperlipidemia [51].

Treatment of AIDS

Hemp's terpeno-phenolic chemicals, which provide the "high" associated with marijuana smoking. There has been a lot of interest in using it to treat a variety of illnesses in the United States, including glaucoma, AIDS wasting, neuropathic pain and treatment of spasticity associated with multiple sclerosis, and chemotherapy-induced nausea. Cannabis has been demonstrated to be effective as a cancer and AIDS treatment in clinical trials [62]. The loss of weight is found to be a common problem in many cases of HIV-AIDS and CBD used in treatments leads to further decrease in weight and loss of appetite so caution should be taken and the drug should be taken under prescribed quantities [21]. In terms of antiviral activity, medical Cannabis has been found to be taken as a supplement by HIV/AIDS patients to relieve neuropathic pain, wasting, nausea, and vomiting [92]. The limited and clinically approved dosage routines of cannabidioil may not show the direct effect on HIV-AIDS but it can lead to suppressing the other diseases occurring due to HIV-AIDS.

Treatment of Alzheimer

The loss of neurons and synapses in the cerebral cortex and certain subcortical regions is a hallmark of Alzheimer's disease. affects memory, thinking and behavior and becomes so critical that they start creating a barrier in daily tasks. There were numerous attempts to figure out the treatment, therapy, remedy, or even antidote of this disease but none has shown promising results except for this recent clinical trial that found evidence that cannabinoids (CBD) could help reduce dementia. The study conducted by Watt and Karl [115], provides "proof of principle" exploring the medicinal potential of CBD and maybe CBD-THC combos in the treatment of Alzheimer's disease. "The phytocannabinoid cannabidiol (CBD) is a strong option for this novel therapy technique," the researchers stated. *In vitro* studies have shown CBD to be neuroprotective [39], to prevent hippocampal and cortical neurodegeneration [49], to have anti-inflammatory and antioxidant properties [83], to reduce hyperphosphorylation [38], and to regulate microglial cell migration [80, 111]. CBD has also been demonstrated to protect against A-mediated neurotoxicity and microglial-activated neurotoxicity [58], to reduce A production by inducing APP ubiquitination [103], and to increase cell survival [15]. These characteristics Although the study has been tested clinically for animals, it provides promising preliminary data and factual translation for any future studies that will be provided for human respondents. Furthermore, it has been a proven science that CBD can be readily available, has limited to no side effects, and is safe for human consumption. The results of different analogous of cannabidiol have been recently studied by Li *et al.* [75], the CBD acts as an inhibitor and generates a protective response and interacts with the endocannabidiol system

(ECS). Also, 2-AG and AEA are the two endocannabinoids that have relevant actions.

5.8 Processing Technologies (Methods and Effects)

Herbal weeds like hemp which contains lots of phytochemicals associated with different parts of the plant need critical analysis and processing methods. Processing is done to improve nutrient bioavailability and functionality, remove dangerous anti-nutrients, and make the product more appealing to consumers. Various methods, like drying, crushing, freezing, cleaning, grading, sorting, storage, are always considered as the basic processing methods employed in the past and even now. The traditional thermal processing processes for the preservation of the extracted compounds increase their shelf life and are made available for human consumption. Many modern technologies, like pulse electric field, radiolysis, high-pressure processing, etc., are majorly incorporated in the processing of critical weeds and have shown high efficiency as compared to traditional methods. Many companies are working since 2001 on processing hemp and making hemp-based raw materials and products like hemp fiber for paper and cloth making, hemp seed for protein and oil extraction and hemp hurds for construction [64]. A little usage of processed hemp has also been seen for animal bedding, nesting, and feeding.

Hemp seed

Hemp seeds can be consumed as raw food by just roasting them in a pan but industrial processing gives it a uniformity by cleaning, grading, sorting, and removes the seed coat, which is hard in texture and inedible in nature for human consumption by dehulling. Further processing includes the concentration of hemp protein done by various methods and extraction of hemp seed oil by hot and cold pressing. The methodology and types of equipment used in these operations depend upon the requirement of the ultimate product. There is no published work found on dehulling of hemp seed but traditionally it is done by using other available dehullers, which gives low efficiency for hemp seed. The seeds of the Hliana cultivar of hemp was used to study the storage quality of hulled seeds for 12 months. The device used for dehulling of seeds consists of a horizontal disc type impeller which gave a maximum yield of 41% and maximum oil yield from hulled hemp seed kernels 19.3% was reported from the screw press [85]. Wang and Xiong [113] described the hemp seed protein into three types which are hemp protein meal

Table 5.5 Different types of hemp protein.

Type of hemp protein	Hemp protein (%)
Hemp Protein Meal	30 to 50%
Hemp Protein Concentrate	65%
Hemp Protein Isolate	90%

Wang and Xiong [113].

(HPM), hemp protein concentrate (HPC), hemp protein isolate (HPI). The extraction of hemp protein begins with deoiled cakes left over from hemp seed oil extraction. HPI > HPC > HPM represents the protein digestibility of hemp protein products (Table 5.5). HPI is commonly extracted using an alkaline technique followed by isoelectric precipitation. Different parameters, such as pH, temperature, extraction time, and centrifugal force, can all affect the HPI's percent yield [95]. Hemp seed protein has low functional qualities when compared to nutritional properties, according to Leonard *et al.* [74]. HPI had significantly poorer emulsifying activity, emulsion stability, and water-holding capacity than SPI (Soy Protein Isolate) [107]. The protein solubility was also low at a pH of less than 8. The functional properties are found to be highly dependent on the structural composition, which varied according to the pH of the ambient environment [79]. But the incorporation of processing technologies has shown immense improvement in the functional properties of hemp protein. The simple fractionation processing technique used to concentrate different compounds available in hemp seed meal found as a by-product from hemp oil extraction gave significant results [90]. Microwave treatment of hemp seed increased the oil yield, carotenoid, and other pigment content, and decreased the p-anisidine value, according to a study [84]. The only part in which the hemp seed oil lacks behind is its oxidative stability due to the availability of unsaturated fatty acids [81]. The pre-mature harvest of hemp seed leads to increased chlorophyll content in oil, which decreased its oxidative stability so complete maturing of seeds is considerable for oil processing.

Hemp extracts/oil processing method

The study of the processing of hemp oil should not be mixed with hemp seed oil as both are two different oils constituting different properties and extraction methods. Hemp oil is extracted mostly from the hemp plant's leaves and flowers and contains a higher ratio of THC than hemp seed oil, making it unfit for human consumption. Winterization, extraction, phase separation, and vacuum distillation are all typical methods for processing hemp oil. The investigation of the extraction of cannabinoid, terpene, and nonpolar contaminating solutes using various solvents such as CO2, ethanol, water, propane, or butane and triglyceride-based oils. for infusion in different food products was done [66]. The flowchart diagram of extraction process of hemp oil is shown in Figure 5.9.

Hemp fiber processing methods and their effects

The extraction of clean and without injuring the length of natural fiber is an important task. These are mainly done by removal of noncellulosic substances like pectin and lignin which acts as binding agents in between the cell wall of plant cells. In past, the retting of fiber is done by using warm water and spinning called the traditional method after that enzyme retting also became another popular method. In the water retting method, the dried hemp stalks are packed into a barrel containing water for approximately 2 weeks with a little addition of bleaching powder into it for the killing of molds. Nowadays chemical or mechanical retting of hemp bast fiber is mainly done by application of applying acidic and alkaline solutions on natural fiber [1, 64]. The removal of pectin from hemp bast is found to be easier as compared to the removal of lignin from the alkaline boiling combined with the chemical process. However, the increased concentration of sodium hydroxide can also lead towards the removal of the pectin layer easily [112]. Dioecious Hungarian Tisza variety of hemp was studied for frost retting with the structural analysis of it was done and the result showed the high concentration of pectin on leaf surfaces and lower lignin content. Placet *et al.* [89] characterized the processing effects on mechanical characteristics of hemp stalks and hemp silver laps (Hemp combed fibers) for Fedora 17 variety. In processing of silver laps, the nontwisted yarns made were not able to make minimum requirements of load to failure capacity which is 15 N for the manufacture of woven fabrics. Other samples made were named as low twisted untied and bound with a natural adhesive yarn, nontwisted yarns bound with natural adhesive yarn with load to failure capacity were, respectively, 19.2, 19.7, and 68.6 N. The field retted hemp for 39 days was found to have the maximum apparent rigidity and tensile strength with mean values of 17.5 and 660, respectively. And minimum shive content of 49.2%. In secondary processing, the low twisted bound ones are found to have a maximum tensile strength of 395 MPa. Flowchart for removal of hemp fiber from hemp stem is shown in Figure 5.10.

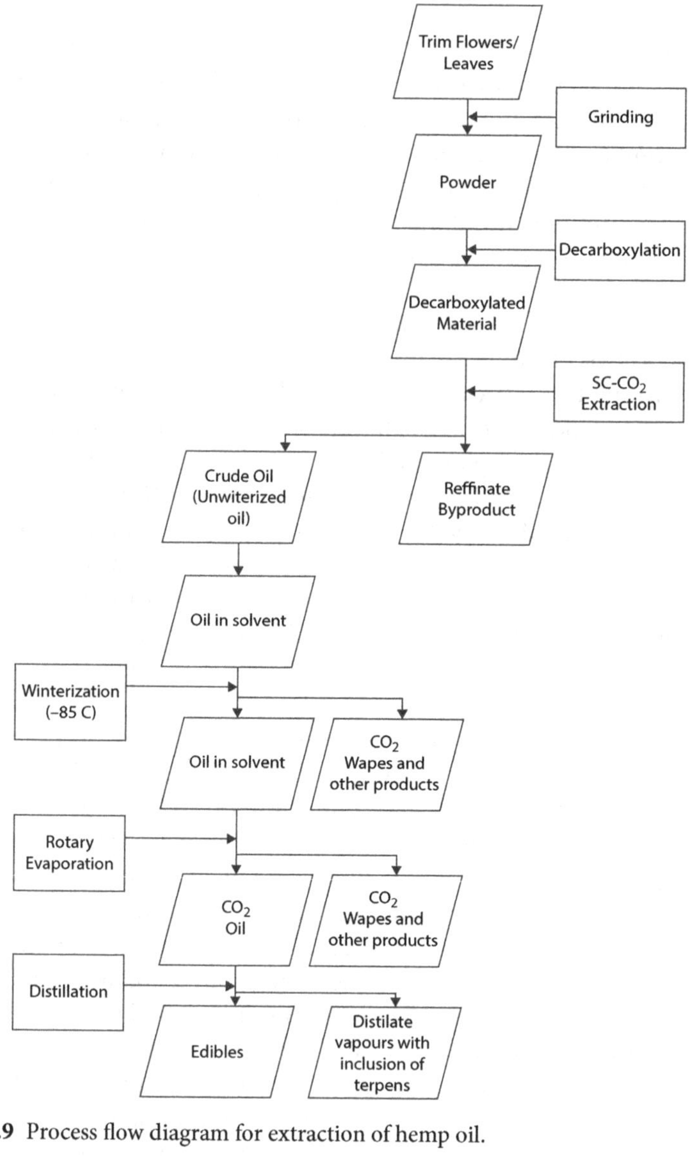

Figure 5.9 Process flow diagram for extraction of hemp oil.

Industrial Applications

Hemp has made a great place in food, pharmaceuticals, textiles, etc industries in small time. Multinational companies are finding interest in cannabis as an alternative solution to many raw materials, which are not so sustainable in nature. These products otherwise find a great market capture all over the world. Further, the research on hemp-based

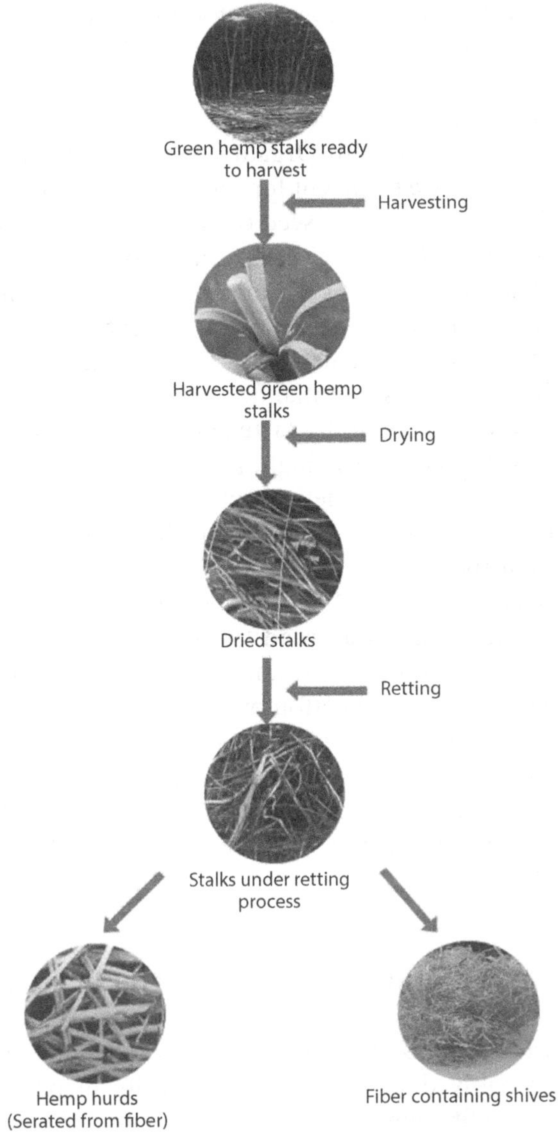

Figure 5.10 Process flow diagram for hemp fiber processing.

new products is also under consideration. The commercialization and retailing of these products need critical analysis about the addition of phytochemicals extracted from cannabis and the process of manufacturing cannabis-infused products. The advancement in acceptance of cannabis-infused foods can be done by mentioning the extraction method of particular phytochemical infused in food on the product label [66]. Hemp is the only plant on the planet that includes all of the human body's necessary fatty acids and amino acids. In comparison to its nutritious composition, cannabis seed has traditionally been underutilized. According to the World Health Organization (WHO), hemp seed protein products (HSPP) are sufficient for the protein requirements of newborns and children [107]. Hemp seed oil has been discovered to be high in omega 3 and omega 6 fatty acids, i.e., the polyunsaturated fatty acids (PUFAs) beneficial for human consumption. The recommended quantity for daily consumption (2–2.5% of caloric intake/day) can provide many nutritional benefits and also helps in reducing tumor growth and cancer possibilities as it has essential fatty acids content of 75% [73]. The virgin hemp seed oil gives a nutty taste and a little bitter aftertaste [81]. Also, the tocopherols content is high, i.e., between 80 and 110 mg/100 g depending on hemp seed varieties, with γ-tocopherol and α-tocopherol as the main tocopherols. Hemp seed oil is found useful in treating atopic dermatitis by changing the fatty acids profiles in plasma and improved clinical symptoms [23]. The absorbance results of UV-B and UV-C is seen which make it a broad spectrum UV protectant and can be used in cosmetics products like sunscreens, etc [84]. Many countries especially India are relying on cotton or flax for their fiber requirements. Alternative solutions to them are the need of the hour and Hemp can fit best into it. Hemp consists of great fiber properties which can cover the demand of the textile market and also the plant fiber continuous reinforcements (PFCs) like tapes, roving, textiles, and other materials for composites [89].

5.9 Conclusion and Prospects for the Future

The diverse positive effects of hemp and its acknowledged mode of action have been thoroughly documented. The majority of prior research on the nutritional benefits, phytochemical makeup, and therapeutic uses of hemp has been evaluated in this paper to determine its current status. The tremendous increase is seen in the demand for products obtained from natural sources as it is also leading towards consumer demand for hemp too. The hemp plant is an emerging oil and protein-rich plant. These types of plant material are becoming an important part of the food, pharmaceutical, FMCG, and nutraceutical industries. Hemp has traveled a long way and is also said to be a beneficial weed from the past. In many areas, people are found to worship hemp and take it as a gift of God, which shows its religious values and beliefs and makes it more valuable. But it has always been challenging to part to make hemp legalize for commercial production. Although some countries have come in front to take this step to legalize hemp and have set an example for others many countries are still thinking of it. The hemp product-making industries have captured a large market and gain huge profits. Although research has progressed in recent years in terms of understanding agronomic principles, nutritional and health advantages, processing technologies, and functional behavior of various components of the hemp plant, there is still more to learn. There is a need to increase the knowledge of consumers through R&D to increase consumer acceptance, which can be done by removing the barriers among the research centers, universities, industries, and public itself.

References

1. Adamsen, A.P.S., Akin, D.E., Rigsby, L.L., Chemical retting of flax straw under alkaline conditions. *Text. Res. J.*, 72, 9, 789–794, 2002.
2. Adesina, I., Bhowmik, A., Sharma, H., Shahbazi, A., A review on the current state of knowledge of growing conditions, agronomic soil health practices and utilities of hemp in the United States. *Agriculture*, 10, 4, 129, 2020.
3. Aizpurua-Olaizola, O., Soydaner, U., Öztürk, E., Schibano, D., Simsir, Y., Navarro, P., Etxebarria, N., Usobiaga, A., Evolution of the cannabinoid and terpene content during the growth of *Cannabis sativa* plants from different chemotypes. *J. Nat. Prod.*, 79, 324–331, 2016.
4. Alexander, S.P.H., Therapeutic potential of cannabis-related drugs. *Prog. Neuropsychopharmacol. Biol. Psychiatry*, 64, 157–166, 2016.
5. Ali, E.M., Almagboul, A.Z., Khogali, S.M., Gergeir, U.M., Antimicrobial activity of *Cannabis sativa* L. *Chin. Med.*, 3, 1, 1–12, 2012.
6. Allegret, S., *The history of hemp. Hemp: Industrial production and uses*, pp. 4–26, 2013, doi: 10.1079/9781845937935.0004.
7. Allegrone, G., Pollastro, F., Magagnini, G., Taglialatela-Scafati, O., Seegers, J., Koberle, A., Werz, O., Appendino, G., The bibenzyl canniprene inhibits the production of pro-inflammatory eicosanoids and selectively accumulates in some *Cannabis sativa* strains. *J. Nat. Prod.*, 80, 731–734, 2017.
8. Amaducci, S., Scordia, D., Liu, F.H., Zhang, Q., Guo, H., Testa, G., Cosentino, S.L., Key cultivation techniques for hemp in Europe and China. *Ind. Crops Prod.*, 68, 2–16, 2015.
9. Andre, C.M., Hausman, J.F., Guerriero, G., *Cannabis sativa*: The plant of the thousand and one molecule. *Front. Plant Sci.*, 7, 1–17, 2016.
10. Audu, B.S., Ofojekwu, P.C., Ujah, A., Ajima, M.N.O., Phytochemical, proximate composition, amino acid profile and characterization of Marijuana (*Cannabis sativa* L.). *J. Phytopharm.*, 3, 1, 35–43, 2014.

11. Bailoni, L., Bacchin, E., Trocino, A., Arango, S., Hemp (*Cannabis sativa* L.) seed and co-products inclusion in diets for dairy ruminants: A review. *Animals*, 11, 3, 856, 2021.
12. Bakro, F., Jedryczka, M., Wielgusz, K., Sgorbini, B., Inchingolo, R., Cardenia, V., Simultaneous determination of terpenes and cannabidiol in hemp (*Cannabis sativa* L.) by fast gas chromatography with flame ionization detection. *J. Sep. Sci.*, 43, 14, 2817–2826, 2020.
13. Baron, E.P., Medicinal properties of cannabinoids, terpenes, and flavonoids in cannabis, and benefits in migraine, headache, and pain: An update on current evidence and cannabis science. *Headache: J. Head Face Pain*, 58, 7, 1139–1186, 2018.
14. Bautista, J.L., Yu, S., Tian, L., Flavonoids in *Cannabis sativa*: Biosynthesis, bioactivities, and biotechnology. *ACS Omega*, 6, 8, 5119–5123, 2021.
15. Beccano-Kelly, D. and Harvey, J., Leptin: A novel therapeutic target in Alzheimer's disease? *Int. J. Alzheimer's Dis.*, 2012, 1–7, 2012.
16. Bertoli, A., Tozzi, S., Pistelli, L., Angelini, L.G., Fibre hemp inflorescences: From crop-residues to essential oil production. *Ind. Crops Prod.*, 32, 329–337, 2010.
17. Biesalski, H.K., Dragsted, L.O., Elmadfa, I., Grossklaus, R., Müller, M., Schrenk, D., Weber, P., Bioactive compounds: Definition and assessment of activity. *Nutrition*, 25, 11–12, 1202–1205, 2009.
18. Borchardt, J.R., Wyse, D.L., Sheaffer, C.C., Kauppi, K.L., Ehlke, R. G. F. N. J., Biesboer, D.D., Bey, R.F., Antimicrobial activity of native and naturalized plants of Minnesota and Wisconsin. *J. Med. Plants Res.*, 2, 5, 098–110, 2008.
19. Bourjot, M., Zedet, A., Demange, B., Pudlo, M., Girard-Thernier, C., *In vitro* mammalian arginase inhibitory and antioxidant effects of amide derivatives isolated from the hempseed cakes (*Cannabis sativa*). *Planta Med. Int. Open*, 3, e64–e67, 2017.
20. Brenneisen, R., Chemistry and analysis of phytocannabinoids and other Cannabis constituents, in: *Forensic Science and Medicine: Marijuana and the Cannabinoids*, 1st ed., M.A. ElSohly (Ed.), pp. 17–49, Humana Press Inc, Totowa, NJ, USA, 2007, ISBN 978-1-58829-456-2.
21. Brown, J.D. and Winterstein, A.G., Potential adverse drug events and drug-drug interactions with medical and consumer cannabidiol (CBD) use. *J. Clin. Med.*, 8, 7, 989, 2019.
22. Callaway, J.C., Hempseed as a nutritional resource: An overview. *Euphytica*, 140, 1, 65–72, 2004.
23. Callaway, J., Schwab, U., Harvima, I., Halonen, P., Mykkänen, O., Hyvönen, P., Järvinen, T., Efficacy of dietary hempseed oil in patients with atopic dermatitis. *J. Dermatol. Treat.*, 16, 2, 87–94, 2005.
24. Canadian Hemp Trade Alliance (CHTA), Available online: http://www.hemptrade.ca/eguide/production/fertility-in-organic-production.
25. Cao, J., Hao, L., Zhang, L., Xu, M., Ge, H., Kang, C., Wang, Z., Optimization of microwave-assisted extraction of total flavonoids from China-hemp leaves

and evaluation of its antioxidant activities, in: *International Conference on Applied Biotechnology*, pp. 555–567, Springer, Singapore, November, 2016.
26. Cecchini, M., Sassi, F., Lauer, J.A., Lee, Y.Y., Guajardo-Barron, V., Chisholm, D., Tackling of unhealthy diets, physical inactivity, and obesity: Health effects and cost-effectiveness. *Lancet*, 376, 9754, 1775–1784, 2010.
27. Cerino, P., Buonerba, C., Cannazza, G., D'Auria, J., Ottoni, E., Fulgione, A., Gallo, A., A review of hemp as food and nutritional supplement. *Cannabis Cannabinoid Res.*, 6, 1, 19–27, 2020.
28. Chen, T., He, J., Zhang, J., Li, X., Zhang, H., Hao, J., Li, L., The isolation and identification of two compounds with predominant radical scavenging activity in hempseed (seed of *Cannabis sativa* L.). *Food Chem.*, 134, 1030–1037, 2012.
29. Chhikara, N. and Panghal, A., Overview of functional foods. In Chhikara, Panghal, Chaudhary (Eds.), *Functional Foods*, pp. 1–20, Wiley and Scrivener, USA, 2022.
30. Chhikara, N., Abdulahi, B., Munezero, C., Kaur, R., Singh, G., Panghal, A., Exploring the nutritional and phytochemical potential of sorghum in food processing for food security. *Nutr. Food Sci.*, 49, 2, 318–332, 2018.
31. Christison, A., On the natural history, action, and uses of Indian hemp. *Mon. J. Med. Sci.*, 4, 19, 26, 1851.
32. Corsi, L., Pellati, F., Brighenti, V., Plessi, N., Benvenuti, S., Chemical composition and *in vitro* neuroprotective activity of fibre-type *Cannabis sativa* L. (hemp). *Curr. Bioact. Comp.*, 15, 2, 201–210, 2019.
33. Desanlis, F., Cerruti, N., Warner, P., Hemp agronomics and cultivation, in: *Hemp Industrial Production and Uses*, P. Bouloc (Ed.), pp. 98–124, CABI, Wollingford, UK, 2013.
34. Devinsky, O., Marsh, E., Friedman, D., Thiele, E., Laux, L., Sullivan, J., Cannabidiol in patients with treatment-resistant epilepsy: An open-label interventional trial. *Lancet Neurol.*, 15, 3, 270–8, 2016, https://doi.org/10.1016/s1474-4422(15)00379-8.
35. Di Bari, V., Campi, P., Colucci, R., Mastrorilli, M., Potential productivity of fibre hemp in southern Europe. *Euphytica*, 140, 25–32, 2004.
36. Dioguardi, F.S., Clinical use of amino acids as dietary supplement: Pros and cons. *J. Cachexia Sarcopenia Muscle*, 2, 2, 75–80, 2011.
37. Elsohly, M.A., Turner, C.E., Phoebe Jr., C.H., Knapp, J.E., Schiff Jr., P.L., Slatkin, D.J., Anhydrocannabisativine, a new alkaloid from *Cannabis sativa* L. *J. Pharm. Sci.*, 67, 1, 124–124, 1978.
38. Esposito, G., De Filippis, D., Carnuccio, R., Izzo, A.A., Iuvone, T., The marijuana component cannabidiol inhibits β-amyloid-induced tau protein hyperphosphorylation through Wnt/β-catenin pathway rescue in PC12 cells. *J. Mol. Med.*, 84, 253–258, 2006a, doi: 10.1007/s00109-005-0025-1.
39. Esposito, G., De Filippis, D., Maiuri, M.C., De Stefano, D., Carnuccio, R., Iuvone, T., Cannabidiol inhibits inducible nitric oxide synthase protein expression and nitric oxide production in β-amyloid stimulated PC12

neurons through p38 MAP kinase and NF-κB involvement. *Neurosci. Lett.*, 399, 91–95, 2006b, doi: 10.1016/j.neulet.2006.01.047.

40. Fadel, D., Assaad, N., Alghazal, G., Hamouche, Z., Lazari, D., "Finola" Cannabis cultivation for cannabinoids production in Thessaloniki-Greece. *J. Agric. Sci.*, 12, 7, 172–181, 2020.
41. Faugno, S., Piccolella, S., Sannino, M., Principio, L., Crescente, G., Baldi, G.M., Pacifico, S., Can agronomic practices and cold-pressing extraction parameters affect phenols and polyphenols content in hempseed oils? *Ind. Crops Prod.*, 130, 511–519, 2019.
42. Faux, A.M., Draye, X., Lambert, R., d'Andrimont, R., Raulier, P., Bertin, P., The relationship of stem and seed yields to flowering phenology and sex expression in monoecious hemp (*Cannabis sativa* L.). *Eur. J. Agron.*, 47, 11–22, 2013.
43. Frassinetti, S., Moccia, E., Caltavuturo, L., Gabriele, M., Longo, V., Bellani, L., Giorgetti, L., Nutraceutical potential of hemp (*Cannabis sativa* L.) seeds and sprouts. *Food Chem.*, 262, 56–66, 2018.
44. Galic, M., Perčin, A., Zgorelec, Ž., Kisić, I., Evaluation of heavy metals accumulation potential of hemp (*Cannabis sativa* L.). *J. Cent. Eur. Agric.*, 20, 2, 700–711, 2019.
45. Goldberg, E.M., Gakhar, N., Ryland, D., Aliani, M., Gibson, R.A., and House, J.D., Fatty acid profile and sensory characteristics of table eggs from laying hens fed hempseed and hempseed oil. *J. Food Sci.*, 77, 4, S153–S160, 2012.
46. Guggisberg, A. and Hesse, M., Putrescine, spermidine, spermine, and related polyamine alkaloids, in: *The Alkaloids: Chemistry and Pharmacology*, vol. 22, pp. 85–188, Academic Press, Cambridge, Massachusetts, 1984. https://doi.org/10.1016/S0099-9598(08)60178-9
47. Guo, T.-T., Zhang, J.-C., Zhang, H., Liu, Q.-C., Zhao, Y., Hou, Y.-F., Bai, N.-S., Bioactive spirans and other constituents from the leaves of *Cannabis sativa* f. sativa. *J. Asian Nat. Prod. Res.*, 19, 8, 793–802, 2016, doi: 10.1080/10286020.2016.1248947.
48. Hakansson, I., Myrbeck, Å., Etana, A., A review of research on seedbed preparation for small grains in Sweden. *Soil Tillage Res.*, 64, 1–2, 23–40, 2002.
49. Hamelink, C., Hampson, A., Wink, D.A., Eiden, L.E., Eskay, R.L., Comparison of cannabidiol, antioxidants, and diuretics in reversing binge ethanol-induced neurotoxicity. *J. Pharmacol. Exp. Ther.*, 314, 780–788, 2005, doi: 10.1124/jpet.105.085779.
50. Hartsel, J.A., Eades, J., Makriyannis, A., Cannabis sativa and hemp, in: *Nutraceuticals: Efficacy, Safety and Toxicity*, 1st ed., R.C. Gupta (Ed.), pp. 735–754, Elsevier Inc., San Diego, CA, USA, 2016, ISBN 978-0-12-802147-7.
51. Hernandèz-Vazquez, E., Ocampo-Montalban, H., Cerón-Romero, L., Cruz, M., Gómez-Zamudio, J., Hiriart-Valencia, G., Estrada-Soto, S., Antidiabetic, antidyslipidemic and toxicity profile of ENV-2: A potent pyrazole derivative against diabetes and related diseases. *Eur. J.*, 803, 159–166, 2017.

52. House, J.D., Neufeld, J., Leson, G., Evaluating the quality of protein from hemp seed (*Cannabis sativa* L.) products through the use of the protein digestibility-corrected amino acid score method. *J. Agric. Food Chem.*, 58, 22, 11801–11807, 2010.
53. Hunter, D., Oldfield, G., Tich, N., Messenheimer, J., Sebree, T., Synthetic transdermal cannabidiol for the treatment of knee pain due to osteoarthritis. *Osteoarthr. Cartil.*, 26, S26, 2018, https://doi.org/10.1016/j. joca.2018.02.067.
54. Irwin, N., Hunter, K., Frizzell, N., Flatt, P.R., Antidiabetic effects of sub-chronic administration of the cannabinoid receptor (CB1) antagonist, AM251, in obese diabetic (ob/ob) mice. *Eur. J. Pharmacol.*, 581, 1–2, 226–233, 2008.
55. Islam, M.S., The influence of fibre processing and treatments on hemp fibre/epoxy and hemp fibre/PLA composites, Doctoral dissertation, The University of Waikato, Hamilton, New Zealand, 2008.
56. Izzo, A.A., Borrelli, F., Capasso, R., Di Marzo, V., Mechoulam, R., Non-psychotropic plant cannabinoids: New therapeutic opportunities from an ancient herb. *Trends Pharmacol. Sci.*, 30, 515–527, 2009.
57. Jadoon, K.A., Ratcliffe, S.H., Barrett, D.A., Thomas, E.L., Stott, C., Bell, J.D., Efficacy and safety of cannabidiol and tetrahydrocannabivarin on glycemic and lipid parameters in patients with type 2 diabetes: A randomized, double-blind, placebo-controlled, parallel group pilot study. *Diabetes Care*, 39, 10, 1777–1786, 2016, https://doi.org/10. 2337/dc16-0650.
58. Janefjord, E., Mååg, J.L., Harvey, B.S., Smid, S.D., Cannabinoid effects on β amyloid fibril and aggregate formation, neuronal and microglial-activated neurotoxicity *in vitro*. *Cell. Mol. Neurobiol.*, 34, 31–42, 2014, doi: 10.1007/s10571-013-9984-x.
59. Jin, D., Dai, K., Xie, Z., Chen, J., Secondary metabolites profiled in cannabis inflorescences, leaves, stem barks, and roots for medicinal purposes. *Sci. Rep.-UK*, 10, 1, 1–14, 2020, https://doi.org/10.1038/s41598-020-60172-6.
60. Joshi, S., An introduction of hemp cultivation in Uttarakhand: A historical and economic perspective. *Studies in Indian Place Names-UGC Care Journal*, 40, 3, 6533–6543, 2020.
61. Joy, J.E., Watson, S.J., Benson, J.A., *Marijuana and medicine. Assessing the science bàse*, National Academy, Washington DC, 1999.
62. Kalita, P., Deka, S., Saharia, B.J., Chakraborty, A., Basak, M., Deka, M.K., An overview and future scope on traditionally used herbal plants of assam having antidiabetic activity. *Int. J. Adv. Pharm. Biol. Chem.*, 3, 2, 299–304, 2014.
63. Karimi, I. and Hayatghaibi, H., Effect of Cannabis sativa L. seed (Hempseed) on serum lipid and protein profiles of rat. *Pak. J. Nutr.*, 5, 6, 585–588, 2006.
64. Karus, M. and Vogt, D., European hemp industry: Cultivation, processing and product lines. *Euphytica*, 140, 1, 7–12, 2004.
65. Khan, B.A., Wang, J., Warner, P., Wang, H., Antibacterial properties of hemp hurd powder against E. coli. *J. Appl. Polym. Sci.*, 132, 10, 41588 (1 to 6), 2015.

66. King, J.W., The relationship between cannabis/hemp use in foods and processing methodology. *Curr. Opin. Food Sci.*, 28, 32–40, 2019.
67. Kleinhenz, M.D., Magnin, G., Ensley, S.M., Griffin, J.J., Goeser, J., Lynch, E., Coetzee, J.F., Nutrient concentrations, digestibility, and cannabinoid concentrations of industrial hemp plant components. *Appl. Anim. Sci.*, 36, 4, 489–494, 2020, doi: 10.15232/aas.2020-02018.
68. Kolodinsky, J. and Lacasse, H., Consumer response to hemp: A case study of Vermont residents from 2019 to 2020. *GCB Bioenergy*, 13, 4, 537–545, 2021.
69. Krajewska, M., Ślaska-Grzywna, B., Andrejko, D., Physical properties of seeds of the selected oil plants. *Agric. Eng.*, 20, 69–77, 2016.
70. Lafontan, M., Piazza, P.V., Girard, J., Effects of CB1 antagonist on the control of metabolic functions in obese type 2 diabetic patients. *Diabetes Metab.*, 33, 2, 85–95, 2007.
71. Latter, H.L., Abraham, D.J., Turner, C.E., Knapp, J.E., Schiff Jr., P.L., Slatkin, D.J., Cannabisativine, a new alkaloid from *Cannabis sativa* L. root. *Tetrahedron Lett.*, 16, 33, 2815–2818, 1975.
72. Lavrieux, M., Jacob, J., Disnar, J.R., Bréheret, J.G., Le Milbeau, C., Miras, Y., Andrieu-Ponel, V., Sedimentary cannabinol tracks the history of hemp retting. *Geology*, 41, 7, 751–754, 2013.
73. Leizer, C., Ribnicky, D., Poulev, A., Dushenkov, S., Raskin, I., The composition of hemp seed oil and its potential as an important source of nutrition. *J. Nutraceuticals Funct. Med. Foods*, 2, 4, 35–53, 2000.
74. Leonard, W., Zhang, P., Ying, D., Fang, Z., Hempseed in food industry: Nutritional value, health benefits, and industrial applications. *Compr. Rev. Food Sci. Food Saf.*, 19, 1, 282–308, 2020.
75. Li, H., Liu, Y., Tian, D., Tian, L., Ju, X., Qi, L., Liang, C., Overview of cannabidiol (CBD) and its analogues: Structures, biological activities, and neuroprotective mechanisms in epilepsy and Alzheimer's disease. *Eur. J. Med. Chem.*, 192, 112163, 2020.
76. Lisson, S.N., Mendham, N.J., Carberry, P.S., Development of a hemp (*Cannabis sativa* L.) simulation model 2. The flowering response of two hemp cultivars to photoperiod. *Aust. J. Exp. Agric.*, 40, 3, 413–417, 2000.
77. Liu, L., Cheng, Y., Zhang, H., Phytochemical analysis of anti-atherogenic constituents of Xue-Fu-Zhu-Yu-Tang using HPLC-DAD-ESI-MS. *Chem. Pharm. Bull.*, 52, 11, 1295–1301, 2004.
78. Liu, M., Fernando, D., Daniel, G., Madsen, B., Meyer, A.S., Ale, M.T., Thygesen, A., Effect of harvest time and field retting duration on the chemical composition, morphology and mechanical properties of hemp fibers. *Ind. Crops Prod.*, 69, 29–39, 2015.
79. Malomo, S.A., He, R., Aluko, R.E., Structural and functional properties of hemp seed protein products. *J. Food Sci.*, 79, 8, C1512–C1521, 2014.
80. Martín-Moreno, A.M., Reigada, D., Ramírez, B.G., Mechoulam, R., Innamorato, N., Cuadrado, A., de Ceballos, M.L., Cannabidiol and other

cannabinoids reduce microglial activation *in vitro* and *in vivo*: Relevance to Alzheimer's disease. *Mol. Pharmacol.*, 79, 6, 964–973, 2011.
81. Matthaus, B. and Bruhl, L., Virgin hemp seed oil: An interesting niche product. *Eur. J. Lipid Sci. Technol.*, 110, 7, 655–661, 2008.
82. McPartland, J.M., Cutler, S., McIntosh, D.J., Hemp production in Aotearoa. *J. Ind. Hemp*, 9, 1, 105–115, 2004.
83. Mukhopadhyay, P., Rajesh, M., Horváth, B., Bátkai, S., Park, O., Tanchian, G., Cannabidiol protects against hepatic ischemia/reperfusion injury by attenuating inflammatory signaling and response, oxidative/nitrative stress, and cell death. *Free Radic. Biol. Med.*, 50, 1368–1381, 2011, doi: 10.1016/j.freeradbiomed.2011.02.021.
84. Oomah, B.D., Busson, M., Godfrey, D.V., Drover, J.C., Characteristics of hemp (*Cannabis sativa* L.) seed oil. *Food Chem.*, 76, 1, 33–43, 2002.
85. Oseyko, M., Sova, N., Petrachenko, D., Mykolenko, S., Technological and chemical aspects of storage and complex processing of industrial hemp seeds. *Ukr. Food J.*, 9, 3, 545–560, 2020.
86. Panghal, A., Kumar, N., Kumar, S., Kumari, A., Chhikara, N., Food function and health benefits of functional foods. In Chhikara, Panghal, Chaudhary (Eds.), *Functional Foods*, pp. 419–441, Wiley and Scrivener, USA, 2022.
87. Pellati, F., Brighenti, V., Sperlea, J., Marchetti, L., Bertelli, D., Benvenuti, S., New methods for the comprehensive analysis of bioactive compounds in *Cannabis sativa* L. (hemp). *Molecules*, 23, 10, 2639, 2018.
88. Piotrowski, S. and Carus, M., *Ecological benefits of hemp and flax cultivation and products*, vol. 5, pp. 1–6, Nova Institute, Germany, 2011.
89. Placet, V., François, C., Day, A., Beaugrand, J., Ouagne, P., Industrial hemp transformation for composite applications: Influence of processing parameters on the fibre properties, in: *Advances in Natural Fibre Composites*, pp. 13–25, Springer, Cham, 2018.
90. Pojic, M., Mišan, A., Sakač, M., Dapčević Hadnađev, T., Šarić, B., Milovanović, I., Hadnađev, M., Characterization of byproducts originating from hemp oil processing. *J. Agric. Food Chem.*, 62, 51, 12436–12442, 2014.
91. Pollastro, F., Minassi, A., Fresu, L.G., Cannabis phenolics and their bioactivities. *Curr. Med. Chem.*, 25, 1160–1185, 2018.
92. Prentiss, D., Power, R., Balmas, G., Tzuang, G., Israelski, D.M., Patterns of marijuana use among patients with HIV/AIDS followed in a public health care setting. *JAIDS J. Acquir. Immune Defic. Syndr.*, 35, 1, 38–45, 2004.
93. Pryce, G. and Baker, D., Control of spasticity in a multiple sclerosis model is mediated by CB1, not CB2, cannabinoid receptors. *Br. J. Pharmacol.*, 150, 4, 519–525, 2007.
94. Radwan, M.M., Chandra, S., Gul, S., ElSohly, M.A., Cannabinoids, Phenolics, Terpenes and Alkaloids of Cannabis. *Molecules*, 26, 9, 2774, 2021.
95. Raikos, V., Duthie, G., Ranawana, V., Denaturation and oxidative stability of hemp seed (*Cannabis sativa* L.) protein isolate as affected by heat treatment. *Plant Foods Hum. Nutr.*, 70, 3, 304–309, 2015.

96. Das, R., Nutraceutical from hemp (*Cannabis sativa* L.) and functional evaluation. *Int. J. Nutr. Agric. Res.*, 3, 1, 23–29, 2016.
97. Ross, S.A., Mehmedic, Z., Murphy, T.P., ElSohly, M.A., GC-MS analysis of the total δ9-thc content of both drug-and fiber-type cannabis seeds. *J. Anal. Toxicol.*, 24, 8, 715–717, 2000.
98. Russo, E.B. and Taming, T.H.C., Potential cannabis synergy and phytocannabinoid-terpenoid entourage effects. *Br. J. Pharmacol.*, 163, 1344–1364, 2011.
99. Sacilik, K., Öztürk, R., Keskin, R., Some physical properties of hemp seed. *Biosyst. Eng.*, 86, 2, 191–198, 2003.
100. Sakamoto, K., Akiyama, Y., Fukui, K., Kamada, H., Satoh, S., Characterization; genome sizes and morphology of sex chromosomes in hemp (*Cannabis sativa* L.). *Cytologia*, 63, 4, 459–464, 1998.
101. Schafer, T. and Honermeier, B., Effect of sowing date and plant density on the cell morphology of hemp (*Cannabis sativa* L.). *Ind. Crops Prod.*, 23, 1, 88–98, 2006.
102. Scheen, A.J. and Paquot, N., Use of cannabinoid CB1 receptor antagonists for the treatment of metabolic disorders. *Best Pract. Res. Clin. Endocrinol. Metab.*, 23, 1, 103–116, 2009.
103. Scuderi, C., Steardo, L., Esposito, G., Cannabidiol promotes amyloid precursor protein ubiquitination and reduction of β amyloid expression in SHSY5YAPP+ cells through PPARγ involvement. *Phytother. Res.*, 28, 1007–1013, 2014, doi: 10.1002/ptr.5095.
104. Stith, S.S., Diviant, J.P., Brockelman, F., Keeling, K., Hall, B., Lucern, S., Vigil, J.M., Alleviative effects of Cannabis flower on migraine and headache. *J. Integr. Med.*, 18, 5, 416–424, 2020.
105. Struik, P.C., Amaducci, S., Bullard, M.J., Stutterheim, N.C., Venturi, G., Cromack, H.T.H., Agronomy of fibre hemp (*Cannabis sativa* L.) in Europe. *Ind. Crops Prod.*, 11, 107–118, 2000.
106. Taheri-Garavand, A., Nassiri, A., Gharibzahedi, S.M.T., Physical and mechanical properties of hemp seed. *Int. Agrophys.*, 26, 2, 211–215, 2012.
107. Tang, C.H., Ten, Z., Wang, X.S., Yang, X.Q., Physicochemical and functional properties of hemp (*Cannabis sativa* L.) protein isolate. *J. Agric. Food Chem.*, 54, 23, 8945–8950, 2006.
108. Van der Werf, H.M., Hemp production in France. *J. Ind. Hemp*, 7, 2, 105–109, 2002.
109. Van der Werf, H.M.G., Agronomy and crop physiology of fibre hemp: A literature review. Centre for Agrobiological Research (CABO) Report 142, pp. 1–12, 1991.
110. Vonapartis, E., Aubin, M.P., Seguin, P., Mustafa, A.F., Charron, J.B., Seed composition of ten industrial hemp cultivars approved for production in Canada. *J. Food Compos. Anal.*, 39, 8–12, 2015.
111. Walter, L., Franklin, A., Witting, A., Wade, C., Xie, Y., Kunos, G., ... Stella, N., Nonpsychotropic cannabinoid receptors regulate microglial cell migration. *J. Neurosci.*, 23, 4, 1398–1405, 2003.

112. Wang, H.M., Postle, R., Kessler, R.W., Kessler, W., Removing pectin and lignin during chemical processing of hemp for textile applications. *Text. Res. J.*, 73, 8, 664–669, 2003.
113. Wang, Q. and Xiong, Y.L., Processing, nutrition, and functionality of hempseed protein: A review. *Compr. Rev. Food Sci. Food Saf.*, 18, 4, 936–952, 2019.
114. Wang, S., Luo, Q., Zhou, Y., Fan, P., CLG from hemp seed inhibits LPS-stimulated neuroinflammation in BV2 microglia by regulating NFkB and Nrf-2 pathways. *ACS Omega*, 4, 16517–16523, 2019.
115. Watt, G. and Karl, T., *In vivo* evidence for therapeutic properties of cannabidiol (CBD) for Alzheimer's disease. *Front. Pharmacol.*, 8, 20, 2017, doi: 10.3389/fphar.2017.00020.
116. Yan, X., Zhou, Y., Tang, J., Ji, M., Lou, H., Fan, P., Diketopiperazine indole alkaloids from hemp seed. *Phytochem. Lett.*, 18, 77–82, 2016.
117. Yan, X., Tang, J., dos Santos Passos, C., Nurisso, A., Simões-Pires, C.A., Ji, M., Lou, H., Fan, P., Characterization of lignanamides from hemp (*Cannabis sativa* L.) seed and their antioxidant and acetylcholinesterase inhibitory activities. *J. Agric. Food Chem.*, 63, 10611–10619, 2015.
118. Zhou, Y., Wang, S., Ji, J., Lou, H., Fan, P., Hemp (*Cannabis sativa* L.) seed phenylpropionamides composition and effects on memory dysfunction and biomarkers of neuroinflammation induced by lipopolysaccharide in mice. *ACS Omega*, 3, 15988–15995, 2018.
119. Zhou, Y., Wang, S., Lou, H., Fan, P., Chemical constituents of hemp (*Cannabis sativa* L.) seed with potential anti-neuroinflammatory activity. *Phytochem. Lett.*, 23, 57–61, 2018.

6

Ocimum Species

Deep Shikha[1]* and Piyush Kashyap[2]

[1]*Department of Food Engineering & Technology, Sant Longowal Institute of Engineering & Technology, Longowal, Punjab, India*

[2]*Department of Food Technology and Nutrition, School of Agriculture, Lovely Professional University, Phagwara – Punjab, India*

Abstract

Ocimum species (Basil or Tulsi) belonging to family of Lamiaceae comprises more than 150 species, distributed through temperate regions of the world. Maximum species of is known to be found in Africa and India. Ocimum is an aromatic and medicinal plant group under a genus of aromatic annual and perennial herbs with a characteristic bushy growth pattern, and leaves are the commercial parts of plant used for essential oil extraction. It also encompasses of anti-oxidant, anti-proliferative, anti-diabetic anti-hypersensitive, anti-bacterial and anti-inflammatory properties due to presence of natural bioactive components including flavonoids, polyphenols, essential oils and saponins with non-volatile compounds. Ocimum species has been in cultivation for a long time and has been known to mankind for its medicinal and aromatic value. It has been a part of therapy in many cultures. In Indian Ayurveda, it has a special place and is a prominent constituent of many ayurvedic medicines. Basils are commonly used for advanced Thai flavors. Industrially plant uses in preparation of soups, sauces and salads as rich in essential oils. This chapter highlighted the origin, history, nutritional value, bioactive profile, and ends up with health benefits and its industrial uses. Considering health benefits and industrial uses of species, are proven to be beneficial for scientific studies and research.

***Keywords*:** Ocimum (Basil or Tulsi) species, production, pharmacology, bioactive components, antioxidant activity

**Corresponding author*: bansaldeepshikha789@gmail.com

Prerna Gupta, Navnidhi Chhikara and Anil Panghal (eds.) Harvesting Food from Weeds, (183–216)

6.1 Origin and History

The plant species of *Ocimum* belongs to the family of *Lamiaceae*, encompassing more than 150 species. It cultivates extensively and is dispersed all over temperate sections of the world with the utmost species in Africa. The best-known species are the intensely aromatic herbs *Ocimum basilicum* (Thai basil) and *Ocimum gratissimum* (African basil), the remedial herb *Ocimum tenuiflorum*, also recognized as *Ocimum sanctum* (tulsi or holy basil in Hindi). In the state's environmental weed approach, sweet basil (*Ocimum basilicum*) is generally regarded as a minor ecological weed. It is commonly found in Western Australia and desert plateaus bioregion in internal central Queensland and is primarily of concern in the northern portions of the state (e.g., on Kimberley Downs and near Broome Station). This species is commonly grown in gardens as a cookery herb with various cultivars and varieties available. However, cultivation from it has been fled, chiefly in the semi-arid and tropical areas of Northern Australia. Most commonly, the varieties grow in waste areas and disturbed sites, along drains and roadsides, but also capture intermittent open woodlands, riparian areas, and natural vegetation. For example, it is a common weed of the ground layer in gidgee woodlands on the floodplain of Dyllingo Creek near Clermont [1].

Ocimum is an aromatic and medicinal plant group under a genus of aromatic annual and perennial herbs. An important member of the *Lamiaceae* family (mints), it is mainly native to warm and tropical regions. *Ocimum* is the genus of basil (syn. Great basil) and is known to most as a culinary herb. *Ocimum* species has been in cultivation for a long time and has been known to mankind for its medicinal and aromatic value. It has been a part of therapy in many cultures. In Indian Ayurveda, it has a special place and is a prominent constituent of many ayurvedic medicines. In our culture, basil plant has been given a divine importance and is worshipped by many as a sacred plant. A common belief is that having 4-5 fresh leaves in the morning on an empty stomach helps to boost immunity and keeps you healthy.

The earliest known records about the spread of basil reveal it to be brought to Europe by Charlemagne, where basil cultivation was done as a medicinal herb began in gardens and monasteries. In the pre-Christian cultures and Middle east region, basil is considered a sacred plant used for religious purposes [2]. The exact center of origin for basil is not clear, but it is believed to have been originated in areas like India, China, Sri Lanka, Asia and Africa. Cultivation of basil is done in many parts of the world. As per a Greek Legen, basil is a plant that heals wounds. Because of its medicinal properties are considered sacred and believed to bring prosperity and healthiness to the soul [2]. The name basil is mentioned in the Greek basileus, which means King or anything royal because of the unique flavor and aroma. Basil is known by different names around the world *viz.* Basilika, Vasiliki, Bosilok, Bazsalikom, Bazalka, Basel, Basilica, Basil, Vasilikos, etc.

6.2 Botanical Distribution

There have been different reports regarding species classification under the genus *Ocimum*. Some reports suggest nearly 65 species [3], while we also find nearly 150 species under the genus [4]. *Ocimum* is distributed all along with Asia, Africa, and Central and South America, but this genus's maximum diversity is found in Africa [5]. Basils have branching stems, black or brown seeds, square or opposite leaves, and flower spikes as physical characteristics. There is much variation in growth patterns, leaf color, and aromatic constituents among Ocimum plants due to the variety of species. Most commercial Ocimum cultivars are from the Ocimum basilicum species, which has a bushy growth habit. There are species with purple foliage, "Dark Opal, "primarily ornamental. The variability in *Ocimum* species forms the basis of its classification into seven purple types *viz.* tall, slender type, large-leafed and robust types, dwarf types, compact types, purple types and lemon-flavored basil [6, 7]. Ocimum species are extensive in India and Africa [8]. *O. gratissimum* and *O. basilicum* have been described for their antiemetic actions [9, 10]. *O. basilicum* Linn, with antiemetic flowers [10], is a spice dispersed in Africa and other regions [11–13]. The major chemical components of *O. basilicum* found in Africa (Cameroon, Egypt, Guinea, Mali, Nigeria, and Rwanda) are as follows: linalool, 1,8-cineole, limonene, eugenol, methyl chavicol, methyl cinnamate, (E)-α-bergamotene, thymol, and methyl eugenol [11]. *O. gratissimum* L. is a significant herbal remedial plant in sub-Saharan African countries and Kenyan communities [14]. *O. gratissimum* is usually found around gardens and village huts in West Africa, [15] and cultured for culinary and therapeutic purposes. The strong aromatic odor of the plant leaves makes them popular for use as flavor in soups and meat [16, 17].

Ahmed *et al.*, 2014 [9] reported antiemetic activity of the whole plant. The main components of the essential oil of African *O. gratissimum* include limonene, p-cymene, β-phellandreneγ-terpinene, eugenol and thymol [18, 19]. The main components extracted from various Ocimum oils include linalool, 1,8-cineol, pinene, ocimene, methyl chavicol, eugenol, camphor, and limonene terpinene [11]. As it is investigated that Ocimum species are of around 150 types worldwide, some of them are shown here in the figure that is found in India and commonly known with names such as *Krishna Tulsi* and *Rama Tulsi* etc. Figure 6.1 depicts some Ocimum species.

(**a**) *Ocimum tenuiflorum* (**b**) *Ocimum adscendens* (**c**) *Ocimum basilicum* (**d**) *Ocimum americanum* (**e**) *Ocimum tenuiflorum* (**f**) *Ocimum gratissimum* (**g**) *Ocimum africanum* (**h**) *Ocimum kilimandscharicum*

Figure 6.1 *Ocimum* species from India.

6.3 Production

Basil is adaptable to a variety of climatic conditions and due to this, cultivation of basil is extended over all the continents of the earth. However, it is primarily cultivated between temperature regimes of 7°C to 27°C, with an annual rainfall range of 0.6 to 4.2 mm. The basil species are highly susceptible to frost, and cold climatic conditions and chilling injury are often observed in plants when cultivated in temperate areas. Being a natural tropical plant, the basil species gives the best results under long-day conditions having plenty of sunlight.

Basil is an herb/shrub that is primarily propagated through seeds. Direct sowing of seeds can be done in the late spring season. Transplanting can also be taken up of already raised nursery seedlings after the cessation of the frost period. The seed of basil is relatively small and should be tested for viability before transferring the seedlings to the main field. Seeds with a germination percentage less than 70 % should be discarded as the source of planting material and it is recommended that the germination percentage be at least 80-95% [20]. The basil seeds are friable delicate and thus require a well-prepared seedbed for optimum germination and subsequent plant establishment. Seedling emergence can be observed between 8 and 14 days after sowing the seeds, which are 0.3-0.6 cm deep. The plant length reaches around 4 inch and can grow in shaded zones and full sun for the finest flowering. This plant becomes attractive to a variety of bees due to its beautiful pink flowers and this plant is drought resistant. Although this basil is a perennial, it is still susceptible to extreme cold. This is a pronounced landscape plant and essential in your herb garden for its fragrance and color.

6.4 Development and Maturation

Basil has a characteristic bushy growth pattern, and leaves are the commercial parts of the plant used for essential oils extraction. Thus growing basil for commercial purposes should aim at more vegetative growth, which is achieved by topping/pinching the leader shoot of the seedling prior to field planting [21]. This encourages the growth of lateral shoots and gives the plant a bushy appearance having a sizable number of leaves. The maturity of basil depends upon the intended use of the plant. Generally, basil is sold as fresh or dried leaves in open markets. Leaves are harvested, leaving the bottom two tiers of leaves. When basil is grown for leaves and extraction of essential oils, the plant is not allowed to bloom and is continuously harvested for leaves. Repeated cuttings/harvestings of leaves also induce more leaf growth, and on average, three to five harvestings per year are possible with the basil plant [20, 21]. In temperate growing conditions, the harvestings are usually low, with only one to three per year. Method of harvesting affects the yield and quality of methyl eugenol, cinnamyl acetate, eugenol and element as investigated in *Ocimum tenuiflorum* [22]. The secondary branches yield higher total essential oil than other branches; however, the biomass yield is poor. *Ocimum* species are reported to sustain in tropical and sub-tropical climates well under various soil and environmental conditions. Well-drained loamy-rich or laterite soil with alkaline to moderate acidic conditions are required by Ocimum species for their favorable vegetative growth. High temperatures have been beneficial for plant growth and essential oil production, and these species thrive in high rainfall and humid conditions [23]. Different species of *Ocimum* require variable conditions for their growth and development, such as *O. sanctum* (holy basil or tulsi) can also be cultivated in partially shaded conditions but, in turn, does not yield a good quantity of eugenol (essential oil).

It is reported that the content of essential oils increases with higher temperature, higher daily light and conditions of water stress, whereas it decreases with the shady atmospheric condition [24]. The plantation time is the third week of February in the nursery, and transplantation is done in mid-April. The plant's irrigation needs vary depending on the season; in the summer, the water supply is increased to keep the soil moist. *Ocimum* species can be propagated through seeds only [25]. The genus Ocimum, member of *Kilimandscharicum, O. minimum, O. lamiifolium, O. selloi, Lamiaceae* family, included almost 200 species and shrubs [26] rated high amongst some of the bewildering herbs for having remarkable medicinal properties. There are vast numbers of diverse varieties and species under this genus [27–29]. Ocimumis widespread in Asia, Central and Southern America and Africa. Ocimum's extraordinary medicinal and therapeutic potentialities make its cultivation more popular. Besides this, the essential oils of this plant also showed remarkable uses in the culinary, aromatherapy treatment and perfumes of herbal toiletries and flavoring agents. The genus comprises a wide range of species, making it difficult to identify among the species; therefore identification process can be optimized with combined analysis of molecular markers, morphological traits, essential oil composition and biological activity [30].

6.5 Nutritional Profile

There are over 150 species of *Ocimum,* and each species has its nutritional value when consumed. Ocimum species are reported to comprise a good amount of proteins, fats, carbohydrates, fibers and other valuable nutrients such as calcium, potassium, iron, phosphorous, vitamin C, vitamin A, and sodium required by our body's daily diet [31]. The composition depends on the Ocimum species variety and climatic condition they are grown. The phytochemical study of *Ocimum basilicum* has revealed that it contains alkaloids, terpenoids, saponins, phenolics, flavonoids, tannins, steroids, cardiac glycosides, reducing sugars, glycosides, carbohydrate, oils, protein, many vitamins and minerals. A recent study has revealed the physiochemical composition of *O. grattissimum* and *O. basilicum* (Table 6.1).

Table 6.1 Physiochemical composition of O. *grattissimum and* O. *basilicum* [32].

S. no.	Nutrients	O. grattissimum	O. basilicum.
1	Ash content (%)	12.18 ± 0.00	15.73 ± 0.00
2	Crude fiber (%)	14.96 ± 0.00	11.31 ± 0.00
3	Carbohydrate (%)	20.8 ± 0.01	14.50 ± 0.00
4	Moisture content (%)	6.93 ± 0.00	5.72 ± 0.00
5	Calories (%)	287.3 ± 0.06 kcal	281.5 ± 0. 02kcal
6	Calcium (mg)	540	460
7	Potassium (mg)	462	483
8	Iron (mg)	11.4	10.5
9	Phosphorous (mg)	26.9	35.9
10	Sodium (mg)	149	159

Studies have also shown the presence of non-volatile composition of essential oils of a few species of northern India, including *Ocimum gratissium L.*, *Ocimum sanctum* (CIM-Ayu), *Ocimum basilicum L.* (Vikarsudha) and *Ocimum kilimandscharicum* Guerke through GC and GC-MS. The process evaluated around 56 different compounds responsible for the composition of essential oils in these species. Among these compounds, phenylpropanoids were discovered to be major constituent in the first three above listed species, i.e. around 65.2-77.6% while in Ocimum, *kilimands* charicum had oxygenated monoterpenes (72%) at the highest [33].

6.6 Bioactive Compounds

Basils (*Ocimum* sp.) are commonly used for the advanced Thai flavors or used as a potherb and act as an anti-oxidant due to the numerous chemical compounds present in it. The species contain some bioactive phytochemical constituents such as alkaloids, flavonoids, phenolics, essential oils, anthocyanins, tannins and saponins [34], which are extracted and utilized in pharmaceutical as well as in food industries. Essential oils such as 1, 8-cineole, anthocyanins, eugenol (71%), estragole, linalool, pinene, methyl chavicol, camphor, terpinene, ocimene, limonene, carvacrol (3%) have shown large-scale application in pharmaceutical industries. Leaves are rich in eugenal, urosolic acid, carvacol, linalool, caryophyllene, limatrol, caryophyllene, methyl carvicol anthocyanins, whereas stem and seeds are good sources of romarinic acid, apigenin, cirsimaritin, isothymusin, isothymonin, xylose, and polysaccharides [35]. The main chemical constituents of *Ocimumsanctum* are ursolic acid, oleanolic acid, eugenol, rosmarinic acid, linalool, carvacrol, and β-caryophyllene, which have been extracted for many years and are being used in food industries, pharmaceutical and other valuable products [36].

Vasudevan *et al.*, (1999) [37] also discussed bioactive compounds of *Ocimum Sp.* This study discussed different aspects, including fungicidal, pesticidal, medicinal and bactericidal effects of plants used in ancient times. In the analytical study of bioactive compounds, using the colorimetry method, it was found that *O. basilicum* sp. consisted of ursolic acid, triterpenes, β-sitosterol, and oleic acid at different stages of plant growth, and some of the compounds such as acidic triterpene were increased in flowers and leaves of the plant whereas decreased in seeds. Another research of French basil species cultivated in Jammu was investigated and found rich in euganol, linalool, ocimene, sesquiterpenes, limonene, etc.

In the case of holy basil, i.e., Ocimum sanctum studies, the presence of ursolic acid and sterols, β-carotene and fatty acids in the Et_2O and petrol extract [38]. Another species, such as *Ocimum gratissimum* was analyzed for its bioactive composition and extraction by using petroleum ether as a solvent. It was reported that sitosterol and ocimol were present in the roots and in the plant. The compounds were indicated in percentage as linalool (0.2%) in the least amount and thymol (47.6%) maximum among other compounds. Similarly, a study investigated for *O. kilimandscharicum* revealed the presence of various compounds such as β-sitosterol (0.02%), camphor (0.04%), ursolic acid (0.18%) and oleanolic acid (0.86%) [39].

The pesticidal activity of plant species was studied for *O. sanctum,* and extracted components were analyzed using Gas chromatography (GC), GC-mass spectrometry and NMR and positive activity against inhibition of β-galactosidase of adult filarial nematode. Similarly, O. basilicum was also shown its repellent activity against *Allacophora foveicollis* insects [40]. Fungicidal activity against food destroying fungi such as *Alternaria tenuis, Curvularia penniseti,* and *Helminthosporium* sp. and favorable results were reported with essential oils of *O. sanctum* [41]. Bactericidal activity of *Ocimum sp.* showed antibiotic effects of plant that helps in inhibition of *Micrococcus pyrogenes var. aureus* and *Mycobacterium tuberculosis* at two different doses of 10 and 100μ/ml [42].

Bioactive and phytochemical compounds in *Ocimum* species are responsible for pharmacological benefits (Table 6.2). In one study of identification with

Table 6.2 Bioactive components of *Ocimum* species.

Sr. no.	Class	Compounds	Structure	References
1	Monoterpene hydrocarbons	α-pinene	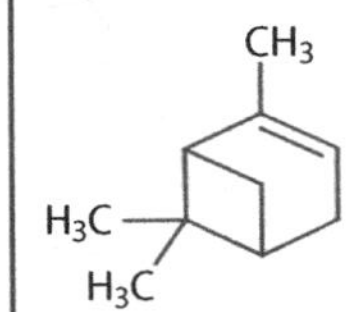	[46–51]
		α-terpinolene	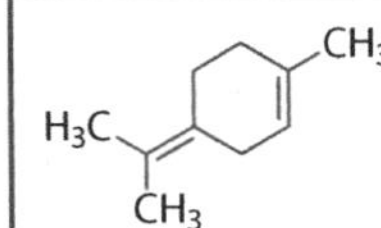	[47]
		Myrcene	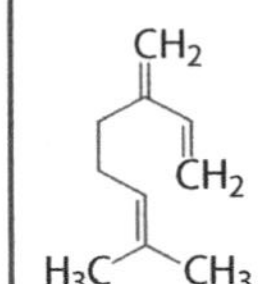	[48–51]
		Limonene	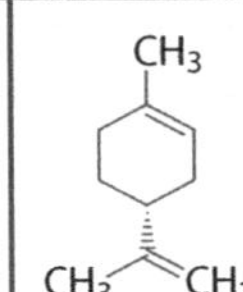	[44–50, 52]

(*Continued*)

Table 6.2 Bioactive components of *Ocimum* species. (*Continued*)

Sr. no.	Class	Compounds	Structure	References
		α-terpinene	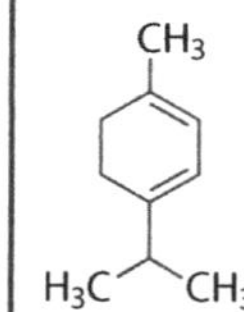	[43, 46–48]
		ρ-cymene	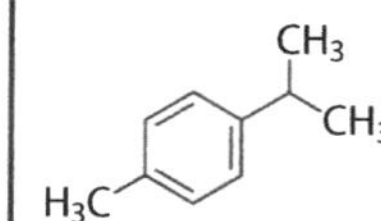	[44, 48]
2	**Sesquimonoterpenes**	sesquiphellandrene	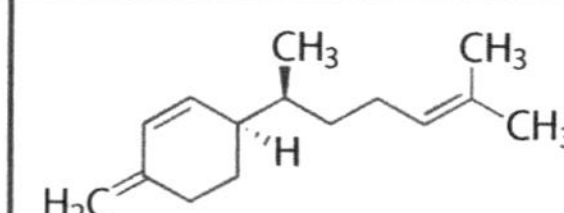	[56–60]
		α-zingiberene	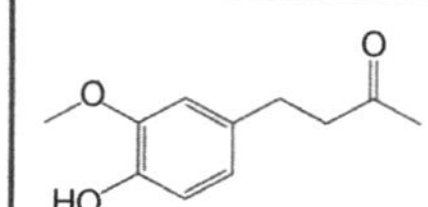	[50, 51]
		Bicyclogermacrene	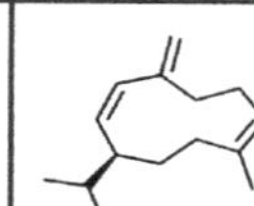	[48–51, 53–55, 62]

(*Continued*)

Table 6.2 Bioactive components of *Ocimum* species. (*Continued*)

Sr. no.	Class	Compounds	Structure	References
		α-Guaiene	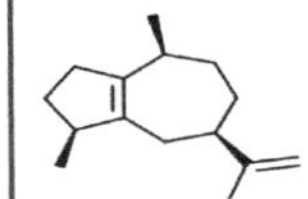	[48–51, 54]
		γ-Gurjunene	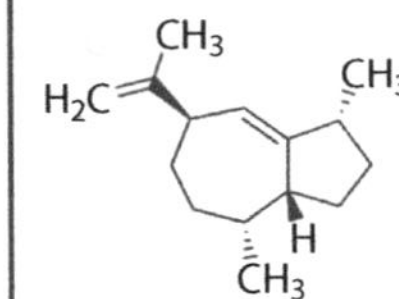	[48]
		Cadinene	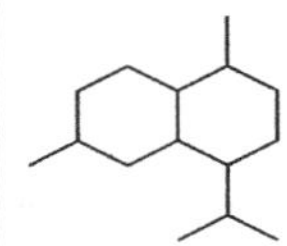	[63]
		Germacrene-A	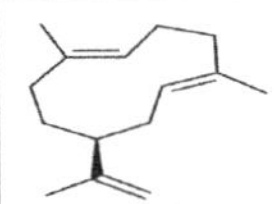	[44, 54, 61]
3	**Monosaccharides**	L-Arabinose	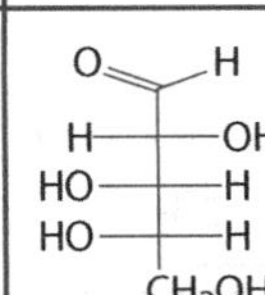	[64]

(*Continued*)

Table 6.2 Bioactive components of *Ocimum* species. (*Continued*)

Sr. no.	Class	Compounds	Structure	References
		L-Rhamnose	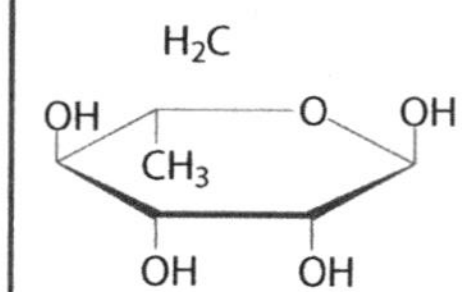	[64]
4	**Aromatic Compounds**	Estragole	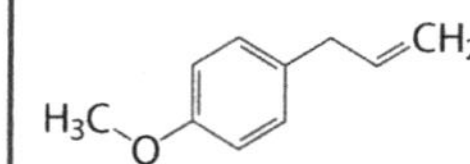	[53]
		Anethole	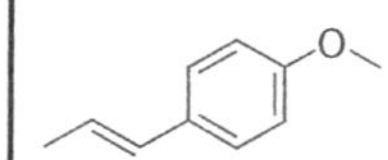	[53]
		Anisaldehyde	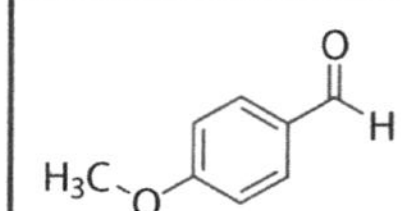	[53]
		Cuminaldehyde	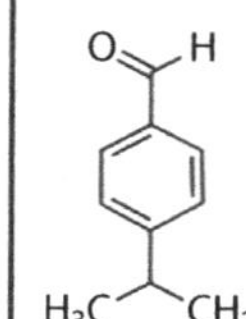	[53]

(*Continued*)

Table 6.2 Bioactive components of *Ocimum* species. (*Continued*)

Sr. no.	Class	Compounds	Structure	References
		Ethyl cinnamate	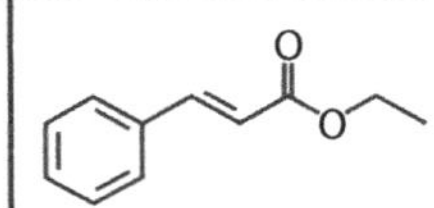	[53]
		Safrole	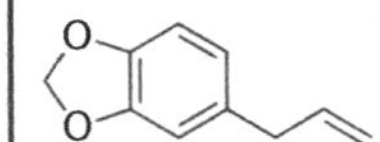	[53]
		Benzaldehyde	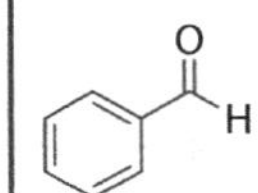	[47]
		Methyl salicylate	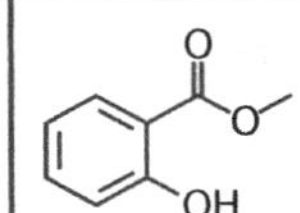	[53]
		Phenyl acetaldehyde	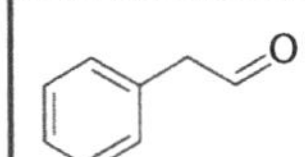	[53]

(*Continued*)

Table 6.2 Bioactive components of *Ocimum* species. (*Continued*)

Sr. no.	Class	Compounds	Structure	References
5	**Flavonoids**	Quercetin	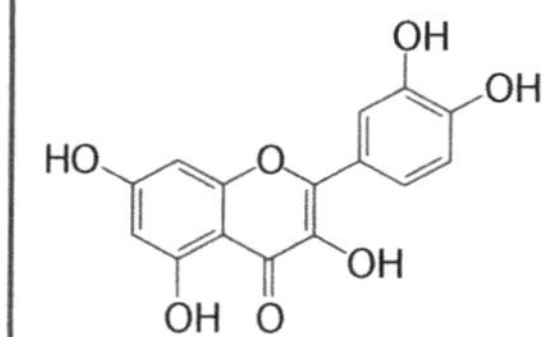	[64]
		Isoquercetrin	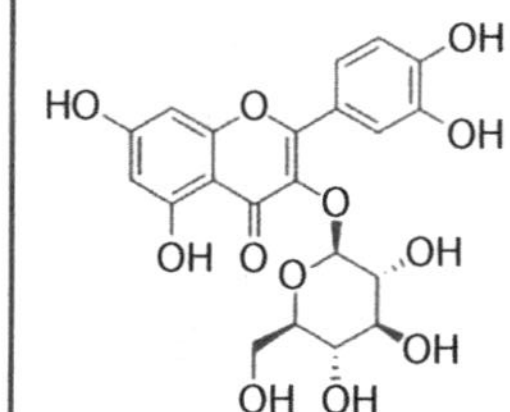	[64]
		Quercetin-3-O-B-D-glucoside	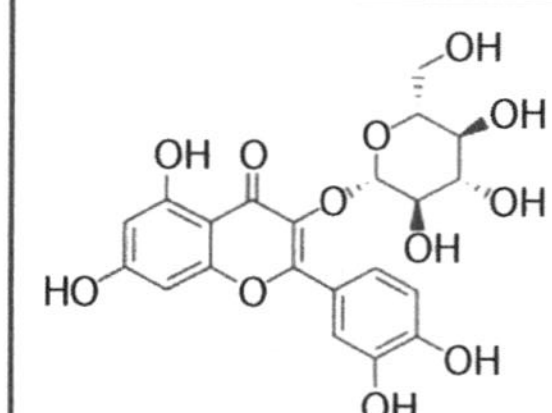	[64]

(*Continued*)

Table 6.2 Bioactive components of *Ocimum* species. (*Continued*)

Sr. no.	Class	Compounds	Structure	References
		Quercetin-3-O-diglycoside	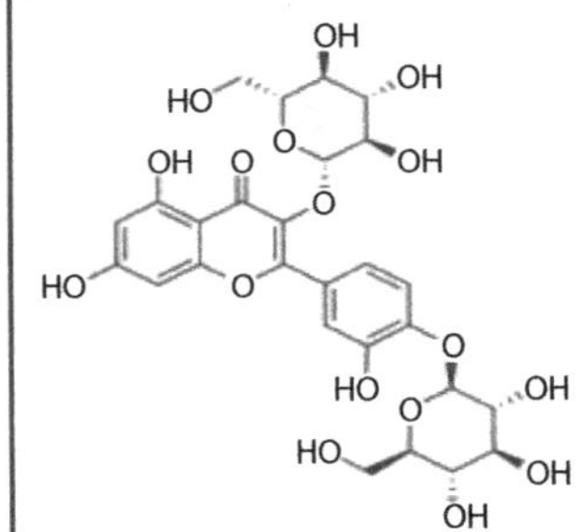	[64]
		Quercetin-3-O-B-D-glucoside-2”-gallate	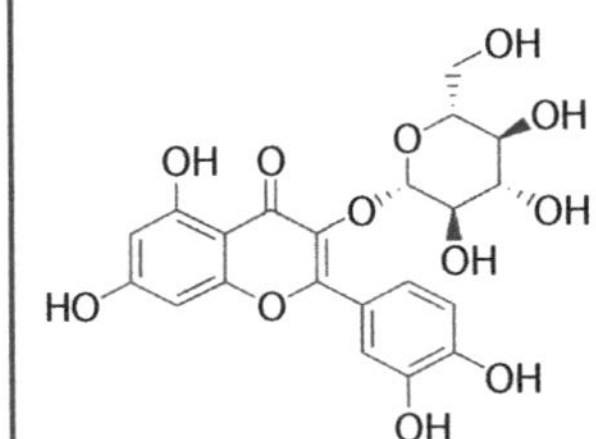	[64]
		Kaempferol	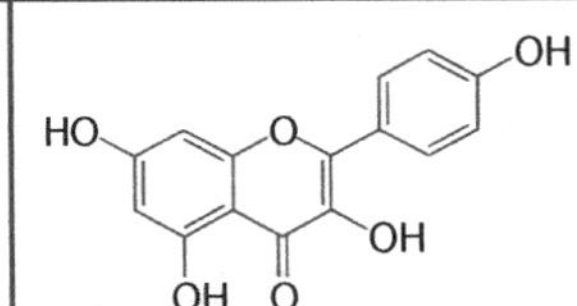	[64]

(*Continued*)

Table 6.2 Bioactive components of *Ocimum* species. (*Continued*)

Sr. no.	Class	Compounds	Structure	References
		Rutin	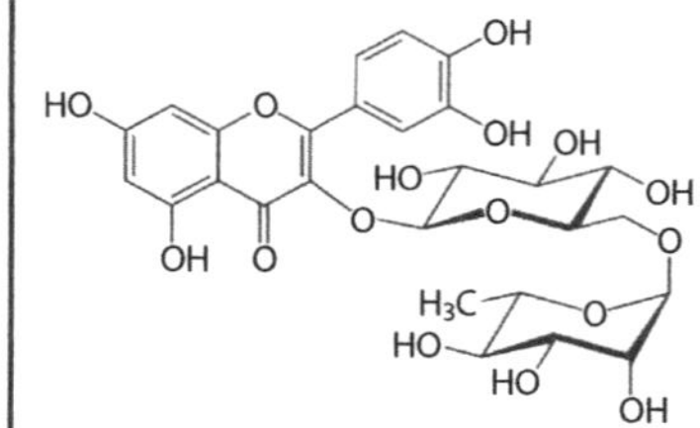	[64]
		2,6-dimethyldiethyl ester	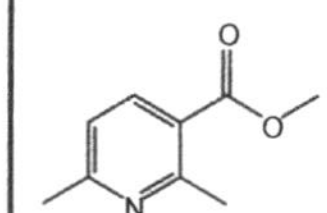	[63]
6	**Oxygenated monoterpene**	Carvone	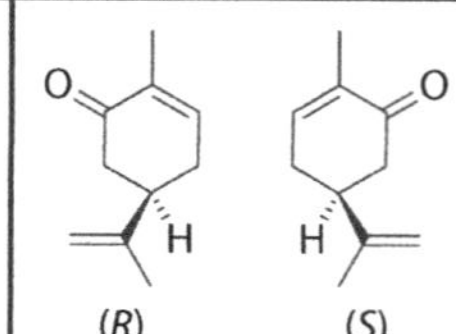	[48]
		Carvacrol	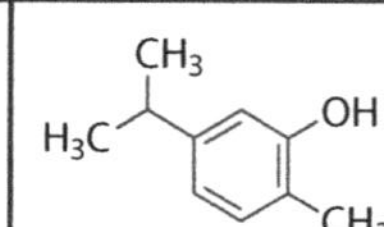	[48]

(*Continued*)

Table 6.2 Bioactive components of *Ocimum* species. (*Continued*)

Sr. no.	Class	Compounds	Structure	References
		Eugenol	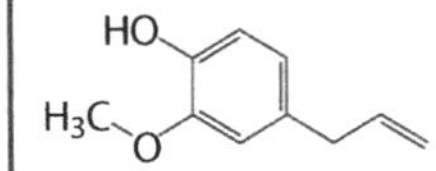	[49–51, 54, 62]
		Geraniol	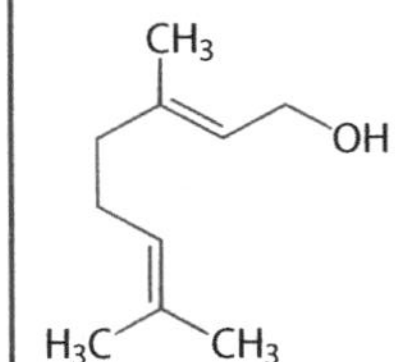	[48–51, 53, 54]
		1,8-Cineole	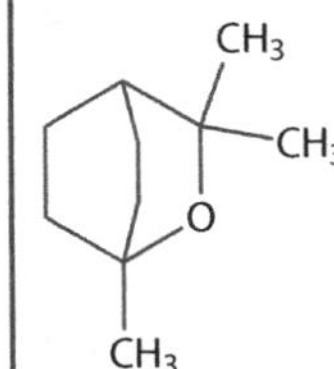	[47, 49–51, 53–55, 61, 62, 65]
		Camphor	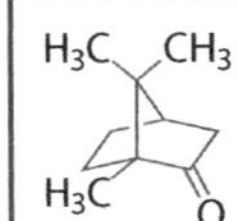	[47, 49–51, 53, 54, 61, 65]

(*Continued*)

Table 6.2 Bioactive components of *Ocimum* species. (*Continued*)

Sr. no.	Class	Compounds	Structure	References
		Menthol	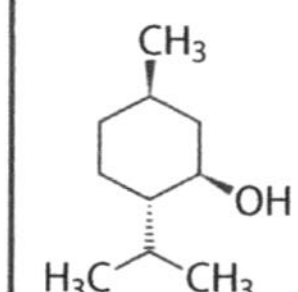	[54]
		Linalool	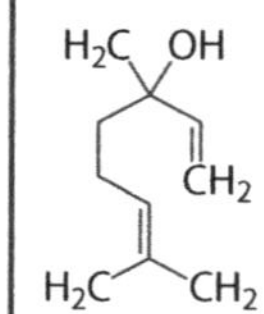	[43–51, 53–55, 62, 65]

the use of extracts of leaves with various solvents including aqueous extracts, ethanol, butanol, and methanol extracts; showing maximum retention of compounds with methanol extracts of plant leaves as well as its active fractions *i.e.* butanol and ethyl acetate including caffeic acid, luteolin, rosmarinic acid, etc. [43]. Borahr *et al.*, [44] reported in their studies about the presence of secondary metabolic compounds in plant species of *Ocimum sanctum*; including 1.6 to 7.6 % of total phenolic compounds. The study also showed that metabolites have flavonoid content ranging between 1.56 to 2.24% among other constituents. Another study on phytochemical compounds and pharmacological benefits have also proven the presence of more than 200 compounds in *Ocimum* basilicum comprising of hydrocarbons, flavonoids, monoterpenes, aromatic compounds etc. and also revealed anti-bacterial, insecticidal, antiproliferative, anti-oxidant, and anti-inflammatory etc. [45].

6.7 Pharmacological Aspect

Plants are the primary source of medicines, and the therapeutic properties of plants are thought to be due to the presence of essential oils and secondary metabolites. The main benefits of medicinal plants include their safety and lower cost, global availability, and efficacy [66]. It's a long-held belief that these plants can be used to assess medicinal properties. First, the Chinese used these plants for therapeutic purposes before 4000 to 5000 B.C. Plants were first used as medicine in India in Rigveda, written between 3500 and 1600 B.C. Ayurveda studied medicinal properties in-depth and found all medical sciences [67].

Tulsi, belongs to the family Lamiaceae, the most sacred plant of India is also considered as most important species for its medicinal and other values. The usage of *Ocimumsanctum* (Tulsi) as an aromatic plant has been well documented in Ayurveda. It is grown in subtropical and tropical including India [68]. It is Omnipresent in all Indian fields. It is a sweet-scented, erect herb. In Sanskrit, "Tulsi" means "the incomparable one". This whole plant is used for medication [69]. In India two forms of Tulsi are more common – light or Rama Tulsi and dark or Shyama (Krishna). The latter is also used commonly for worship besides its medicinal value. Various other species are usually found in India like *O. basilicum, O. canum, O. ammericanum, O. kilimandscharicum, O. micranthum*, and *O. camphora* [70, 71].

This plant's anti-microbial, anti-inflammatory, immunomodulatory, anti-stress, antipyretic, hypoglycemic, anti-asthmatic, analgesic, and hypotensive activities have been studied pharmacologically. Tulsi plant has also been effective in various animal models [72]. The essential oil of leaves contains eugenol, carvacrol, caryophylline, limatrol, and methylchavicol. The seed oil is mainly composed of sitosterol and fatty acids. The roots contain sitosterol and three triterpenes A, B, and C. The leaves also contain n-triacontanol and ursolic. Nerol, eugenol, its methyl ether,terpinen 4-decylaldehyde, caryophyllene, selinene, pinenes, a-pinene and camphene have been identified in essential oil. It also contains linalool, rosmarinic acid, thymol, citral and methyl chavicol etc. [73].

Ocimum species have a relatively high rating in terms of pharmacological activities. Research have concluded that *Ocimum* species show antimicrobial, immunomodulatory, anti-inflammatory, antiulcer, antidiabetic, Hepatoprotective, antihyperlipidemic, cardio protective, anti-oxidant, radio protective, memory enhancing, antihypertensive, anticoagulant, anti-cataract, anthelmintic and anti-nociceptive activities [74]. The biological properties of essential oils extracted from Ocimum species have been the subject of increasing research. Extracted essential oils are more in focus because of their distinctive properties. Essential oils of Basil contain several compounds and their composition depends on growing conditions such as light, temperature and irrigation [75]. *Ocimum* oils show anti-microbial, anti-oxidant, insect repellent, insecticidal, anti-inflammatory, antiulcer, analgesic, anti-carcinogenic, skin permeation enhancer, immunomodulatory, cardio-protective and anti-lipidemic properties [76].

The main bioactive compound found in Basil essential oil is estragole, and linalool, a phenylpropene used in manufacturing perfumes, acts as a food additive, whereas linalool is used as a scent in cleaning agents [77]. Compounds such as eugenol, citronellol, linalool, citral, limonene, and terpineol are well known for their anti-inflammatory and antibacterial properties. This activity of eugenol in Basil makes the plant more valuable for treating individuals with arthritis or any other inflammatory responses. Sweet basil (*O. basilicum*) is proven to contain the highest eugenol compared to other species and hence shows the highest anti-oxidant activity. Brewed basil water helps relieve nausea and headache while its oil helps inhibit the growth of pathogenic bacteria such as Staphylococcus, Enterococci, Shigella, and *Pseudomonas*.

Allopathic medications are known to cause significant toxicity in normal tissues, thus they possess side effects to human body [78, 79]. As a result, researchers are scouring the globe for the most effective antitumor agents from various sources. Recent pharmacological research has focused on the need for safe and effective chemotherapeutic drugs to treat malignancies [79, 80]. Several medicinal plants with diverse anticancer qualities exist in the indigenous system of medicine (Traditional Indian Medicine), and they require extensive investigation in order to generate antitumor herbal medications. Ethnomedicinal plants have made a significant contribution to discovering new drugs for diseases such as cancer, hypertension, diabetes, and others. India is also called "Botanical garden of the World" because of the highest production of medicinal plants. According to medical information contained in old Indian literature, certain therapeutic plants have been utilized for thousands of years in some form or another under the indigenous system of medicine. Out of 45,000 plant species present in India, 15,000 to 20,000 are medically important. On the other hand, traditional communities only use about 7000-7500 plants for medicinal purposes. Only a few medicinal plants have piqued scientist's interest in researching them as a tumor treatment [79, 81]. Natural products account for more than half of all modern drugs in clinical use, and many of them have been shown to induce apoptosis in various tumor cells [82]. Traditional medicine is used by more than 80% of people for primary health services in underdeveloped nations, according to WHO estimates [81, 83]. Some medicinal plants and their products can help to prevent cancer. The consumption of many vegetables and fruits can help to avoid getting cancer. Doctors advise people who want to lower their cancer risk to eat several fruits and vegetables each day. Several plant-derived products have been shown to have potent antitumor activity against various cancer cell lines [84].

6.7.1 Analgesic Activity

Ocimum species are reported to show analgesic activities when carried on mice. Methanolic extract of *Ocimum* basilicum was investigated for its analgesic activity. The tail immersion method was carried out on Swiss mice, and it was noted that 200 mg/kg of the extract was enough to be compared with standard aspirin dosage [85]. In one study of anti-nociceptive action of basil species, results revealed that analgesic activity of leaves with alcoholic extract at two different doses 50 and 100 mg/kg was significant [86]. Hannan *et al.*, 2011 [87] reported in their studies about *Ocimum* sanctum and showed that activity was dose-dependent (250 mg/kg and 500 mg/kg of body weight) in ethanolic extracts of the plant.

6.7.2 Anti-Microbial Activity

Ocimum species are well noted for their anti-microbial activity and Ocimum sanctum is one of the most important species. It shows the anti-microbial activity as it is revealed that ethanolic, methanolic, and organic extracts of *Ocimum* sanctum L. exhibit a wide range of inhibition against various microbes such as *Escherichia coli,* Staphylococci, Shigella, *Staphylococcus aureus and Enterobacteria species* [88]. The results have shown that *O. Sanctum* acts as an effective anti-microbial agent. The anti-bacterial activity of extracts from *O. gratissimum* leaves was also tested against *Staphylococcus aureus, Escherichia coli, Salmonella typhi* and *Salmonella typhimurium,* and concluded the effectiveness of the extract to act against microbes.

6.7.3 Anti-Oxidant Activity

Ocimum sanctum L. aqueous and methanolic extracts show significant anti-oxidant activity in both *in vivo* and *in vitro* assay. The Phytochemical investigations of *Ocimum sanctum* L. leaf show phenols (eugenol, cirsilineol, isothymucin, apigenin, and vosamarinic acid) and flavonoids (orientin and vicenin). *Ocimum sanctum,* known for heart, contains eugenol, a phenolic compound that can inhibit copper-dependent LDL oxidation and reduce iron-mediated lipid peroxidation [74]. Eugenol is an important essential oil present in different *Ocimum* species, which acts against cyclooxygenase. Cyclooxygenase is an enzyme that activates the inflammatory reactions inside the human body, and eugenol acts opposite to it [89].

6.7.4 Hepatoprotective Activity

Ocimum species are reported to have significant hepatoprotective effects. It is concluded that the ethanolic extract of *O. basilicum* leaves shows hepatoprotective effects and acts against liver damage induced by hydrogen peroxide and carbon tetrachloride. The extract also revealed the significant effect on anti-lipid peroxidation *in vitro* studies [89]. Chattopadhyay, *et al.* [90] reported the hepatoprotective effect of Ocimum sanctum species against paracetamol induced hepatic damage in rats and results were remarkable as doses of Ocimum sp. showed reduced in raised enzyme serum levels.

6.7.5 Anti-Diabetic Activity

Ocimum species have hypoglycemic properties, with *Ocimum basilicum* and *Ocimum sanctum* being the most effective against diabetes. The oral administration of aqueous extract of *O. basilicum* and sanctum lowered the blood sugar level of hyperglycemic streptozotocin-diabetic rats [71] without affecting the basal plasma insulin levels. Another study found that *Ocimum sanctum* L. reduced cortisol and glucose levels in the blood, suggesting that it may help to regulate corticosteroid-induced diabetes [88].

The leaves of the extract of *O. gratissimum* were also reported to possess some antidiabetic activity when treated on streptozocin-induced in diabetic rats [91].

6.7.6 Anti-Inflammatory Activity

Ocimum species have been reported to contain several anti-inflammatory compounds. Sieboldogenin, a compound extracted from ethyl acetate extract of *Ocimum sanctum* has shown *in vitro* and *in vivo* inflammatory activities. Several compounds, including eugenol, isothymusin, apigenin, cirsimaritin, and acid obtained from the acetone extract of *Ocimum* sanctum leaves show COX-1 inhibitory activity [92]. Manaharan *et al.*, [93] showed the activity of Ocimum sanctum essential oil against inflammatory action by suppressing the expression of MMP-9 in LPS-induced inflammatory cells, which triggered the anticipation of cancer cell migration.

6.7.7 Antifungal Activity

Ocimum species have shown a significant effect against fungus, and aqueous, hexane, chloroform, n-butanol extracts of *Ocimum sanctum* showed antifungal activity. *Ocimum sanctum* works by acting against the biodeterioration of food that occurs during storage. The aqueous and acetone extract of *Ocimum sanctum* was also effective against plant fungi such as *Alternaria tenuis, Helminthosporium*. Essential oils of *O. sanctum* (eugenol) were tested on various fungi such as *Alternaria solani, Candida guillermondii, Colletotricum capsici*, Curvularia species, and it has shown positive effects. Therefore, *Ocimum sanctum* essential oils can act against the fungal spoilage of food during storage [88].

6.8 Health Benefits

Whether leaf or seed, holy basil (*Ocimum sanctum*) is regarded as the natural tonic for each part of our body. Scientific research has concluded that *Ocimum sanctum*, native to Southeast Asia, is worshiped and full of health benefits as well. Basil leaves contain many important natural chemical compounds known to be disease-preventing. Tulsi has a unique combination of pharmacological actions that have been shown to help with physical, chemical, metabolic, and psychological stress [94]. It is also used in ancient times (Ayurveda) and in traditional Chinese medicine to treat digestive tract disorders, like stomach ache, diarrhea or kidney complications and other infections [92]. It is also used as African traditional medicine, where they use Basil to treat whooping cough and fever.

In West Africa, a Basil leaf decoction is generally used for treating whooping cough [92]. Basil protects visceral organs and tissues against degradation caused by heavy metals or industrial pollutants [94]. It has positive effects on lowering blood glucose levels and blood pressure and normalizing a person's lipid profile. The plant is used to treat nausea, diarrhea, and vomiting, whereas its fresh flowers and leaves are used for bronchitis and if mixed with black pepper, it is effective against malaria. Some of the *Ocimum* species, such as *O. sanctum* and *O. grattisimum* are rich in Vitamin K that is essential for the production of clotting factors in the blood. Leaves of such species are low in calories but full of essential minerals and vitamins for which it is regarded as an immunity booster. A total reduction in cholesterol, phospholipids, total lipids and triglycerides was observed on adding *O. sanctum* leaf powder to the diet of diabetic as well as non-diabetic rats [95] and concluded that *O. sanctum* can control the above when tests were performed on liver, heart and kidney [96].

6.9 Industrial Utilization

The primary ingredient in pesto sauce is basil but can also be used for flavoring various soups and sauces. The whole plant is useful. After soaking in water, the seeds turn mucilaginous and edible. The leaves can be used in salads. It is commonly used in Ayurveda and traditional Chinese medicine to treat digestive problems like stomach aches and diarrhea, kidney complaints, and infections. Basil is used to treating whooping cough and various fevers in African traditional medicine. In West Africa, a leaf decoction is used to treat coughs. Various Ocimum species have been studied for their anti-inflammatory and anti-nociceptive properties. These include *O. gratissimum* [97–99]; *O. basilicum* [100–103]; *O. sanctum* [104–106]; *O. americanum* [107]; O. *suave* [108]; *O. micranthum* [109]; and *O. lamiifolium* [110]. The majority of the studies that have been published so far have used rat and mouse models to investigate the activity of essential oils. Kapewangolo *et al.* [111] found that *Ocimum labiatum* ethanolic extract significantly inhibited pro-inflammatory cytokines such as IL-2, IL-4, IL-6, and IL-17.

Ocimum species have yielded several anti-inflammatory compounds. Sieboldogenin, isolated from the ethyl acetate fraction of *O. sanctum*, inhibited carrageenan-induced hind paw edema *in vitro* and *in vivo*, with doses of 10 and 50 mg/kg showing significant LOX inhibition (IC50 of 38 mM) [112]. Many compounds, including eugenol, cirsilineol, isothymonin, isothymusin, apigenin, cirsimaritin, and rosmarinic acid obtained from the acetone extract of the leaves of O. sanctum also showed good COX-1 inhibitory activity. Eugenol was discovered to be the most effective COX-1 inhibitor, with a 97 % inhibition at 1 mM [113].

6.10 Conclusion and Future Prospectus

Ocimum species (Tulsi)are found in a large number of every region of the world and with greater phytochemicals, bioactive compounds such as linalool, α-terpinolene, Myrcene, Limonene, α-terpinene, etc. and also pharmacological benefits including anti-oxidant, anti-diabetic, anti-inflammatory and hepato-protective etc. activities. Religious concern about "Rama tulsi" and "Shyama tulsi" made it more famous for everyone. This plant is used based on recent studies and used in recent times for fever, cough, cold, and infections to protect people from major diseases. Therefore, from this content, we concluded that increased intake of Ocimum species in the form of essential oils and extracted juices would be helpful and will fortify our immune system to fight against body aches and chronic diseases.

References

1. Anonymous, *Weeds of Australia. Identification and Diagnostic Tools*, CSIRO, Collingwood, 2016.
2. Borloveanu, M., *Leacuri Mănăstiresti. Terapii Pentru Trup si Suflet*, Lumea Credintei, Bucuresti, 2014.
3. Paton, A., Harley, R.M., Harley, M.M., Ocimum-an overview of relationships and classification, in: *Medicinal and Aromatic Plants-Industrial Profiles*, Y. Holm and R. Hiltunen (Eds.), pp. 1–8, Harwood Academic, Amsterdam, The Netherlands, 1999.
4. Sahoo, Y., Pattnaik, S.K., Chand, P.K., *In-vitro* clonal propagation of anaromatic medicinal herb*Ocimum* basilicumL. *(sweet basil) by axillary shoot proliferation. In Vitro Cell. Dev. Biol. Plant*, 33, 293–296, 1997.
5. Paton, A. and Putievsky, E., Taxonomic problems and cytotaxonomic relationships between varieties of *Ocimum* basilicumand related species (labiatae). *Kew Bull.*, 5, 1–16, 1996.
6. Darrah, H.H., Investigation of the cultivars of the Basils (*Ocimum*). *Econ. Bot.*, 28, 63–67, 1974.

7. Darrah, H.H., Cultivated basils, Buckeye printing. Agris.fao.org. 37–40, 1980.
8. Mbakwem-Aniebo, C., Onianwa, O., Okonko, I.O., Effects of *Ocimum* gratissimum leaves on common dermatophytes and causative agent of Pityriasis versicolor in rivers state, Nigeria. *J. Microbiol. Res.*, 2, 4, 108–113, 2012.
9. Ahmed, S., Hasan, M.M., Ahmed, S.W., Natural antiemetics: An overview. *Pak. J. Pharm. Sci.*, 27, 1583–1598, 2014.
10. De Albuquerque, U.P., De Medeiros, P.M., De Almeida, A.L.S., Monteiro, J.M., De Freitas Lins Neto, E.M., De Melo, J.G., Dos Santos, J.P., Medicinal plants of the caatinga (semi-arid) vegetation of NE Brazil: A quantitative approach. *J. Ethnopharmacol.*, 114, 325– 354, 2007.
11. Pandey, A.K., Singh, P., Tripathi, N.N., Chemistry and bioactivities of essential oils of some *Ocimum* species: An overview. *Asian Pac. J. Trop. Biomed.*, 4, 9, 682–694, ISSN 2221-1691, 2014.
12. Abdullahi, M., Biopotency role of culinary spices and herbs and their chemical constituents in health and commonly used spices in Nigerian dishes and snacks. *Afr. J. Food Sci.*, 5, 3, 111–124, 2011.
13. Azhar, A.F., Sreeramu, B.S., Srinivasappa, K.N., *Cultivation of Spice Crops*, Universities Press, India, 2005.
14. Matasyoh, L.G., Matasyoh, J.C., Wachira, F.N., Kinyua, M.G., Muigai, A.W.T., Mukiama, T.K., Antimicrobial activity of essential oils of *Ocimum gratissimum* L. from different populations of Kenya. *Afr. J. Tradit. Complement. Altern. Med.*, 5, 2, 187–193, 2008.
15. Iwu, M.M., *African Medicinal Plants*, pp. 109–110, CRC Press, Maryland, 1993.
16. Ijeh, I., II, Omodamiro, O.D., Nwanna, I.J., Anti-microbial effects of aqueous and ethanolic fractions of two spices, *Ocimum* gratissimum and Xylopia aethiopica. *Afr. J. Biotechnol.*, 4, 9, 953–956, 2005.
17. Okoli, C.O., Ezike, A.C., Agwagah, O.C., Akah, P.A., Anticonvulsant and anxiolytic evaluation of leaf extracts of *Ocimum* gratissimum, a culinary herb. *Pharmacognosy Res.*, 2, 1, 36, 2010.
18. Tchoumbougnang, F., Zollo, P.A., Dagne, E., Mekonnen, Y., *In vivo* antimalarial activity of essential oils from Cymbopogon citratus and *Ocimum* gratissimum on mice infected with Plasmodium berghei. *Planta Med.*, 71, 01, 20–23, 2005.
19. Ntezurubanza, L., Scheffer, J.J.C., Svendsen, A.B., Composition of the essential oil of *Ocimum* gratissimum grown in rwanda1. *Planta Med.*, 53, 05, 421–423, 1987.
20. Simon, J.E., Morales, M.R., Phippen, W.B., Vieira, R.F., Hao, Z., Basil: A source of aroma compounds and a popular culinary and ornamental herb, in: *Perspectives on New Crops and New Uses*, vol. 16, pp. 499–505, 1999.
21. Dalton, C.C., Application of gas analysis to continuous culture, in: *Primary and Secondary Metabolism of PlantCell Cultures*, K.H. Neumann, W. Barz, E. Reinhard (Eds.), pp. 85–65, Springer, Berlin, Germany, 1985.

22. Kothari, S.K., Bhattacharya, A.K., Ramesh, S., Essential oil yield and quality of methyl eugenol rich *Ocimum* tenuiflorum Lf (syn. O. sanctum L.) grown in south India as influenced by method of harvest. *J. Chromatogr. A*, 1054, 1-2, 67–72, 2004.
23. Pushpangadan, P. and George, V., Basil, in: *Handbook of Herbs and Spices*, pp. 55–72, Woodhead Publishing, Sawston, Cambridge, 2012.
24. Chang, X., Alderson, P.G., Wright, C.J., Effect of temperature integration on the growth and volatile oil content of basil (*Ocimum* basilicum L.). *J. Hortic. Sci. Biotechnol.*, 80, 593–598, 2005.
25. Ekren, S., Sönmez, Ç., Özçakal, E., Kurttaş, Y.S.K., Bayram, E., Gürgülü, H., The effect of different irrigation water levels on yield and quality characteristics of purple basil (*Ocimum* basilicum L.). *Agric. Water Manage.*, 109, 155–161, 2012.
26. Charles, D.J. and Simon, J.E., Comparison of extraction methods for the rapid determination of essential oil content and composition of basil. *J. Am. Soc. Hortic. Sci.*, 115, 458–462, 1990.
27. Mondello, L., Zappia, G., Cotroneo, A., Bonaccorsi, I., Chowdhury, J.U., Yusuf, M., Dugo, G., Studies on the essential oil-bearing plants of Bangladesh. Part VIII. Composition of some *Ocimum* oils O. basilicum L. var. purpurascens; O. sanctum L. purple, O. americanum L. citral type; O. americanum L. camphor type. *Flavour Fragr. J.*, 17, 335–340, 2002.
28. Labra, M., Miele, M., Ledda, B., Grassi, F., Mazzei, M., Sala, L.F., Morphological characterization, essential oil composition and DNA genotyping of *Ocimum* basilicum cultivars. *Plant Sci.*, 167, 725–731, 2004.
29. Klimankova, E., Holadová, K., Hajšslová, J., Èajka, T., Poustka, J., Koudela, M., Aroma profiles of five basil (*Ocimum* basilicum L.) cultivars grown under conventional and organic conditions. *Food Chem.*, 107, 464–472, 2008.
30. Kashyap, C.P., Ranjeet, K., Vikrant, A., Vipin, K., Therapeutic potency of *Ocimum* kilimandscharicum Guerke-A review. *Glob. J. Pharmacol.*, 5, 3, 191–200, 2011.
31. Pattanayak, P., Behera, P., Das, D., Panda, S.K., *Ocimum* sanctum Linn. A reservoir plant for therapeutic applications: An overview. *Pharmacogn. Rev.*, 4, 7, 95–105, 2010.
32. Shuaib, Musah, M., Yerima, H., Sani, H., Amusat, K., Comparative nutritional and anti-nutritional analysis of *Ocimum* grattissimum and *Ocimum basilicum*. Academia Arena, 7, 7, 77–81, 2015.
33. Padalia, R.C. and Verma, R.S., Comparative volatile oil composition of four *Ocimum* species from northern India. *Nat. Prod. Res.*, 25, 6, 569–575, 2011.
34. Tchatchouang, S., Beng, V.P., Kuete, V., Chapter 11-Antiemetic African medicinal spices and vegetables, in: *Medicinal Spices and Vegetables from Africa*, V. Kuete, (Ed.), pp. 299–313, Academic Press, Cambridge, ISBN 9780128092866, 2017.
35. Jaggi, R.K., Madaan, R., Singh, B., Anticonvulsant potential of holy basil, *Ocimum* sanctum Linn. and its cultures. *Indian J. Exp. Biol.*, 41, 1329–33, 2003.

36. Panchal, P. and Parvez, N., Phytochemical analysis of medicinal herb (*Ocimum* sanctum). *Int. J. Nanomater. Nanotechnol. Nanomed.*, 5, 2, 008–011, 2019.
37. Vasudevan, P., Kashyap, S., Sharma, S., Bioactive botanicals from basil (*Ocimum* sp.). *Journal of Scientific & Industrial Research (JSIR)*, 58, 332–338, 1999.
38. Skaltsa, H., Couladi, M., Philianos, S., Singh, M., Phytochemical study of the leaves of *Ocimum* sanctum. *Fitoterapia*, 8, 13, 286, 1987.
39. Mookherjea, S., Chemical examination of the inflorescence of *Ocimum kilimandscharicum Gurke. Indian Chem. Soc. J.*, 50, 1, 69, 1973.
40. Dube, S., Upadhyay, P.D., Tripathi, S.C., Antifungal, physicochemical, and insect-repelling activity of the essential oil of *Ocimum* basilicum. *Can. J. Bot.*, 67, 7, 2085–2087, 1989.
41. Shekhawat, P.S. and Prasada, R., Antifungal properties of some plant extracts: 2. Growth inhibition studies. *Sci. Cult.*, 37, 40, 1971.
42. Gupta, K.C. and Viswanathan, R., Combined action of streptomycin and chloramphenicol with plant antibiotics against tubercle bacilli. Part I: streptomycin and chloramphenicol with cepharanthine. Part II: Streptomycin and allicin. *Antibiot. Chemother.*, 5, 1, 24–7, 1955.
43. Mousavi, L., Salleh, R.M., Murugaiyah, V., Phytochemical and bioactive compounds identification of *Ocimum* tenuiflorum leaves of methanol extract and its fraction with an anti-diabetic potential. *Int. J. Food Prop.*, 21, 1, 2390–2399, 2018.
44. Borah, R. and Biswas, S.P., Tulsi (*Ocimum* sanctum), excellent source of phytochemicals. *Int. J. Environ. Agric. Biotech.*, 3, 5, 265258, 2018.
45. Marwat, S.K., Khan, M.S., Ghulam, S., Anwar, N., Mustafa, G., Usman, K., Phytochemical constituents and pharmacological activities of sweet Basil-Ocimum basilicum L.(Lamiaceae). *Asian J. Chem.*, 23, 9, 3773, 2011.
46. Ismail, M., Central properties and chemical composition of *Ocimum* basilicum. essential oil. *Pharm. Biol.*, 44, 8, 619–626, 2006.
47. Chang, X., Alderson, P.G., Wright, C.J., Variation in the essential oils in different leaves of basil (*Ocimum* basilicum L.) at day time. *Open Horticulture J.*, 2, 1, 13–16, 2009.
48. De Martino, L., De Feo, V., Nazzaro, F., Chemical composition and *in vitro* anti-microbial and mutagenic activities of seven Lamiaceae essential oils. *Molecules*, 14, 10, 4213–4230, 2009.
49. Ozcan, M. and Chalchat, J.C., Essential oil composition of *Ocimum* basilicum L. *Czech J. Food Sci.*, 20, 6, 223–8, 2002.
50. Hassanpouraghdam, M.B., Hassani, A., Shalamzari, M.S., Menthone- and estragole-rich essential oil of cultivated *Ocimum* basilicum L. from Northwest Iran. *Chemija*, 21, 1, 59–62, 2010.
51. Hassanpouraghdam, M.B., Gohari, G.R., Tabatabaei, S.J., Dadpour, M.R., Inflorescence and leaves essential oil composition of hydroponically grown *Ocimum* basilicum L. *J. Serb. Chem. Soc.*, 75, 10, 1361–1368, 2010.

52. Lawrence, B., Mookheyee, B., Wills, B., Developments in food sciences, flavors and fragrances: A world perspective, in: *Proceedings of the 10th International Congress of Essential Oils, Fragrances, and Flavors*, vol. 848, Elsevier Inc., Washington DC, 1988.
53. Lee, S.J., Umano, K., Shibamoto, T., Lee, K.G., Identification of volatile components in basil (*Ocimum* basilicum L.) and thyme leaves (Thymus vulgaris L.) and their anti-oxidant properties. *Food Chem.*, 91, 1, 131–137, 2005.
54. Bassolé, I.H.N., Lamien-Meda, A., Bayala, B., Tirogo, S., Franz, C., Novak, J., Dicko, M.H., Composition and anti-microbial activities of Lippia multiflora Moldenke, Mentha x piperita L. and *Ocimum* basilicum L. essential oils and their major monoterpene alcohols alone and in combination. *Molecules*, 15, 11, 7825–7839, 2010.
55. Hussain, A., II, Anwar, F., Sherazi, S.T.H., Przybylski, R., Chemical composition, anti-oxidant and anti-microbial activities of basil (*Ocimum* basilicum) essential oils depends on seasonal variations. *Food Chem.*, 108, 3, 986–995, 2008.
56. Sims, C.A., Juliani, H.R., Mentreddy, S.R., Simon, J.E., Essential oils in holy basil (*Ocimum* tenuiflorum L.) as influenced by planting dates and harvest times in North Alabama. *J. Med. Act. Plants*, 2, 3, 33–41, 2014.
57. Pino, J.A., Rosado, A., Rodriguez, M., Garcia, D., Composition of the essential oil of *Ocimum* tenuiflorum L. grown in Cuba. *J. Essent. Oil Res.*, 10, 4, 437–438, 1998.
58. Brophy, J.J., Goldsack, R.J., Clarkson, J.R., The essential oil of *Ocimum* tenuiflorum L.(Lamiaceae) growing in Northern Australia. *J. Essent. Oil Res.*, 5, 4, 459–461, 1993.
59. Yamani, H.A., Pang, E.C., Mantri, N., Deighton, M.A., Anti-microbial activity of Tulsi (*Ocimum* tenuiflorum) essential oil and their major constituents against three species of bacteria. *Front. Microbiol.*, 7, 681, 2016.
60. Raina, A.P., Kumar, A., Dutta, M., Chemical characterization of aroma compounds in essential oil isolated from "Holy Basil" (*Ocimum* tenuiflorum L.) grown in India. *Genet. Resour. Crop Evol.*, 60, 5, 1727–1735, 2013.
61. Sajjadi, S.E., Analysis of the essential oils of two cultivated basil (*Ocimum* basilicum L.) from Iran. *DARU J. Pharm. Sci.*, 14, 3, 128–130, 2006.
62. Vani, S.R., Cheng, S.F., Chuah, C.H., Comparative study of volatile compounds from genus. *Ocimum. Am. J. Appl. Sci.*, 6, 3, 523, 2009.
63. Zhang, J.W., Li, S.K., Wu, W.J., The main chemical composition and *in vitro* antifungal activity of the essential oils of *Ocimum* basilicum Linn. var. pilosum (Willd.) Benth. *Molecules*, 14, 1, 273–278, 2009.
64. Bhatti, H.A., *Isolation and Structure Elucidation of the Chemical Constituents of Ocimum Basilicum L. and Centella Asiatica and Synthesis of Lawsonin and its Relative Dhydrobenzofuran Derivatives*, pp. 1–217, Ph.D. Thesis, International Center for Chemical and Biological Sciences H.E.J. Research Institue of Chemistry University of Karachi, Pakistan, 2008.

65. Ismail, M., Central properties and chemical composition of *Ocimum* basilicum. essential oil. *Pharm. Biol.*, 44, 8, 619–626, 2006.
66. Ahmed, M., Ahamed, R.N., Aladakatti, R.H., Ghosesawar, M.G., Reversible anti-fertility effect of benzene extract of *Ocimum* sanctum leaves on sperm parameters and fructose content in rats. *J. Basic Clin. Physiol. Pharmacol.*, 13, 1, 51–9, 2002.
67. Pandey, M.M., Rastogi, S., Rawat, A.K.S., Indian traditional ayurvedic system of medicine and nutritional supplementation. *Evid. Based Complementary Altern. Med.*, 2013, 1–12, 2013.
68. Banerjee, S., Parashar, R., Kumar, A., Rao, A.R., Modulatory influence of alcoholic extract of Ocimum leaves on carcinogen-metabolizing enzyme activities and reduced glutathione levels in mouse. *Nutr. Cancer*, 25, 2, 205–217, 1996.
69. Bansod, S. and Rai, M., Antifungal activity of essential oils from Indian medicinal plants against human pathogenic Aspergillus fumigatus and A. niger. *World J. Med. Sci.*, 3, 2, 81–88, 2008.
70. Bhargava, K.P., Antistress activity of Ocimum sanctum Linn. *Indian J. Med. Res.*, 73, 443–451, 1981.
71. Chattopadhyay, R.R., Hypoglycemic effect of O. sanctum leaf extract in normal and streptozotocin diabetic rats. *Indian J. Exp. Biol.*, 31, 891–893, 1993.
72. Chiang, L.C., Ng, L.T., Cheng, P.W., Chiang, W., Lin, C.C., Antiviral activities of extracts and selected pure constituents of Ocimum basilicum. *Clin. Exp. Pharmacol. Physiol.*, 32, 10, 811–816, 2005.
73. Dhar, M.L., Dhar, M.M., Dhawan, B.N., Mehrotra, B.N., Ray, C., screening of Indian plants for biological activity: Part I. *Indian J. Exp. Biol.*, 6, 4, 232–235, 1968.
74. Collin, H., Herbs, spices and cardiovascular disease, in: *Handbook of Herbs and Spices*, pp. 126–137, Woodhead Publishing, 126–137, 2006.
75. Aldarkazali, M., Rihan, H.Z., Carne, D., Fuller, M.P., The growth and development of sweet basil (*Ocimum basilicum*) and bush basil (*Ocimum minimum*) grown under three light regimes in a controlled environment. *Agronomy*, 9, 743, 2019.
76. Mahajan, N., Rawal, S., Verma, M., Poddar, M., Alok, S., A phytopharmacological overview on Ocimum species with special emphasis on Ocimum sanctum. *Biomed. Prev. Nutr.*, 3, 2, 185–192, 2013.
77. Smith, R.L., Adams, T.B., Doull, J., Feron, V.J., Goodman, J., II, Marnett, L.J., Sipes, I.G., Safety assessment of allylalkoxybenzene derivatives used as flavouring substances—methyl eugenol and estragole. *Food Chem. Toxicol.*, 40, 7, 851–870, 2002.
78. Pandey, G.P., *Hepatogenic Effect of Some Indigenous Drugs on Experimental Liver Damage*, Doctoral Dissertation, College of Veterinary Science & Animal Husbandry, 1990.

79. Somkuwar, A.P., *Studies on Anticancer Effects of Ocimum Sanctum and Withania Somnifera on Experimentally Induced Cancer in Mice*, Doctoral dissertation, Jawaharlal Nehru Krishi Viswavidyalaya, Jabalpur, 2003.
80. Govind, P. and Madhuri, S., Medicinal plants: Better remedy for neoplasm. *Indian Drugs*, 43, 11, 869–874, 2006.
81. Madhuri, S. and Pandey, G., Some dietary agricultural plants with anticancer properties. *Plant Arch.*, 8, 1, 13–16, 2008.
82. Rosangkima, G. and Prasad, S.B., Antitumour activity of some plants from Meghalaya and Mizoram against murine ascites Dalton's lymphoma, *Indian Journal of Experimental Biology (IJEB)*, 42, 981–988, 2004.
83. Sivalokanathan, S., Ilayaraja, M., Balasubramanian, M.P., Efficacy of Terminalia arjuna (Roxb.) on N-nitrosodiethylamine induced hepatocellular carcinoma in rats, *Indian Journal of Experimental Biology (IJEB)*, 43, 264–267, 2005.
84. Sharma, M. and Govind, P., Ethnomedicinal plants for prevention and treatment of tumours. *Int. J. Green Pharm.*, 3, 1, 2, 2009.
85. Ch, M.A., Naz, S.B., Sharif, A., Akram, M., Saeed, M.A., Biological and pharmacological properties of the sweet basil (Ocimum basilicum). *J. Pharm. Res. Int.*, 7, 5, 330–339, 2015.
86. Khanna, N. and Bhatia, J., Anti-nociceptive action of Ocimum sanctum (Tulsi) in mice: Possible mechanisms involved. *J. Ethnopharmacol.*, 88, 2-3, 293–296, 2003.
87. Hannan, J.M.A., Das, B.K., Uddin, A., Bhattacharjee, R., Das, B., Chowdury, H.S., Mosaddek, A.S.M., Analgesic and anti-inflammatory effects of Ocimum sanctum (Linn) in laboratory animals. *Int. J. Pharm. Sci. Res.*, 2, 8, 2121, 2011.
88. Siva, M., Shanmugam, K.R., Shanmugam, B., Venkata, S.G., Ravi, S., Sathyavelu, R.K., Mallikarjuna, K., Ocimum sanctum: A review on the pharmacological properties. *Int. J. Basic Clin. Pharmacol.*, 5, 558–565, 2016.
89. Bilal, A., Jahan, N., Makbul, S., Bilal, S.N., Habib, S., Hajra, S., Phytochemical and pharmacological studies on Ocimum basilicum. *Int. J. Curr. Res. Rev.*, 4, 73–83, 2012.
90. Chattopadhyay, R.R., Sarkar, S.K., Ganguly, S., Medda, C., Basu, T.K., Hepatoprotective activity of Ocimum sanctum leaf extract against paracetamol included hepatic damage in rats, *Indian J. Pharmacol.*, 24, 163–165, 1992.
91. Mohammed, A., Tanko, Y., Okasha, M.A., Magaji, R.A., Yaro, A.H., Effects of aqueous leaves extract of Ocimum gratissimum on blood glucose levels of streptozocininduced diabetic wistar rats. *Afr. J. Biotechnol.*, 6, 18, 2087–2090, 2007.
92. Dzoyem, J.P., Melong, R., Tsamo, A.T., Maffo, T., Kapche, D.G., Ngadjui, B.T., Eloff, J.N., Cytotoxicity, anti-oxidant and anti-bacterial activity of four compounds produced by an endophytic fungus Epicoccum nigrum associated with Entada abyssinica. *Rev. Bras. Farmacogn.*, 27, 251–253, 2017.

93. Manaharan, T., Thirugnanasampandan, R., Jayakumar, R., Ramya, G., Ramnath, G., Kanthimathi, M.S., Antimetastatic and anti-inflammatory potentials of essential oil from edible Ocimum sanctum leaves. *Sci. World J.*, 2014, 1–5, 2014.
94. Cohen, M.M., Tulsi-ocimum sanctum: A herb for all reasons. *J. Ayurvedaand Integr. Med.*, 5, 4, 251–259, 2014.
95. Dini, I., Chapter 14-Spices and herbs as therapeutic foods, in: *Handbook of Food Bioengineering, Food Quality: Balancing Health and Disease*, A.M. Holban and A.M. Grumezescu (Eds.), pp. 433–469, Academic Press, Cambridge, 2018.
96. Suanarunsawat, T., Ayutthaya, W.D., Songsak, T., Rattanamahaphoom, J., Anti-lipidemic actions of essential oil extracted from *Ocimum san*ctum L. leaves in rats fed with high cholesterol diet. *J. Appl. Biomed. Sci.*, 7, 45–53, 2009.
97. Tarkang, P.A., Okalebo, F.A., Siminyu, J.D., Ngugi, W.N., Mwaura, A.M., Mugweru, J., Guantai, A.N., Pharmacological evidence for the folk use of Nefang: antipyretic, anti-inflammatory and anti-nociceptive activities of its constituent plants. *BMC Complement. Altern. Med.*, 15, 1, 1–11, 2015.
98. Ajayi, A.M., Tanayen, J.K., Ezeonwumelu, J.O.C., Dare, S., Okwanachi, A., Adzu, B., Ademowo, O.G., Anti-inflammatory, anti-nociceptive and total polyphenolic content of hydroethanolic extract of Ocimum gratissimum L. leaves. *Afr. J. Med. Med. Sci.*, 43, Suppl 1, 215, 2014.
99. Tanko, Y., Magaji, G.M., Yerima, M., Magaji, R.A., Mohammed, A., Anti-nociceptive and anti-inflammatory activities of aqueous leaves extract of Ocimum Gratissimum (Labiate) in Rodents. *Afr. J. Tradit. Complement. Altern. Med.*, 5, 2, 141–146, 2008.
100. Singh, S., Mechanism of action of anti-inflammatory effect of fixed oil of Ocimum basilicum Linn, *Indian Journal of Experimental Biology (IJEB)*, 248-252, 1999.
101. Min, S.S., Han, S.H., Yee, J., Kim, C., Seol, G.H., Im, J.H., Lee, M.J., Anti-nociceptive effects of the essential oil of Ocimum basilicum in mice. *Korean J. Pain*, 22, 3, 206–209, 2009.
102. Venâncio, A.M., Onofre, A.S.C., Lira, A.F., Alves, P.B., Blank, A.F., Antoniolli, A.R., de Araujo, B.S., Chemical composition, acute toxicity, and anti-nociceptive activity of the essential oil of a plant breeding cultivar of basil (Ocimum basilicum L.). *Planta Med.*, 77, 08, 825–829, 2011.
103. Raina, P., Deepak, M., Chandrasekaran, C.V., Agarwal, A., Wagh, N., Kaul-Ghanekar, R., Comparative analysis of anti-inflammatory activity of aqueous and methanolic extracts of Ocimum basilicum (basil) in RAW264. 7, SW1353 and human primary chondrocytes in respect of the management of osteoarthritis. *J. Herb. Med.*, 6, 1, 28–36, 2016.
104. Godhwani, S., Godhwani, J.L., Vyas, D.S., Ocimum sanctum: An experimental study evaluating its anti-inflammatory, analgesic and antipyretic activity in animals. *J. Ethnopharmacol.*, 21, 2, 153–163, 1987.

105. Singh, S., Majumdar, D.K., Rehan, H.M.S., Evaluation of anti-inflammatory potential of fixed oil of Ocimum sanctum (Holybasil) and its possible mechanism of action. *J. Ethnopharmacol.*, 54, 1, 19–26, 1996.
106. Manaharan, T., Thirugnanasampandan, R., Jayakumar, R., Ramya, G., Ramnath, G., Kanthimathi, M.S., Antimetastatic and anti-inflammatory potentials of essential oil from edible Ocimum sanctum leaves. *Sci. World J.*, 2014, 1–5, 2014.
107. Yamada, A.N., Grespan, R., Yamada, Á. T., Silva, E.L., Silva-Filho, S.E., Damião, M.J., Cuman, R.K.N., Anti-inflammatory activity of Ocimum americanum L. essential oil in experimental model of zymosan-induced arthritis. *Am. J. Chin. Med.*, 41, 04, 913–926, 2013.
108. Masresha, B., Makonnen, E., Debella, A., *In vivo* anti-inflammatory activities of Ocimum suave in mice. *J. Ethnopharmacol.*, 142, 1, 201–205, 2012.
109. de Pinho, J.P.M., Silva, A.S.B., Pinheiro, B.G., Sombra, I., de Carvalho Bayma, J., Lahlou, S., Magalhães, P.J.C., Antinociceptive and antispasmodic effects of the essential oil of Ocimum micranthum: Potential anti-inflammatory properties. *Planta Med.*, 78, 07, 681–685, 2012.
110. Mequanint, W., Makonnen, E., Urga, K., *In vivo* anti-inflammatory activities of leaf extracts of Ocimum lamiifolium in mice model. *J. Ethnopharmacol.*, 134, 1, 32–36, 2011.
111. Kapewangolo, P., Omolo, J.J., Bruwer, R., Fonteh, P., Meyer, D., Anti-oxidant and anti-inflammatory activity of Ocimum labiatum extract and isolated labdane diterpenoid. *J. Inflamm.*, 12, 1, 1–13, 2015.
112. Mallick, P., Anti-inflammatory activities of sieboldogenin from *Ocimum sanctum*: Experimental and computational studies. *World J. Pharm. Pharm. Sci.*, 3, 1459–1465, 2014.
113. Kelm, M.A., Nair, M.G., Strasburg, G.M., DeWitt, D.L., Anti-oxidant and cyclooxygenase inhibitory phenolic compounds from Ocimum sanctum Linn. *Phytomedicine*, 7, 1, 7–13, 2000.

7

Role of Bioactive Compounds of *Bauhinia variegata* and their Benefits

Deepika Kaushik[1], Mukul Kumar[2]*, Ravinder Kaushik[3] and Ashwani Kumar[4]

[1]Department of Biotechnology, Faculty of Applied Sciences and Biotechnology, Shoolini University, Solan (H.P.), India
[2]Food Technology and Nutrition, School of Agriculture, Lovely Professional University, Phagwara, Punjab, India
[3]School of Health Sciences, University of Petroleum and Energy Stuides, Deharadun, Uttrakhand, India
[4]Department of Post Harvest Technology, College of Horticulture and Forestry, Jhansi, India

Abstract

Bauhinia variegata belongs to Fabaceae family and is a deciduous tree of medium-sized that grows naturally in Asia (India, China, Pakistan, Mayanmar, East Indies, China, Thailand, Combida, Vietnam, and Loas) and North America (Jamaica). It has roots, brown bark, green spread leaves, white to purplish flowers, and seed. The flowers are used as a vegetable in many parts of Asia, and the other parts, like bark, are used in medicinal preparations. All these parts are rich in phytochemicals like bark contains kaempferol 7, pyranoside, hesperidin, kaempferol-3-o-β-D-gluco, stem contains isoqucercitroside, rutoside, myricetol glycoside, kaempferol, glycoside, bauhinione, 2,7-dimethoxy-3-methyl-9, leaves contain heptatriacontane, B-diol 7 dotetracont-15-en-9-ol, catechol, tannins, ellagic acid, sterol, the flower contains kaempferol-3-galactoside, kaempferol-3-rhamno glucoside, cynidin-3-glucoside, malvidin-3-glucoside, malvidin-3-diglucoside, peonidin-3-glucoside, which show effective properties in human health. These phytochemicals exhibit therapeutic properties like anticarcinogenic activity and antimicrobial activity (root bark, stem bark, and leaves), antidiabetic action (stem bark, leaves, flower), hemagglutinating activity (crude seeds), hypolipidemic activity (root, stem bark), hematinic activity and immunomodulatory activity

**Corresponding author*: mukulkumar@gmail.com

Prerna Gupta, Navnidhi Chhikara and Anil Panghal (eds.) Harvesting Food from Weeds, (217–266)

(stem bark), nephroprotective activity (root), neural activity, and antioxidant effects (root and flower).

Keywords: *Bauhinia variegate*, bioactive compounds, therapeutic value, traditional uses

7.1 Introduction

Bauhinia variegata is the fast-growing flowering tree that belongs to the Leguminosae family. It is a medium-sized tree that grows all over the world at an altitude of 1300 m [1]. The genus of Bauhinia contains more than 200 species [2]. Worldwide, it has been known by various names, such as *Bwaycheng* and *Bwechin* (Burmese), Ayata, Karalabhogi, Mandara, Irkubalitu, Bilikanjivala, Ulipe Arisinantige, Kanjivala, Kempukanjivala, and Kempumandara (Canarese) and Bauhinie panache and Arbe de saint Thomas (French) [3, 4]. It is commonly called Mountain Ebony in English. It also has various vernacular names like *Kachnar*, *kattaki*, *Phalgu*, and *Bodanta* in India (Figure 7.1). The different parts of the *Bauhinia variegata viz.*, bark, seeds, bud, flower, and fruit are used traditionally for consumption. It showed higher therapeutic value, such as antidiabetic, antiobesity,

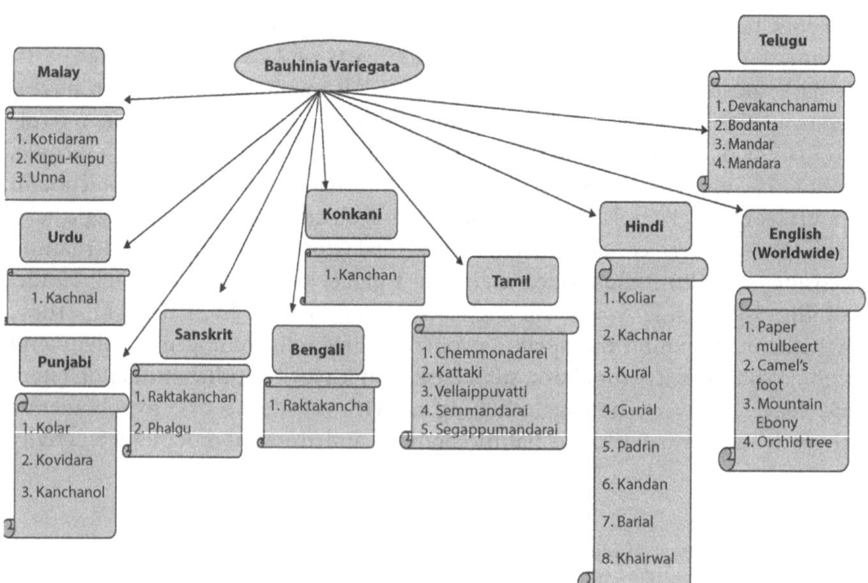

Figure 7.1 Local name of *Bauhinia variegate* in India [4, 14, 15].

anti-inflammatory, antidiarrheal, antioxidant activity, wound healing, anti-ulcer, hematinic, immunomodulatory, hemagglutinating, antimicrobial, hepatoprotective [5]. Medicinal plants can be poisonous if wrong plant parts or concentrations are used [6]. Medicinal plants and herbal medicines are assumed to be harmless. According to the Ayurveda *Bauhinia variegate* tree is reported in different fields *viz., Ruksha guna, Katu Vipaka, Kasaya rasa,* and *Shita virya.* In the literature, it was reported that a part of *Bauhinia variegate* tree is used to cured various diseases, such as wounds (vrna), worm infection (krinnroga), cervical lymphadenitis (apaci), and scrofula (gandamala) [7, 8].

7.2 Origin and Distribution of *Bauhinia variegata*

Bauhinia variegata originated from the East Indies and Jamaica [3]. It is mostly found in tropical and subtropical countries, such as India, Burma, Pakistan, Loas, China, North Thailand, North Vietnam, and Combida [4]. In India, the *Bauhinia variegata* grows in several states, like Delhi, Punjab, Himachal Pradesh, Bihar, Madhya Pradesh, Jammu and Kashmir, Manipur, Karnataka, Nagaland, Tripura, Uttar Pradesh, Rajasthan, West Bengal, Tamil Nandu, and Orissa [9]. This tree can grow in the slightly acidic and acidic soil, along the roadside, yards, parks, streets for decoration purposes [10]. For the optimum growth of this tree, the required temperature and rainfall range are 7°C to 44°C and 760 to 1900 mm, respectively [11]. The tree commonly grows up to 33 to 39 ft [2].

7.3 Cultivation

Bauhina variegata tree required complete sun exposure to grow and moderate moisture in the winter. Proper watering is required during the summer and winter seasons [12]. Fertile soil is required to grow *Bauhinia variegate* tree, which has the capability to retain moisture content [13].

7.4 Morphology

Bauhinia variegata is a small size tree that attended the 15-m height with a 50-cm diameter. It contains flowers, leaves, stems, bark, and pods [16]. The size of the *Bauhinia variegata* leaves is 1 to 2 mm, which contains tiny stipules. The leaf blade is 1.5 to 4 × 2.2 to 6 cm orbicular, leathery, pubescent are axial, veined (7-9), apex bifid to ca. 1/3, glabrous axially, and rounded lobes (present at apex). The size of the lamina and petiole is 6 to 16 cm and 3 to 4 cm, respectively [17]. The lamina of a tree is ovate to spherical, broader, and elongated. The inner and outer color of *Bauhinia variegata* bark is pale pink and brown [18]. The bark of the tree is soft, bitter in taste, and fibrous [19]. The bark of the plant is grayish brown externally and cream-colored from the inner side. After cutting of the bark, it gradually turns into red, and on drying, it became brown and smooth. The external surface of the bark always remains grayish-brown and rough due to the presence of fissures, transverse cracks, and exfoliations. After drying, it became curved and channelled [20]. The phloem is represented by sieve tubes, companion cells, phloem parenchyma, phloem fibers, crystal fibers, and stone cells, transverse by uni, and biseriate medullary rays [21]. The tree contains 10 to 15 seeds and pods are hard, dehiscent, flat [17]. Usually, the twigs are light green in color with the zigzag slightly hairs and slender appearance. In the end, the twigs cluster of the flowers is unbranched and racemes. The buds of the tree are light green in color, contain the fairly hairy calyx, and are pointed with five angled buds. These buds are splits open to one side of the tree whose other side remains attached. It contains five petals which are wavy margined, unequal, and close to the base. The color of the flower is white and purplish according to the different varieties of the *Bauhinia variegata*. In the tree, few flowers have stalk-like, stout stalk, shorts, and narrow basal tube. However, the flower of the *Bauhinia variegata* contained the five curved stamens with the stalked, lessened, very slender, and have the green color appearance [3, 22]. Pollen grains are spheroidal and tricolporate, broadly opened with large, thickened, and circular pores. Also, the flower contains the one-celled ovary, dots like stigma and style [15]. The ovary is superior with marginal placentation. The flower buds are apex protruding, obovoid, and puberulent in nature and they contain the hypanthium turbinate. The petals of the flower are subequal, subsessile, 8 to 10 mm, oblanceolate and yellowish in color. It contains the unequal, 10 fertile stamens, ca. 3 mm, another small, and size of filaments is 6 to 7 mm. Microscopic studies of flowers reveal uni and multicellular covering trichomes pointed at the apex and broad at the base and thin-walled multicellular balloon-shaped glandular trichomes. The *Bauhinia variegate* tree contains 15 to 20× 1.8 to 2.2 cm legumes which are linear-cylindric, glabrous, and valves woody. The branches of the *Bauhinia variegate* tree are 3 to 6 m spread outward and the size of the lobed leaves is 10 to 15 cm [3, 17]. *Bauhinia variegate* is an Indian tree that remains leafless from January to April and from the month of November to December the leaf fall is started. The flowering is grown on the tree in the leafless condition at the age of 2 to

3 years. In favorable conditions (light and moisture), the seeds are started to disperse from the pods and the germination process occurred [22].

7.5 Composition

Bauhinia variegata tree contains a higher amount of carbohydrates, fat, fiber, ash, protein, and oil which are available in various parts *viz.*, flower, seeds, buds, and dried leaves as shown in Figure 7.2. It contains several fatty acids such as stearic acid, linolenic acid, oleic acid, margaric acid, palmitic acid, behenic acid, and arachidic acid [23]. These fatty acids help in the metabolic regulatory system and plasma membrane. Linoleic acid is the only fatty acid among others, which consider as a precursor of metabolic regulatory compounds [24]. *Bauhinia variegata* leaves are a rich source of minerals, vitamin C, reducing sugar, and polyphenolic compounds [25]. Flowers are a good source of digestive carbohydrates, fats, and protein [22]. The various part of the tree contains a higher amount of phytotherapeutic compounds. In a study, it was reported that *Bauhinia variegta* oil shows several physicochemical characteristics, such as iodine value (84.5 g of 12/100 g of oil), peroxide value (1.9 meq O_2/kg of oil), refractive index

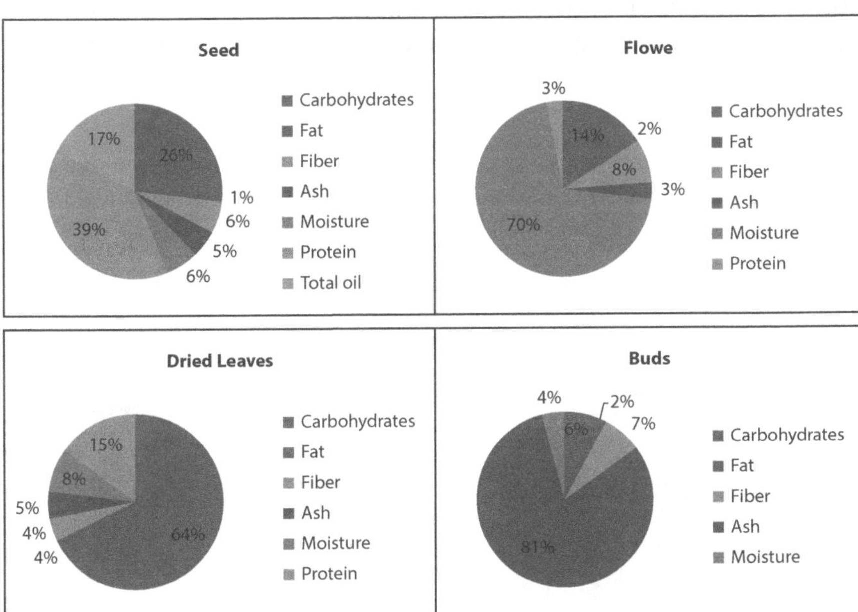

Figure 7.2 Composition of the various part of *Bauhinia variegta* [4, 17].

(1.4589), saponification value (191.3 mg of KOH/g of oil), unsponification matter (0.9 %), moisture content (6.7 %), fiber (6.9 %), ash (4.8 %), and color (2.2 unit of red color and 30.0 unit for yellow color) [4, 26, 27]. In another report, the *Bauhinia variegta* seeds contain significant amount of total proteins, carbohydrates, total oils, and free fatty acids (41.9 %, 28.4 %, 18.0 %, and 0.6 %) respectively . It also includes the component of fatty acids, such as palmitoleic acid (0.4 %), stearic acid (17.5 %), palmitic acid (22.1 %), oleic acid (on the basis of C18:1 (0.5 %) and Cis 9.7 (13.4 %), margaric acid (0.3 %), linolenic acid (0.6 %), linoleic acid (42.1 %), nevonic acid (0.6 %), eicosapentanenoic acid (0.2 %), and behenic acid (0.5 %) [28, 29]. The flower of the *Bauhinia variegta* is white and pale violet. The white flower contains the kaempferol-3-rhamno glucoside and kaempferol-3-rhamno glucoside, whereas cynidin-3-glucoside, malvidin-3-glucoside, malvidin-3-diglucoside, and peonidin-3-glucoside are present in the pale violet flower [30–32].

7.6 Bioactive Compound of *Bauhinia variegta*

Various parts of *Bauhinia variegta viz.*, root, leaves, seed, flower, stem and non-woody aerial part contains a huge number of bioactive compounds as shown in Figure 7.3. The *Bauhinia variegta* plant contains a huge number of a compound, such as ombuin, hesperidin, rugtoside,

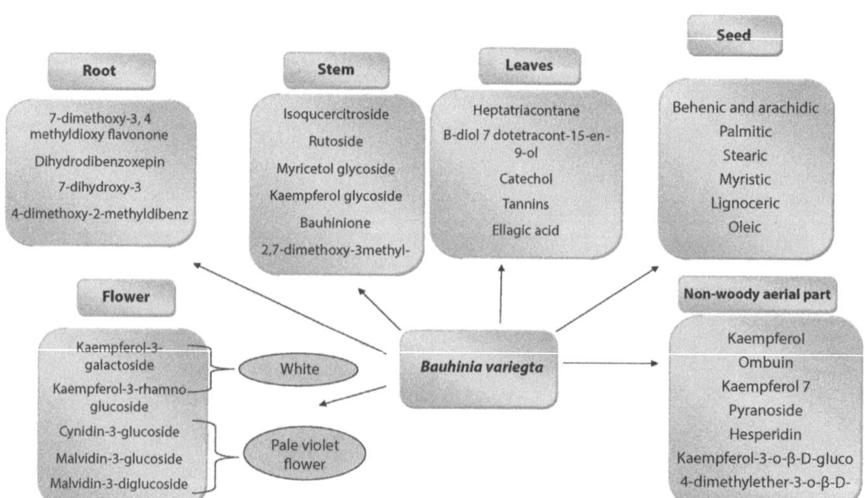

Figure 7.3 Compound present in different part of *Bauhinia variegta* [12, 36–40].

quercitroside, lupeol, catechol, ellagic acid, lysine, valine, and proline [12, 15, 33]. These compounds show several health benefits, such as wound healing (by intake of boiled water), skin prevention (decrease the swelling and maintaining the texture of the skin), reducing the inflammatory activities (show high healing (Ropan), and cold (Sita) properties). The soxhlet technique, reflux technique, solvent extraction technique, and microwave-assisted extraction technique are used for the preparation of extracts to isolate the compounds from the *Bauhinia variegta* [18, 34, 35].

7.7 Role and Structure of Bioactive Compounds of *Bauhinia variegta*

These compounds show several mechanisms, which help in the prevention of several harmful diseases. Compound ombuin helps in the reduction of inflammatory and viral activity by controlling the function of microphage in the inflammatory process [41]. In another report, it was observed that rutoside help in strengthening the blood vessels and circulation of blood [42]. It also reduces the low-density lipoprotein (LDL) level in the body. Rutoside also helps in inhibiting the hepatoxicity level by suppressing of TGF-Beta/Smad signaling pathway [43]. In the mouse model, lupeol intake helps in the inhibition of inflammation and cancer activity due to decreasing the level of Type II cytokines, IL-13, IL-5, and IL-4 [44]. In another study, it was observed that malvidin-3-glucoside protects the endothelial cells by the inhibition of apoptotic proteins and ROS production [45]. The stearic acid is used as saturated fat in daily life due to the presence of hydrogen atoms in the carbon chain [46]. However, lysine intake helps in blocking down the stress response receptors and reduces the level of anxiety [39]. In another report, the intake of valine is responsible for the repairing of muscle tissue by raising the level of glucose in muscles [23].

7.7.1 Cyanidin-3-Glucoside

Cyanidin-3-Glucoside is the natural anthocyanin that shows high therapeutic properties, such as anti-inflammatory and antioxidant properties, due to the presence of two hydroxyl groups on the B-ring [34]. It was absorbed in the stomach and gastrointestinal tract [30, 42, 46, 47]. The Cyanidin-3-Glucoside metabolites show the higher activity in different biological system, i.e., phloroglucinaldehyde (PGA), protocatechnic acid (PCA), vanillic acid (VA), and ferulic acid (FA) [26, 38, 48]. In a study, the scientist revealed that in rats, mice, and macrophage PCA shows the antioxidant properties, which increased the total antioxidant capacity (T-AOC), glutathione peroxidase activity (GPx), superoxide dismutase activity (SOD), and catalase level. However, it decreases the level of hydroperoxides, *malondialdehyde* levels (MDA), and reactive oxygen species (ROS) [29, 43, 46, 49].

7.7.2 Malvidin-3-Glucoside

Malvidin-3-glucoside is a natural dietary anthocyanin that shows higher atherosclerosis, vascular, and antiatherogenic properties [23]. It helps the endothelial cells from apoptotic death by inhibiting the production of apoptotic protein and ROS activity. It also plays a crucial role in degrading the formation of reactive species after the cell aggression [44]. It has the ability to improve the expression of proapoptotic protein in the endothelial cell. The basic mechanism for the cytoprotective is the modulation that occurs on mitochondrial through the apoptotic pathway [50].

7.7.3 Proline

Proline is the type of amino acid present in the *Bauhinia variegta* [51]. It shows the effective properties against wound healing, skin damage, gut lining, connective tissues, joint repairing, and asymmetric catalyst in organic reaction [52]. It also helps in the protection of protein integrity through the molecular chaperone and improves the enzymatic activity [27]. In a study, the scientist revealed cyclic glycine proline is responsible for the regulation of insulin. It observed change in the ratio of the insulin-like growth factor is associated with a change in weight as compared with the normal weight of women. In women, it was observed with an increase in the cyclic glycine proline is also responsible for the weight loss [20].

7.7.4 Arachidic Acid

Arachidic acid is the polyunsaturated fatty acid present in the cell membrane of the body [39]. In the study, it reported arachidic acid is responsible for enhancing the growth in the body during exercise. It also helps in maintaining the physiological response for the exercise [53]. In another report, the scientist revealed that cyclooxygenase (COX) and 5-lipoxygenase are used to metabolize the arachidonic acid and synthesize the leukotrienes. Similarly, it suppresses the prostaglandins by the arachidic acid inhibition [54].

7.7.5 Valine

Valine is the amino acid that plays an important role in repairing the tissue, enhancing muscle growth, stimulant activity, and protein synthesis [31]. The study is reported the basic mechanism of valine is to hydrolase the methacrylic (MC), hydratase, 3-hydroxyisobutyryl-CoA, alpha-keto acid dehydrogenase, and co-enzyme (CoA) with the branched-chain amino-transferase in the human liver [26]. The scientist revealed that L-valine used glucose for energy production and to prevent muscle breakdown [47]. It also helps in the smooth functioning of the nervous system and immune system [55].

7.7.6 Isoleucine

Isoleucine plays a crucial role in growth, immunity, glucose transportation, fatty acid metabolism, wound healing, and regulating the blood glucose level [56]. In the study, it reported that isoleucine help in the detoxification of nitrogenous waste present in the human body, which is excreted by the kidney [57]. It also helps in the formation and production of red blood cells (RBC) and hemoglobin [58].

7.7.7 Palmitic Acid

Palmitic acid is a saturated fatty acid that shows effective properties against inflammation, insulin sensitivity, hemostasis, cholesterol metabolism, and helps in regulating metabolic health [13, 59]. The primary mechanism for the prevention of beta-cell apoptosis is through the glucose induces process [36]. Table 7.1 represent the structure of *Bauhinia variegta* leaves, flower, and non-woody part compounds and their health benefits.

7.8 Traditional Uses as a Food

Traditionally, the flower of *Bauhinia variegata* is used for the preparation of various products, such as pickle, chutney, and curry, due to the presence of crude protein, crude fiber, fats, antioxidants, and total carbohydrates in the higher amount [64]. Buds are also used for the preparation of curry and

Table 7.1 The structure and health benefits of *Bauhinia variegta* compounds which are present in the leaves, flower, non-woody aerial part, buds, bark, seed and stem.

Part	Compound	3D structure	PubChem CID	Health benefits	References
Non woody aerial part	**Ombuin**	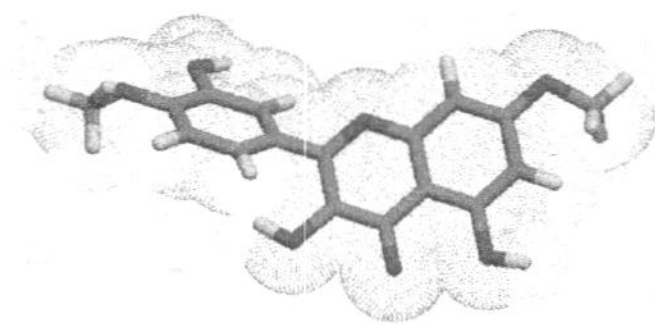	5320287	Reduction in anti-inflammatory and antiviral activity	[12, 33]
	Hesperidin	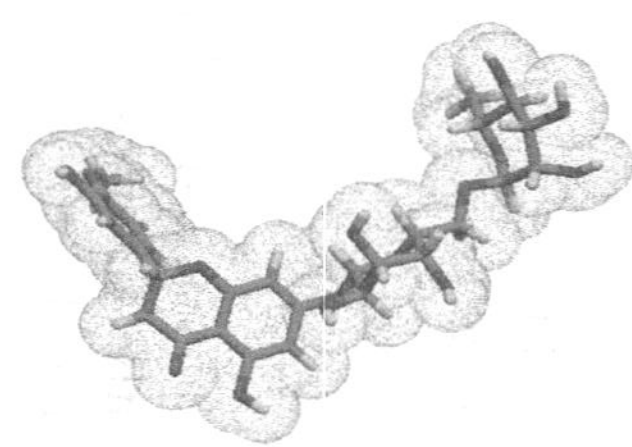	10621	Help in the treatment of skin lightening, anti-inflammation, wound healing, UV protection, antimicrobial and antiskin cancer	[15]
	Kaempferol	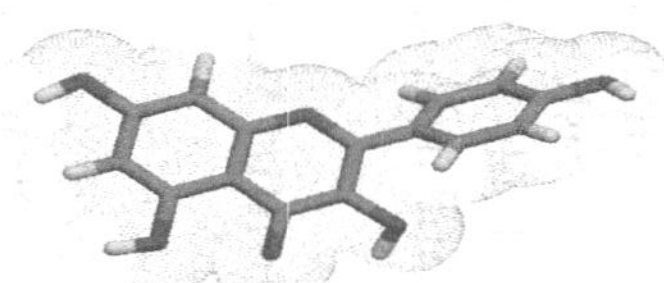	5280863	Play a defense action against the free radicals Reduce the risk of chronic diseases, metastasis, angiogenesis, apoptosis and inflammation	[15, 33]

(*Continued*)

Table 7.1 The structure and health benefits of *Bauhinia variegta* compounds which are present in the leaves, flower, non-woody aerial part, buds, bark, seed and stem. (*Continued*)

Part	Compound	3D structure	PubChem CID	Health benefits	References
	Kaempferol-3-o-β-D-glucopyranoside	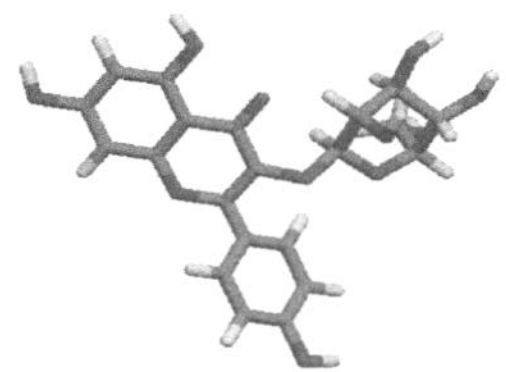	9911508	Shown more anticancerous, antioxidant, antiobesity, antiulcer, neuroprotective, cardioprotective and antidiabetic properties	[15]
Stem	**Quercitroside**	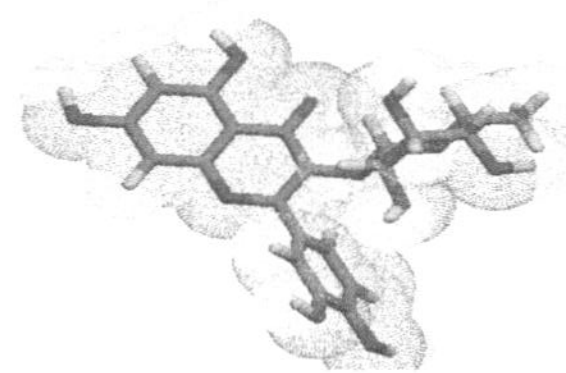	5280459	Help in the reduction of allergy symptoms, inflammation and blood pressure	[39]
	Rutoside	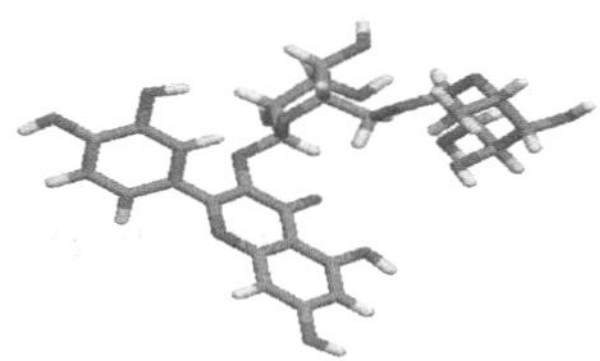	5280805	Reduce the level of cholesterol and arthritis pain Prevent from the blood clotting Improve the blood circulation level	[24, 60]

(*Continued*)

Table 7.1 The structure and health benefits of *Bauhinia variegta* compounds which are present in the leaves, flower, non-woody aerial part, buds, bark, seed and stem. (*Continued*)

Part	Compound	3D structure	PubChem CID	Health benefits	References
	β-sitosterol		222284	Target the cancerous, diabetic, and immunomodulatory activity. Improve the antimicrobial activity Increase the antioxidant activity	[15, 24]
	Lupeol	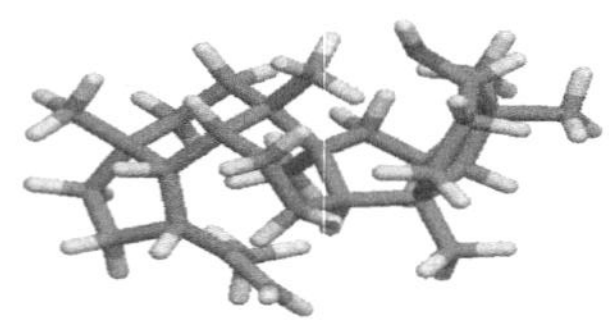	259846	Shown anti-inflammatory and anticancer activity	[15, 33]
Leaves	**Catechol**	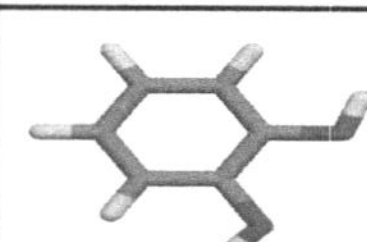	289	Help in inhibiting the ERK2 activity in lung cancer cells by the reduction of c-Myc phosphorylation	[15]

(*Continued*)

Table 7.1 The structure and health benefits of *Bauhinia variegta* compounds which are present in the leaves, flower, non-woody aerial part, buds, bark, seed and stem. (*Continued*)

Part	Compound	3D structure	PubChem CID	Health benefits	References
	Ellagic acid	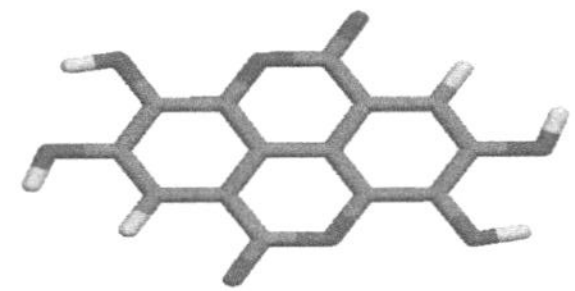	5281855	Used for the treatment of cancer, skin dark patches, memory skill	[33]
	Vitamin C	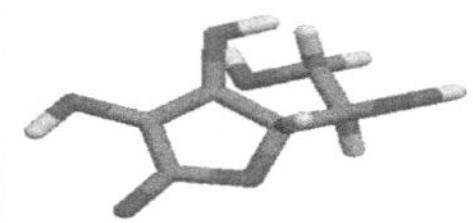	54670067	Help in the treatment of cancer, atherosclerosis, diabetes, neurodegenerative disease and metal toxicity	[24, 39]
Flower	**Cynidin-3-Glucoside**	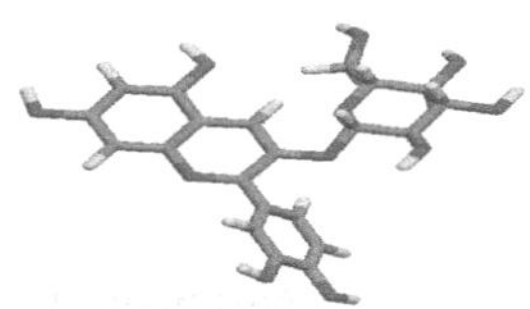	197081	Showed the antitumor, anti-inflammatory, antihypertension, antimutagenic and anticarcinogenic activities.	[39]

(*Continued*)

Table 7.1 The structure and health benefits of *Bauhinia variegta* compounds which are present in the leaves, flower, non-woody aerial part, buds, bark, seed and stem. (*Continued*)

Part	Compound	3D structure	PubChem CID	Health benefits	References
	Malvidin-3-Glucoside	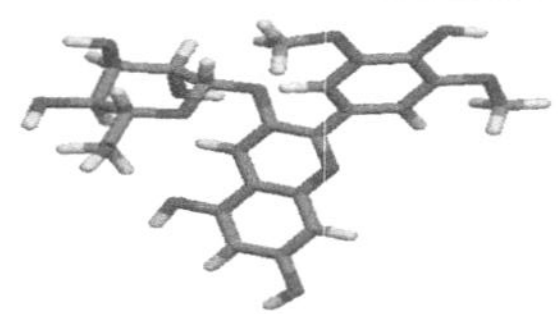	11249520	Shown the vascular protective effects and antiatherogenic properties	[23]
	Peonidin-3-Glucoside	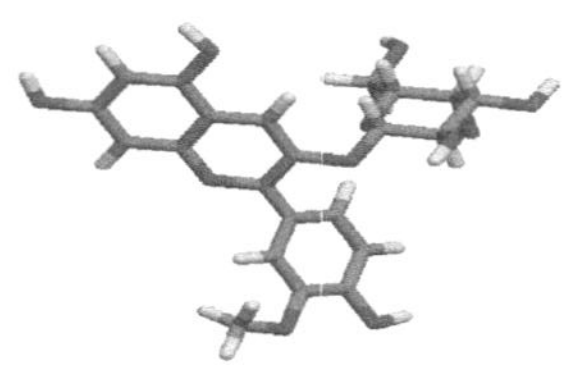	443654	Antioxidant properties	[23]
	Kaempferol-3-Galactoside	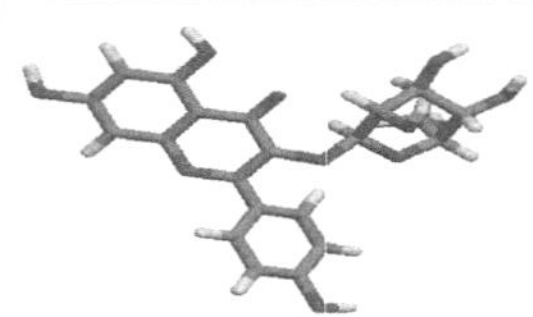	5462193	Increase the antioxidant activity Showed anticancer and anti-inflammatory properties	[15]

(*Continued*)

Table 7.1 The structure and health benefits of *Bauhinia variegta* compounds which are present in the leaves, flower, non-woody aerial part, buds, bark, seed and stem. (*Continued*)

Part		Compound	3D structure	PubChem CID	Health benefits	References
Seed	Fatty acids	**Myristic acid**	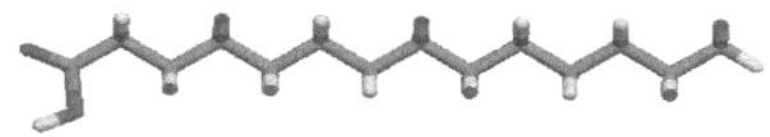	11005	Work like a surfactant, cleansing and opacifying agent Help in texture enhancer	[15, 33]
		Palmitic acid	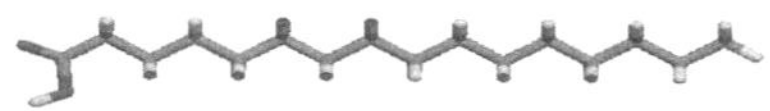	985	Shown the dyslipidemia, hyperglycemia and fat accumulation properties	
		Steric acid	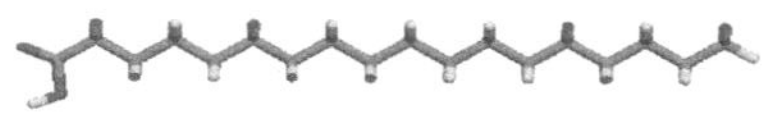	5281	Help in improve the texture and consistency Work like a non-toxic saturated fat which does not affect the human cholesterol levels	
		Arachidic acid		10467	Help in inflammation and wound healing Target the emotion, pain and stress responses Maintain the energy balance level	[61]

(*Continued*)

Table 7.1 The structure and health benefits of *Bauhinia variegta* compounds which are present in the leaves, flower, non-woody aerial part, buds, bark, seed and stem. (*Continued*)

Part	Compound	3D structure	PubChem CID	Health benefits	References
	Lignoceric acid	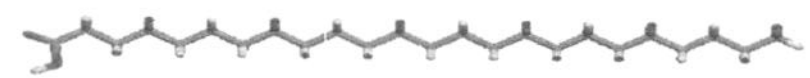	11197	Help in blanching of saturated fatty acids	[15, 33]
	Oleic acid		445639	Reduction in coronary heart disease	
	Linoleic acid	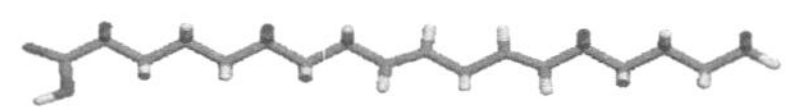	5280450	Shown antiatherogenic and anticarcinogenic activities Help in the reduction of body fat Enhancing the growth cell Help in the reduction of catabolic immune stimulation effects	

(*Continued*)

Table 7.1 The structure and health benefits of *Bauhinia variegta* compounds which are present in the leaves, flower, non-woody aerial part, buds, bark, seed and stem. (*Continued*)

Part		Compound	3D structure	PubChem CID	Health benefits	References
Seed	**Amino acids**	**Lysine**	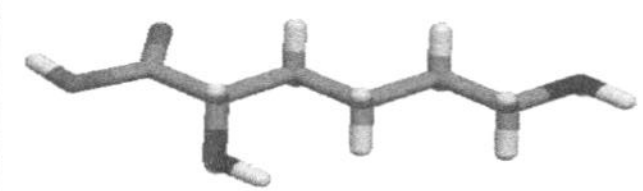	5962	• Target the stress response receptors which help in reduction of anxiety • Improving the property of calcium absorption and retention Help in the wound healing Prevent from the cold sores	[62, 39]
		Threonine	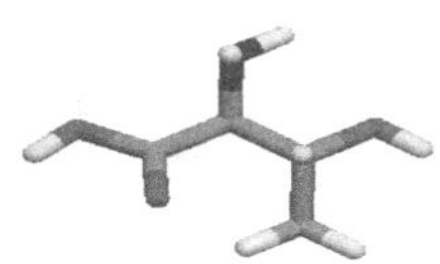	6288	Shown the familial spastic paraparesis amyotrophic lateral sclerosis, spinal spasticity and multiple sclerosis activity	

(*Continued*)

Table 7.1 The structure and health benefits of *Bauhinia variegta* compounds which are present in the leaves, flower, non-woody aerial part, buds, bark, seed and stem. (*Continued*)

Part	Compound	3D structure	PubChem CID	Health benefits	References
	Valine		6287	Help in enhancing the energy and repairing of muscle tissue Increase endurance	
	Methionine	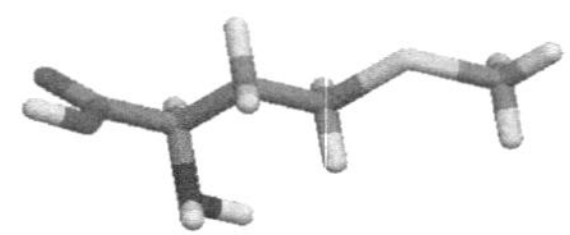	6137	Shown antioxidant properties Help in detoxification of body to prevent from the liver damage	
	Isoleucine	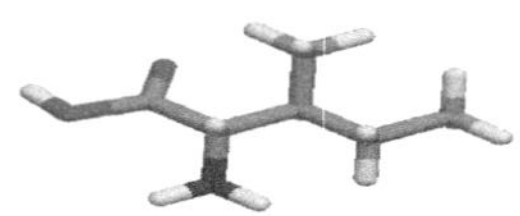	6306	Control the blood sugar level Help in boosting energy and increase the healing of injured muscles	
	Leucine	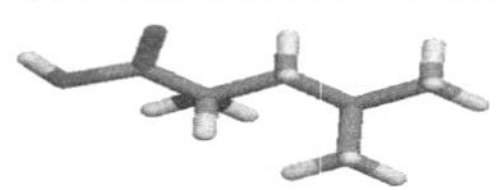	6106	Help in the metabolism, protein synthesis and tissue regeneration	[63]

(*Continued*)

Table 7.1 The structure and health benefits of *Bauhinia variegta* compounds which are present in the leaves, flower, non-woody aerial part, buds, bark, seed and stem. (*Continued*)

Part	Compound	3D structure	PubChem CID	Health benefits	References
	Phenylalanine	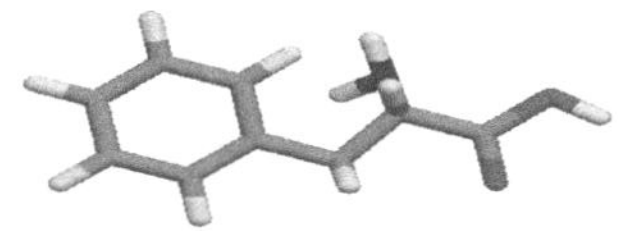	6140	Prevent from the Parkinson's disease, osteoarthritis, rheumatoid arthritis, deficit-hyperactivity disorder (ADHD), skin disease and chronic pain	[39]
	Histidine		6274	Help in the kidney dialysis Prevention from the ulcers, allergic diseases and rheumatoid arthritis	[33, 39]
	Proline	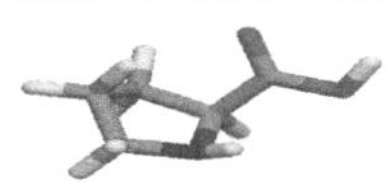	145742	Help in protein synthesis and metabolism process Increase the healing speed of wound Higher antioxidative properties	[15]

(*Continued*)

Table 7.1 The structure and health benefits of *Bauhinia variegta* compounds which are present in the leaves, flower, non-woody aerial part, buds, bark, seed and stem. (*Continued*)

Part	Compound	3D structure	PubChem CID	Health benefits	References
	Glycine		750	Prevent in liver and heart diseases Help in muscle loss and diabetes Improve the sleep	[39]
	Serine	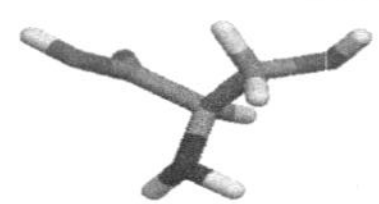	5951	Help in maintaining the neuronal processes growth Also help in increasing the level of L-serine in the brain	[15]
	Glutamic acid	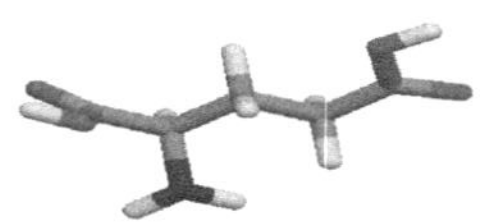	33032	Help in the treatment of low blood sugar level, muscular dystrophy and epilepsy Prevention in the personal behavioural issues and nerve damage	[24]

(*Continued*)

Table 7.1 The structure and health benefits of *Bauhinia variegta* compounds which are present in the leaves, flower, non-woody aerial part, buds, bark, seed and stem. (*Continued*)

Part	Compound	3D structure	PubChem CID	Health benefits	References
	Aspartic acid	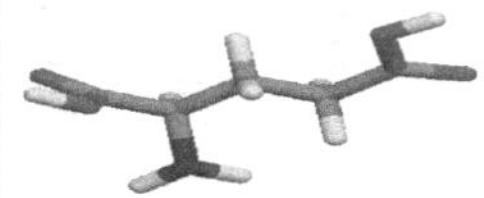	5960	Help in improving the stamina athletes Prevent from the toxins production in the body Improve the function of immune system Prevent from the brain disorder issues	[59]
	Arginine	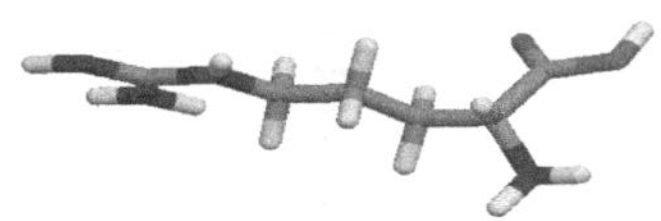	6322	Help in the reduction of blood pressure level Prevent from the heart diseases, inflammation and erectile dysfunction Control the blood sugar level	[15]

Table 7.2 Traditional uses of the various part of *Bauhinia variegta* in food.

Bauhinia variegta		
Parts	**Uses**	**References**
Bud	Used as a flavoring agent Used in the preparation of flour, curry, nuggets, chutney, pakoras, raita and pickle Intake as a fried products	[24, 65]
Leaves	Boiled and eaten as a vegetable Used in curry and pakoras preparation Used in the feed of animals Used as a flavoring agent	[17, 24, 33]
Seeds	Used as the pulses in the various tribal region of India (due to the presence of higher content of amino acids)	[33]
Flower	Used as flavoring agent in fish and meat industry Used in the preparation of nuggets, pickle and pakoras	[33, 64, 66]

pickles. As seen in Table 7.2, *Bauhinia variegata* is traditionally used as a food all over the world.

7.9 Therapeutic Value of *Bauhinia variegata*

Bauhinia variegatata is more popular in the tribal and nontribal people due to its higher medicinal properties [33, 67]. The various part of this tree contains a huge amount of bioactive compound, which help in fighting against several diseases as shown in Table 7.3.

7.9.1 Antidiabetic Activity

Bauhinia variegate leaf, stem, and bark are rich sources of protein (appear like insulin), which help in diabetes [63, 68]. It also helps in lowering the blood glucose level due to the presence of heptatricontan-12, ellagic acid, tannins, peonidin-3-glucoside, and crude protein [58, 69]. In this study, it is reported that an ethanolic extract of *Bauhinia variegate* shows higher antidiabetic activity and hypoglycemic activity

Table 7.3 Therapeutic value of *Bauhinia variegta* and its model studies.

Therapeutic value	Part used	Compound	Models	Mode of action	References
Antidiabetic	Leaves, Flower, Stem bark	Heptatriacontan-12, 13-diol, ellagic acid, tannins, crude protein, peonidin-3-glucoside, Kaempferol glycoside and volatile oil,	Strepto-zotocin induce diabetes model Alloxan-induced diabetic rats	Reduce the level of blood glucose in hyperglycemia due to the presence of a similar amino acid sequence as insulin, which presents in *Bauhinia variegta* tree. The oral dose given to diabetic rats is 200 mg/kg.	[8, 63, 69, 76]
			Alloxan induced hyperglycemic rats	The activity of the pancreatic exocrine is improved through the transferring of blood glucose to peripheral and it enhances the property.	[3, 58, 77, 82]

(*Continued*)

Table 7.3 Therapeutic value of *Bauhinia variegta* and its model studies. (*Continued*)

Therapeutic value	Part used	Compound	Models	Mode of action	References
Anticancerous	Leaves, Stem bark, Root bark	Phenanthra-quinone 5, 7-dihydroxyflavanone-4'-O-a-L-rhamnopyranosylb-glucopyranoside 5, 7-dihydroxy and 5, 7 dimethoxy flavanone-4-O-a-L-rhamnopyranosyl-b-D-glucopyranosides, spathulenol, δ-and γ-cadinene, flavonoids quercetin, rutin, kaempferol, β-sitosterol and 5,7, dimethoxy flavanone-4-o-L	Skin papilloma model in Swiss albino mice	Target to reduce the tumor rate and tumor burden by intake of 500 and 1000 mg/kg dose	[18, 41, 54]
Antimicrobial	Leaves, Stem bark	Tannin, lupeol, glycoside, octacosanol and flavonone glycoside	Against Gram-negative bacteria	By the inhibition of bacterial growth Polar Extract of *Bauhinia variegata* show the effective properties against *E. coli*, Pseudomonas species, and *Klebsiella pneumonia*	[16, 22, 84]
Antifungal	Leaves	Crude protein, catechol and alkaloids	-----	-----	[84]

(*Continued*)

Table 7.3 Therapeutic value of *Bauhinia variegta* and its model studies. (*Continued*)

Therapeutic value	Part used	Compound	Models	Mode of action	References
Analgesic activity	Leaves	β -sitosterol	Nociceptive mouse models	Reduce the level of Prostaglandin E2 and edema in serum, liver homogenate and granuloma	[16, 22, 78]
Anti-inflammatory	Non-woody aerial parts, Root	Flavonol glycoside 5,7,3',4'-tetrahydroxy-3-methoxy-7-O-alpha-L-rhamnopyranosyl(1-->3)-O-beta-galactopyranoside, Kaempferol, ombuin, 7,4-dimethylether-3-o-β-D-glucopyranoside, hesperidin, isorhamnetin-3-o--β-D-glucopyranoside, 3β trans (3,4 dihydroxycinnamoyloxy) olean-12-en-28-oic acid	Nociceptive mouse models Hind paw edema method	Reduction of pulmonary granuloma and hepatic diameter Reduce the level of Prostaglandin E2 and edema in serum, liver homogenate and granuloma Induced the non-immunological carrageenan	[3, 16, 41, 73]
Antihyperepidemic	Flower	kaempferol 3-glucoside, melvidin-3-glucoside	-----	-----	[75]

(*Continued*)

Table 7.3 Therapeutic value of *Bauhinia variegta* and its model studies. (*Continued*)

Therapeutic value	Part used	Compound	Models	Mode of action	References
Neural activity	Flower	acetylcholinesterase	Thin Layer Chromatography	Target to inhibit the acetylcholinesterase and enzyme	[22, 85]
Antidiarrheal	Flower	Peonidin-3-glucoside	-----	-----	[23, 52]
Antioxidant activity	Flower Root Root bark	Tannins, amino acids, isoquercitroside, kaempferol-3-glucoside, melvidin-3-glucoside, glutamic acid, keto acid, rutoside, myricetrol, oleic acid, β-sitosterol, and ellagic acid	Female rats DPPH radical scavenging assay	Level of hyperlipidemic states is reduced due to the presence of oleic acid and β-sitosterol Target the H_2O_2-induced oxidative and prevent damage to pBR322 DNA	[71, 86]
Antiulcer	Stem	Lupeol, β-sitosterol, naringenin 5, and 7 dimethyl-dibenz oxepin	Albino mice	Target the pyloric ligation, decrease the level of free acidity, gastric secretion volume and ulcer index control	[19, 22, 73, 78, 87]

(*Continued*)

Table 7.3 Therapeutic value of *Bauhinia variegta* and its model studies. (*Continued*)

Therapeutic value	Part used	Compound	Models	Mode of action	References
Antitumor	Stem	7 dimethyl-dibenz oxepin	Swiss albino mice against Dalton's ascetic lymphoma (DAL)	Ethanolic extract of *Bauhinia variegta* target to reduce the level of N-nitrosodiethylamine induced. Also, help in elevation of SGOT, SGPT, ALP, GGTP, GST, LPO, and GPX	[18, 22]
Chemoprevention	Stem	Quercertin, 5,7-Dimethoxy and dihydroxy flavonone-4-O-l-rhamnopyronosyl-d-glucopyranosides, flavanone 5-hydroxyl 7,3,4,5-tetramethoxy flavone-5-O-d-xylopyronosyl (1→2) _-l-rhamnopyroanoside, sitosterol, lupeol, 7,12-dimethylbenz (a) anthracene and dihydrodiben-	Wistar male rats model	Hepatocellular carcinogenesis is inhibited with interaction of N-nitrosodiethylamine. It reported liver weight of the rats is increased (7.8±0.12 g) by induce of N-nitrosodiethylamine as compared with normal rats liver weight (4.1±0.10 g/100 g) body weight Lowering the catalase, SOD and antioxidants level by the induce of N-nitrosodiethylamine	[87]

(*Continued*)

Table 7.3 Therapeutic value of *Bauhinia variegta* and its model studies. (*Continued*)

Therapeutic value	Part used	Compound	Models	Mode of action	References
Cytotoxic effect	Stem	sitosterol, lupeol and dihydrodiben	Human breast cancer (HBL-100) cells and human epithelial larynx cancer (HEp2)	The level of catalase antioxidants and SOD is normal by the inducing of N-nitrosodiethylamine. In study it reported HEp2 and HBL-100 cell lines showed significant difference and the IC50 values 250_g/ml for HEp2 and >300 g/ml for HBL-100 cell lines.	[19, 39, 68]
Wound healing	Root	Kaempferol-3-o-β-D-glucopyranoside and cytidine-3-glucoside	-----	By the developing a new skin on wound with the presence of antioxidant properties and anti-inflammatory.	[63]
Nephroprotecting	Root	Flavonone-5, 7-dimethoxy-3, 4 methyl-enedioxyflavonone	*In-vivo* cisplatin-induced nephropathy model in rats	By decreasing the level of urea and creatinine serum Target the body weight with the increased in urine output	[22, 80]
Hemagglutinating	Crude seeds	Palmitic acid, oleic acid, proteins and stearic acid	Wistar rats	Target protein content and prevent from the hemagglutinating activity. Also, detect the viral particle present in it.	[22, 88]

(*Continued*)

Table 7.3 Therapeutic value of *Bauhinia variegta* and its model studies. (*Continued*)

Therapeutic value	Part used	Compound	Models	Mode of action	References
Antiobesity	Root bark	Dihydrodibenzoxepin, 4-dimethoxy-2-methyl-dibenz oxepin, 7 dihydroxy-3, 6b dihydroo-1, β -sitosterol, phenolic and flavonoid compounds	Female rats	By the feeding of hypercaloric diet	[22]
Hepatoprotective	Stem bark	β -sitosterol	Sprague-Dawley rats	Target the level of ALP (alkaline phosphatase), ALT, AST, total lipids, protein and GGT Hepatotoxicity is responsible to increasing the level of protein In study, oral dose of 100 and 200 mg/kg ethanolic extract is given to Sprague-Dawley rats and observed the level of ALP, AST, GGT, ALT is decreased .	[18, 76]
Hematinic	Stem bark	Hentriacontane and stigmasterol	Hemolytic anemic rats	By increasing the level of hemoglobin content in blood	[22, 72, 74, 89]

(*Continued*)

Table 7.3 Therapeutic value of *Bauhinia variegta* and its model studies. (*Continued*)

Therapeutic value	Part used	Compound	Models	Mode of action	References
Hypolipidemic activity	stem bark, root	iso-octyl poly-oxyethylene phenol	Albino rats Triton WR-1339	Lowering down the level of lipoprotein and plasma lipids Target the level of high density lipid in Triton WR-1339 and increased the oxidative marker	[90]
Antibacterial	Stem bark	reducing sugar, octacosonal and stigmasterol	Bacterial strains Agar disc diffusion method and agar well diffusion method	Target the different bacterial strains viz. *Pseudomonas pseudoalcaligenes, Klebsiella pneumonia, Bacillus cereus, Staphyllococcus aureus* and *Escherichia Coli* through the agar disc diffusion method and agar well diffusion method It was reported that antibacterial activity are more in ethanolic extract as compared with the methanolic extract due to presence of higher bioactive compounds	[14, 41, 48, 91]

(*Continued*)

Table 7.3 Therapeutic value of *Bauhinia variegta* and its model studies. (*Continued*)

Therapeutic value	Part used	Compound	Models	Mode of action	References
Immuno-modulatory effect	Stem bark	Tannins	*In-vitro* immunomodulatory activity of *Bauhinia variegata* Linn on human neutrophils	The non-specific immune system is activated and increased the function of phagocytic of human neutrophils. Phagocytosis plays the role of defense mechanism. Acts as chemo-attractant. Target the neutrophils to kill the micro-organisms	[70]

as compared with other solvents [43]. In a report, the scientist revealed glucose level, high-density lipid cholesterol (HDL), triglycerides, and total cholesterol level is reduced in the alloxan-induced diabetic model due to the intake of *Bauhinia variegate* leaf and bark [60]. The study shows that *Bauhinia variegate* contains natural flavonoids and glycosyl flavonoids, which show higher antidiabetic properties [68]. In another study, it reported an ethanolic extract of *Bauhinia variegate to* show higher antidiabetic properties due to the conjunction with the chloroplast protein [55].

7.9.2 Antioxidant Activity

Bauhinia variegate also accounted for quercetin, rutin, apigenin, and epigenin 7-O-glucoside. Quercitin and flavonoids are powerful cancer-preventing agents due to their good collaboration with biomolecules [70]. Aqueous and ethanolic extract of the root of *Bauhinia variegate* has cancer prevention agents. It might be flavonoids and other phytochemicals in the extract, which helps in the prevention of cancer [53]. Among all the concentrates, methanol concentrate was found to be good dissolvable and have great cell reinforcement activity [71].

7.9.3 Antidiarrheal Activity

Bauhinia variegate leaves show an inhibitory effect at different dose levels on animal models castor oil-induced diarrhea in rats and gastrointestinal motility test by using the charcoal meal [72]. The effect of the extract on castor oil-induced diarrhea is more when compared to untreated ones. The effect of extract on gastrointestinal motility of charcoal meal decreased the propulsion as compared to control group [16, 56].

7.9.4 Antiulcer Activity

Bauhinia variegate exhibits antiulcer properties as shown in study, which held on a rat (*in vitro*). The antiulcer activity of leaf extract of this plant was evaluated against gastric ulcer induced by aspirin and pyloric ligation induced ulcer models in rats [42]. The stomach was examined for ulcer. Oral administration of ethanolic extract in rats significantly decreased the gastric secretion volume, total free acidity, and ulcer index, which showed its effectiveness against ulcer treatment to control [19].

7.9.5 Antitumor Activity

Bauhinia variegate ethanol extract also exhibits antitumor activity in the case of Dalton's ascitic lymphoma (DAL) as evaluated in Swiss albino mice. Ethanolic extract of the plant was found responsible for the increase in peritoneal cell count. The extract-treated tumor-bearing mice show enhancement of mean survival time as compared to the control group. After 14 days of inoculation, the ethanolic extract can reverse the changes in hematological parameters and protein [19]. The antitumor activity of ethanol extract was evaluated against Ehrlich ascites carcinoma in Swiss albino mice and found to be toxic against Ehrlich ascites carcinoma tumor cells [73]. It is also effective against N-nitrosodiethylamine induced liver tumors and human cancer lines as chemopreventive and cytotoxic effect of ethanol extract evaluated in N-nitrosodiethylamine induced liver tumor in rat and human cancer cell lines. Oral administration of ethanol extract suppressed the liver tumor induced by N-nitrosodiethylamine as revealed a decrease in elevated levels of serum glutamate pyruvate transaminase (SGPT), alkaline phosphatise (ALP), glutathione peroxidise (GPx), serum glutamate oxaloacetate transaminase (SGOT), total bilirubin, gamma glutamate transpeptidase (GGTP), and glutathione S-transferase (GST). The extract increases the enzymatic antioxidant levels and total proteins as compared to that liver tumor of rats. Ethanol extract of plant was found to be cytotoxic against human epithelial larynx cancer (HEp2) and human breast cancer cells [19].

7.9.6 Antigoitrogenic

Antigoitrogenic is the substance that helps to control the disruption in thyroid hormones production in the body *Bauhinia variegate in vivo* study on rats showed that this plant has a tendency to bring the goitrogenic thyroid to a normal level. They administered neomercazole in rats to induce goiter [40].

7.9.7 Anti-Inflammatory

Bauhinia variegate exhibits anti-inflammatory properties. Researchers carried out an *in vivo* study for the assessment of the anti-inflammatory potential of the leaf, stem, root and bark of *Bauhinia variegate* [18]. They used three animal models viz. Cotton pellet induced granuloma formation, carrageenan-induced rat paw edema, and adjuvant-induced arthritis in rats. Petroleum ether fraction and ethanol extract were tested against the models. The study revealed that both exhibit anti-inflammatory activity but petroleum ether fraction shows more potent activity [29].

7.9.8 Antimicrobial Activity

Bauhinia variegate alcoholic extract was found to be effective against *Bacillus subtilis, Pseudomonas aeruginosa, Salmonella typhi, Staphylococcus aureus, Shigella dysentriae,* and *Vibrio cholera.* The extract showed more potent activity against gram +ve bacteria [74]. It shows the effective properties against gram +ve bacteria *Staphylococcus aureus, Bacillus subtilis,* and *Streptococcus epidermidis* and gram –ve, such as *Escherichia coli, Shigella flexineria, Pseudomonas auriginosa* [75]. Hydromethanolic extract of this plant inhibits the growth of microorganisms. The crude extract of this plant exhibits more effectiveness than ampicillin against 2-g +ve *Staphylococcus aureus, Streptococcus pyogens,* and 2-g –ve *Escherichia coli* and *proteus mirabilis* [45]. The leaf extract of *Bauhinia variegate* also showed toxicity against ringworms causing fungi *Epidermophyton floccosum, Trichophyton mentagrophytes,* and *Microsporum gypseum* [66]. In another study of this plant, bark powder was defatted with petroleum ether. The residue was collected and air-dried and separated into five batches (20 g each). The deffated and non-deffated material was individually extracted in solvents with increasing polarity. The antibacterial activity deffated and non-deffated was determined by agar well diffusion method at three concentrations (10 mg/ml, 5 mg/ml, and 2.5 mg/ml) using Muller Hinton Agar no. 2 (Hi media) [29]. The deffated extracts as well as those non-deffated extracts were diluted in dimethylsulphoxide (DMSO). The experiment was performed thrice. The growth of microorganisms was determined by the measurement of the zone of inhibition and the mean values. The antibacterial activity of deffated extracts was found more effective against non-deffated extracts against tested microorganisms. The polarity of the solvent plays an imperative role in exhibiting potential antibacterial activity [14].

7.9.9 Analgesic Activity

Analgesic activity helps in the control or relieve from the pain, which are caused by the damage in the body *Bauhinia variegate* contain the isoflavonoids, which are responsible for the analgesic activity and mediated via the formation and release of various autocoids [76]. The basic mechanism is not cleared till date, but several studies have shown that flavonoids are responsible to reduce the pain by blocking the cellular regulatory proteins expression.

7.9.10 Antiobesity Effect

Bauhinia variegate methanolic extract of stem and root barks have an antiobesity effect as tested in female rats which are fed with a hypercaloric diet. The rats are fed for 40 days with the hypercaloric diet to the anti-obesity activity [15]. Plant extract shows a hypolipidemic effect and thus reduced obesity due to presence of beta-sitosterol. It also helps in reducing the level of brain serotonin. Treatment of these obese rats with methanolic extract decreases the total cholesterol, triglycerides, and low-density lipoprotein and increased high-density lipoprotein and brain's serotonin level [53].

7.9.11 Anticataract Activity

Bauhinia variegate exhibits anticataract activity which helps in a significant reduction of clouding or opaqueness in the eye's lens. Stem bark shows the presence of Rhamnocitrin, a flavonoid that has a strong antioxidant effect can be used effectively against cataracts. Anticataract activity of stem bark studied in ovine and chick embryo lens model [53, 77].

7.9.12 Antihyperlipidemic Activity

Bauhinia variegate leaves methanol extract shows an antihyperlipidemic property. The extract showed a significant reduction in cholesterol and triglyceride level. Administration of dose of Triton WR-1339 to adult rats induced hyperlipidemia. Studies revealed that methanol extract of the plant significantly reduced cholesterol and plasma triglycerides levels [65].

7.9.13 Antiarthritic

The study reported *Bauhinia variegate* ethanolic extract exhibits antiarthritic properties as tested in the rat. Oral administration of ethanol extracts incomplete Freund's adjuvant (CFA) rat for 15 days simultaneously. On the 15th day, blood was collected from the rats, and then serum was separated. The result shows that the administration of extract significantly altered the biochemical parameters and also the level of various antioxidant enzymes evaluated in the liver and kidney of control and extract treated rats such as catalase, glutathione peroxidise (GPx), superoxide dismutase (SOD), and lipid peroxidise (LPO). Various parameters were also evaluated such as alanine aminotransferase (ALT), alkaline phosphatase (ALP), total cholesterol, and triglycerides [78].

7.9.14 Chelation Action

The *Bauhinia variegate* show the high chelation action to remove the iron, lead, arsenic, and mercury content from the body which are toxic for the human [90]. It works like a drug in the body to cure autism, Alzheimer's diseases, and heart diseases. In the report, the scientist revealed that aqueous extract of *Bauhinia variegate* shows the higher chelation property as compared with the ethyl alcohol, chloroform, acetone, ethyl acetate, and petroleum ether [22].

7.9.15 Cytotoxic Activity

Bauhinia variegate stem shows the presence of flavanone and its structure analyzed by color reactions and spectral analysis [14]. Flavanone tested for cytotoxic activity against many human tumor lines representing leukemia colon, CNS, melanoma, ovarian, renal, prostate, non-small cell lung, and breast cancers [54].

7.9.16 Hepatoprotective Property

Bauhinia variegate bark extract also shows the hepatoprotective property. Bark extract examined by using carbon tetrachloride intoxicated Sprague-Dawley rat liver. Carbon tetrachloride is commonly used as a hepatotoxin and increases the level of total lipids in the liver, which leads to liver damage. Alcoholic stem bark extract was orally administered in male rats in different doses [76]. The study was carried out in female albino Wistar rats to test hepatoprotective property in *Bauhini variegate.* Aqueous extract of bark revealed hepatoprotective property against carbon tetrachloride. Rats treated with extract of *Bauhinia variegate* shows improvement in the functional integrity of the liver cells [30].

7.9.17 Hemagglutination

Bauhinia variegate seeds saline extract shows hemagglutination property. This property evaluated on erythrocytes of vertebrates, such as monkey, man, rabbit, goat, rat, sheep, cow, buffalo, horse, mule, and fowl, where the extract showed this property [12].

7.9.18 Immunomodulatory Activity

Bauhinia variegate stem bark ethanolic extract has immunomodulatory properties [68]. Immunomodulatory agents of the plants enhance the immune response toward the pathogen by activating the immune system [70]. The effect of ethanolic extract of stem on phagocytic activity was carried out by carbon clearance test and neutrophil activation was evaluated by neutrophil adhesion test for a nonspecific immune response. After oral administration of stem ethanolic extract, there is a significant increase in the primary and secondary humoral antibody response [16]. The acetone, water extract of *Bauhinia variegate* stem bark increases the phagocytic function of human neutrophils. Bark extract significantly increases the neutrophil chemotactic movement. Tannins obtained from bark show immunomodulatory properties and help in boosting the immune system [21].

7.9.19 Mosquito Control

Bauhinia variegate aqueous leaf extract also helps in mosquito control. The study revealed that silver nanoparticles (AgNPs) synthesized using aqueous extract by reduction of Ag+ ions from silver nitrate solution. These silver nanoparticles were characterized by UV-visible spectrophotometry, Fourier transform infrared spectroscopy (FTIR), transmission electron microscopy (TEM), scanning electron microscopy (SEM), energy–dispersive X-ray analysis (EDX), and X-ray diffraction analysis (XRD). Aqueous leaf extract and synthesized silver nanoparticles evaluated against the larvae of *Anopheles subpictus, Aedes albopictus,* and *Culex tritaeniorhynchus* [53, 65, 79].

7.9.20 Nephroprotective

Bauhinia variegate ethanolic extract studied *in vivo* in rats. A study was carried out on cisplatin-induced nephropathy in rats. Findings revealed that administration of ethanolic extract for 14 days reversed the changes induced by cisplatin administered to rats. The toxic control group results in acute kidney dysfunction as evidenced by the increase in serum urea and creatinine, and lowered in urine output and body weight with multiple histological damages. Treatment of rats with ethanol extract for 14 days results in the lowering of serum level of creatinine and urea, decreased urine creatinine and albumin with weight gain, and elevated urine output as compared to the toxic group [80].

7.9.21 Proteinase Inhibitor

Bauhinia variegate helps in the inhibition of proteins by reducing the level of trypsin activity by the controlling of protein catalytic reaction [22]. The seeds of *Bauhinia variegate* play important role in the inhibition of blood clotting enzymes [43].

7.9.22 Wound Healing Activity

In a study, scientists reported 200 mg/kg and 400 mg/kg doses of *Bauhinia variegate* of different solvent extract (ethanol and aqueous) are given to albino rats. And it was observed that both doses show the significant wound healing effect on albino rats by excision and incision wound model [18].

7.9.23 *Bauhinia variegate*

Is also shown effective pesticidal properties such as antihelminthic activity, insecticidal activity, and molluscicidal activity.

7.9.23.1 Anthelmintic Activity

Bauhinia variegate bark shows antihelmintic activity due to the presence of rutoside [81]. It helps to destroy or kill the parasitic intestinal worm [43]. Aqueous and chloroform extract of bark investigated for their anthelmintic activity against *Pheretima posthuma* and *Ascardia galli.* Flavonoids help in various biological activities, including analgesic, anti-inflammatory, antioxidant, antiulcer activity, and hepatoprotective activity. Tannins act as a protectant against the helminthic activity. Preliminary phytochemical screening of chloroform extract of bark shows the presence of tannins, steroids, flavonoids, coumarins, reducing sugars, and carbohydrates. The presence of these phytoconstituents may be responsible to show a potent anthelmintic activity [28]. In a study, the scientist revealed chloroform and aqueous extract of *Bauhinia variegate* show high antihelmintic activity as compared with high other solvents. It inhibits the spontaneous motility and the time death of worms. In a result, it was observed that both extracts show maximum inhibition activity at 100 mg/ml [76].

7.9.23.2 Insecticidal Activity

Bauhinia variegate stem extract showed juvenile activity against *Dysdercus cingulatus* nymphs. The plant shows the effective property to control the *Plutella xylostella* and *Candida* activity in the plant. It also helps in inhibiting the growth of Brown root rot [43].

7.9.23.3 Molluscicidal Activity

Bauhinia variegate is a rich source of saponin and quercetin which are the more effective component of molluscicidal [44]. The different solvents, such as ether, chloroform, carbon tetrachloride, acetone, and ethanol, show good soluble properties in *Bauhinia variegate* [53]. In a study, the scientist revealed that molluscicidal activity was higher in *Bauhinia variegate* leaf against the *Indoplanoribis exustus* [29].

7.10 Health Benefits of *Bauhinia variegatata*

7.10.1 Flower

The flower of *Bauhinia variegatata* tree is a rich source of vitamin C, rutoside, myricetrol, apigenin, carotenoid pigments, active biomolecules, and tannins, which help in the treatment of edema, dysentery, snakebites, ulcers, gastrointestinal, genito urinary problems, malaria, laxative properties, anthelmintic, and bleeding piles [14, 30]. Mundas use the 20-ml flower decoction for female for the prevention of galactogogue [61]. It was reported that an aqueous and ethanolic extract of *Bauhinia variegata* flower showed higher free radical scavenging activity and nephroprotective activity. It helps in reducing the level of BUN, urine creatine, serum urea, and serum creatinine due to inducing gentamicin in the nephrotoxicity model [59]. It shows significant antidiabetic, antihyperlipidemic, and antioxidant activity as tested in streptozotocin (STZ)-induced diabetic rats [65]. Flowers are rich in carotenoid pigments and active biomolecules. It contains anthocyanins like malvidin-3-glucoside, cynidin-3-glucoside, peonidin-3-glucoside, and peonidin-3-diglucoside [41, 68]. Anthocyanins have the potential to protect against DNA cleavage, antioxidant activity, and antibacterial activity to stop lipid peroxidation [29].
Bauhinia variegata has two varieties of flowers *viz.*, white and red flower.

White flower is used to cure biliousness, cough, anal troubles, blood diseases, tuberculosis, and burning sensation [40]. It also provides sweet, cooling, appetizing, and astringent to the bowel [62].

Red flower of *Bauhinia variegata* shows astringent, cooling, and sweet properties, also cure the vaginal discharge, headache in malaria, ulcers, leprosy, tuberculosis, biliousness by intake in a lower amount. The root juice of the red flower tree is used as a precaution from snakebite [81].

7.10.2 Buds

In the study, Mali and Dhake reported buds intake in the dried form help in the treatment of piles, eye ailments, liver, diarrhea, vaginal discharge in women, menorrhagia due to the presence of higher antioxidant properties [92]. Traditionally, flower buds and flowers are used as astringent, antibacterial, liver tonic, antihelmintic, antileprotic, and in the treatment of edema, wounds, eye and skin diseases, ulcers, hemorrhoids, and dysmenorrheal [24]. It also intakes with the combination of water and black pepper in vermifuge and for the proper vaginal discharge [62]. Buds are a good source of fiber components and can be beneficial for patients with cardiovascular disease and constipation [81].

7.10.3 Roots

Roots of *Bauhinia variegata* are a rich source of flavonoids that help in curing ulcer, obesity, malaria, inflammation, cancer, snake poison antidote, dyspepsia [79]. Shrikande and shrikande reported that intake of 15 ml *Bauhinia variegata* root juice per day helps in the reduction of indigestion and obesity [14].

7.10.4 Barks

Barks also show higher medicinal properties for the treatment of leprosy, piles, flatulence, dysentery, tumor, skin diseases, ulcer, and malaria [3, 14]. It is a rich source of vitamin C, flavonoids, phenolic compounds, which show effective properties against degenerative diseases with oxidative stress [34]. The powder of bark is used to prevent blood pressure problems [74]. Fresh bark juice with ginger helps in curing the scrofula in the konkan [14]. It was reported that non-tribal people of Assam and Khasi used the paste of *Bauhinia variegata* bark for faster healing of wounds, fractured bone, and cuts [23]. In the study, it was observed that methanolic bark extract of *Bauhinia variegate* helps in the reduction of body weight and food intake due to the presence of β-sitosterol in female rats fed with a hypercaloric diet. Also, help in elevating the serotonin level in the brain and reducing the lipid profile [76]. The bark extract showed higher free radical scavenging activity [65]. The bark, root, and flower mixture with boiled rice water is used as a maturant for boil and abscesses [29].

7.10.5 Leaves

Bauhinia variegata leave juice is more consumed in different parts of India *viz.*, Orissa, South India, Bengal, Madhya Pradesh, Sikkim for the treatment of headache, wounds, stomach tumors, and jaundice [18]. Singh reported leaves to contain a huge amount of quercetin and saponin which show molluscicidal activity [43]. The paste of *Bauhinia variegate* leaves is used for curing the infection of the urinary [62]. The Lodhas use the mixture of pepper paste and leaf paste (1: 2) w/w to get instant relief from the boils [18, 83]. According to Bhils and Andh, leaves infusion is used for the treatment of piles in the form of laxative [72]. The leaves also help in lowering the blood glucose level due to the presence of apigenin-7-o-gluicoside amide [93]. The intake of dried *Bauhinia variegata* leaves helps in curing the cough problem. A study conducted by Kumar and Rajni, on leave extract of *Bauhinia variegata* and observed an extract of *Bauhinia variegata* leave showed the antisecretory activity due to the presence of free radical scavenging activity. It reported the ethanolic extract of leaves showed an analgesic effect due to the presence of flavonoid content. In another study, it observed aqueous and ethanolic extract of *Bauhinia variegata* leaves help in lowering down the blood glucose level in streptozotocin-induced diabetic rats [91]. A survey of the literature revealed that dehydrated leaves powder of this plant are rich in iron, calcium, carbohydrates, energy, and it can be used in the preparation of *Idli* and *Kachori* [17].

7.11 Other Uses

Bauhinia variegata tree showed several other benefits as shown in Table 7.4. It was used in different fields, such as dying industries, equipment industries, animal feed industries, the construction field, and ornamental industries. The young pods and seeds of the plant are cooked and eaten by tribes, such as Kathkors, Gondas of India [82].

7.12 *Bauhinia variegate* Was Used Mythologically

There are several mythologies where the *Bauhinia variegate* is used for the ritual of treatment by the "Ojhai" and "Gunin." They do curing by the invocation of souls with the two phases.

Table 7.4 Other uses of the various part of the *Bauhinia variegta* tree.

Bauhinia variegta		
Parts	**Uses**	**References**
Stem	Used for the rope making due to the weak stems fiber	[41]
Leaves	Used for the animal feeding and control the mosquito	[94]
Roots	Used for the gums and mucilage	[15]
Seeds	Used in the industry for the production of fiber, oil, gum and tannins	[15]
Wood	Used for the manufacturing of different operational tools for the agriculture field	[95]
Flower	Used in perfumes Ornamental plant Intake as a vegetable (cooked products)	[72]
Bark	Play the important role in the production of dyes that are used in dying industries for fabrics	[15]

Phase I: Use the different modes of divination for the diagnoses of the different diseases. In this phase, they take the bark of *Bauhinia variegate* and utter the name of Demons and God for the treatment [21].

Phase II: In this mode of phase "Ojhai" and "Gunin" take the oil of kachnar and use the magical words for the treatment [58].

7.13 Market Product

Nowadays, a huge number of products are prepared from the *Bauhinia variegate*, such as juice, kachnar ki chhal, kanchnar kashaya, kanchnar guggulu, kanchnar churna, kanchnar guggulu capsules, kachnar thyroid capsules, kachnar herbal extract, vitalizing serum prepared for the falling hair problem [22].

7.14 Conclusion and Future Perspectives

Bauhinia variegate is part of the Leguminosae family, which is traditionally used as a medicinal plant, due to the presence of a huge amount of bioactive compounds, such as Kaempferol-3-o-β-D-glucopyranoside, cytidine-3-glucoside, β-sitosterol, phenolic, and flavonoid. In this chapter, we focused on the origin, composition, compound present, traditional uses, other use. Also, we study the role of the bioactive compound of *Bauhinia variegate* to cure harmful diseases *viz.*, anti-inflammatory, immunomodulatory effect antibacterial, antioxidant, antidiabetic, antiulcer, antiobesity, anticancerous hematinic, hepatoprotective, hypolipidemic, nephroprotective, and wound. A brief future prospective and different study of this plant is discussed in this chapter. *Bauhinia variegate* tree is more used in the coming modern era in medicine and the food industry. The fresh and dried part *viz.*, flower, bud, and leaves are used for the preparation of ready serve beverage, chutney, squash, wine, vinegar, appetizer, juice, instant food product mix, flavoring agent, animal feed, pharmaceutical industry, flour, and extraction of the compound for further studies.

References

1. Ghaisas, M.M., Shaikh, S.A., Deshpande, A.D., Evaluation of the immunomodulatory activity of ethanolic extract of the stem bark of Bauhinia variegata Linn. *Int. J. Green Pharm.*, 3, 1, 2009.
2. Samant, S.S., Kishor, K., Upreti, B.M., Bharti, M., Bohra, N., Sharma, P., Tewari, M.L., Diversity, distribution, indigenous uses and conservation prioritization of the economically important floristic diversity in Nadaun Block of Hamirpur District, Himachal Pradesh. *Int. J. Biodivers. Conserv.*, 6, 7, 522–540, 2014.
3. Sahu, G. and Gupta, P.K., A review on Bauhinia variegata Linn. *Int. Res. J. Pharm.*, 3, 1, 48–51, 2012.
4. Sharma, R.K., Rajani, G.P., Sharma, V., Komala, N., Effect of ethanolic and aqueous extracts of Bauhinia variegata Linn. on gentamicin-induced nephrotoxicity in rats. *Indian J. Pharm. Educ. Res.*, 45, 2, 192–198, 2011.
5. Kumar, A., Anand, V., Dubey, R.C., Goel, K.K., Toxicological evaluation of Bauhinia variegata Linn. for estimating the neurological, hematological and physical alterations in the albino mice. *Environ. Conserv. J.*, 17, 3, 97–102, 2016.

6. Guan, J., Mallah, G.S., Liu, K., Thorstensen, E., Shorten, P.R., Mitchell, E.A., Taylor, R., The role for cyclic Glycine-Proline, a biological regulator of Insulin-like Growth Factor-1 in Pregnancy-related Obesity and Weight changes. *J. Biol. Regul. Homeost. Agent*, 32, 3, 465–78, 2018.
7. Halim, A., Hamed, A., Fyiad, A.A.A., Aboulthana, W.M., El-Sammad, N., Youssef, A.M., Ali, M.M., Assessment of the anti-diabetic effect of Bauhinia variegata gold nano-extract against streptozotocin induced diabetes mellitus in rats. *J. Pharm. Allied Sci.*, 10, 05, 077–091, 2020.
8. Kansal, M., Shukla, P., Shukla, P., A boon to human health-bauhinia variegata. *Int. J. Pharmacogn.*, 64, 11-12, 798–08, 2009.
9. Space, J.C., Waterhouse, B.M., Miles, J.E., Tiobech, J., Rengulbai., K., *Report to the Republic of Palau on Invasive Plant Species of Environmental Concern*, p. 179, USDA Forest Service, Honolulu, 2009.
10. Sudheerkumar, K., Seetaramswamy, K.S., Babu, B., Kumar, P.K., Phyto pharmacognostical and isolation of chemical constituents from bauhinia variegata leaf extract. *J. Pharmacogn. Phytochem.*, 4, 1, 189–191, 2015.
11. Singh, K.L., Singh, D.K., Singh, V.K., Multidimensional uses of medicinal plant kachnar (Bauhinia variegata Linn.). *Am. J. Phytomed. Clin. Ther.*, 4, 2, 58–72, 2016.
12. Jash, S.K., Roy, R., Gorai, D., Bioactive constituents from Bauhinia variegata Linn. *Int. J. Pharm. Biomed. Res.*, 5, 2, 51–54, 2014.
13. Raghuveer, I., Kumar, A., Gurjar, H., Gupta, N., Kumar, S., Kumar, M., Plant profile, phytochemistry and pharmacology of Bauhinia variegata Linn. (Kachnar): An overview. *Int. J. Pharmacol.*, 1, 5, 279–287, 2014.
14. Parekh, J., Karathia, N., Chanda, S., Evaluation of antibacterial activity and phytochemical analysis of Bauhinia variegata L. bark. *Afr. J. Biomed. Res.*, 9, 1, 53–56, 2006.
15. Kumar, P., Shaunak, I., Verma, M.L., Biotechnological application of health promising bioactive molecules, in: *Biotechnological Production of Bioactive Compounds*, pp. 165–189, Elsevier, Amsterdam, Netherlands, 2020.
16. Ahmed, A.S., Esameldin, E.E., Moodley, N., McGaw, L.J., Naidoo, V., Eloff, J.N., The antimicrobial, antioxidative, anti-inflammatory activity and cytotoxicity of different fractions of four South African Bauhinia species used traditionally to treat diarrhoea. *J. Ethnopharmacol.*, 143, 3, 826–839, 2012.
17. Dhyani, N. and Gupta, A., Nutritional composition of dehydrated Kachnar leaves (Bauhinia purpurea) powder. *Int. J. Home Sci.*, 2, 2, 363–364, 2016.
18. Al-Snafi, A.E., The pharmacological importance of Bauhinia variegata. *A review. Int. J. Pharma Sci. Res.*, 4, 12, 160–164, 2013.
19. Rajkapoor, B., Jayakar, B., Murugesh, N., Antitumour activity of Bauhinia variegata on Dalton's ascitic lymphoma. *J. Ethnopharmacol.*, 89, 1, 107–109, 2003.
20. Yang, P., Yuan, C., Wang, H., Han, F., Liu, Y., Wang, L., Liu, Y., Stability of anthocyanins and their degradation products from cabernet sauvignon red wine under gastrointestinal pH and temperature conditions. *Molecules*, 23, 354, 2018.

21. Patil, J.K., Patel, M.R., Sayyed, H.Y., Patel, A.A., Pokal, D.M., Suryawanshi, H.P., Ahirrao, R.A., Pharmacognostic and phytochemical investigation of Bauhinia variegata (Linn.) Benth. stem bark. *PSM*, 3, 1, 1–12, 2012.
22. Chandra, T.R., Suresh, C., Sanghamitra, D., Kumar, G.R., Kanchnara (Bauhinia variegatalinn): A critical review. *Int. J. Ayurveda Pharma Res.*, 3, 7, 39–46, 2015.
23. Connor, K.F., Bauhinia variegata L, in: *Tropical Tree Seed Manual Agric. Handbook*, vol. 721, J.A. Vozzo, (Ed.), pp. 332–334, US Department of Agriculture, Forest Service, Washington, DC, 2002.
24. Anwar, S., Jahan, N.S.A., Aslam, S., Microwave-assisted extraction and nitric oxide and superoxide attenuation potential of polyphenolics from Bauhinia variegata. *Asian J. Chem.*, 25, 13, 71–25, 2013.
25. Easdown, W.J., Ravishankar, M., Kaur, D.P., Bhushan, K.B., Traditional leafy vegetables of a tribal community in Jharkhand, India, in: *XXIX Int. Hortic. Congress Horticulture: Sustaining Lives, Livelihoods Landscapes (IHC2014)*, 1102, 43–52, 2014.
26. Taniguchi, K., Nonami, T., Nakao, A., Harada, A., Kurokawa, T., Sugiyama, S., Takagi, H., The valine catabolic pathway in human liver: Effect of cirrhosis on enzyme activities. *Hepatology*, 24, 6, 1395–1398, 1996.
27. Verbruggen, N. and Hermans, C., Proline accumulation in plants: A review. *Amino Acids*, 35, 753–759, 2008.
28. Bairagi, S.M., Aher, A.A., Nimase, P.K., *In vitro* anthelmintic activity of Bauhinia variegata bark (Leguminosae). *Int. J. Pharm. Pharm. Sci.*, 4, 672–674, 2012.
29. Morling, J.R., Yeoh, S.E.N., Kolbach, D.N., Rutosides for prevention of post-thrombotic syndrome. *Cochrane Database Syst. Rev.*, 11, 1–15, 2018.
30. Kulshrestha, P.K., Mishra, A.K., Pal, V.K., Pandey, S., Tripathi, D., Yadav, P., The antimicrobial activity of Bauhinia variegata Linn. flower extract (methanolic). *Asian J. Pharm. Clin. Res.*, 4, 1, 46–47, 2011.
31. Paixao, J., Dinis, T.C.P., Almeida, L.M., Protective role of malvidin-3-glucoside on peroxynitrite-induced damage in endothelial cells by counteracting reactive species formation and apoptotic mitochondrial pathway. *Oxid. Med. Cell. Longev.*, 2012, 1–12, 2012.
32. Saleem, M., Lupeol, a novel anti-inflammatory and anti-cancer dietary triterpene. *Cancer Lett.*, 285, 2, 109–115, 2009.
33. Sahu, T. and Sahu, J., Bauhinia racemosa: A review of its medicinal properties. *World J. Pharm. Res.*, 4, 686–696, 2015.
34. Cheng, Y.T., Lin, J.A., Jhang, J.J., Yen, G.C., Protocatechuic acid-mediated DJ-1/PARK7 activation followed by PI3K/mTOR signaling pathway activation as a novel mechanism for protection against ketoprofen-induced oxidative damage in the gastrointestinal mucosa. *Free Radic. Biol. Med.*, 130, 35–47, 2019.
35. Tan, J., Hu, R., Li, B., Li, Y., Yang, X., He, Z., Wu, S., Protocatechuic acid as a phenolic intermediate to ameliorate non-alcoholic fatty liver disease by inhibiting enterococcus faecalis in mice. *Antioxidants*, 8, 479, 2019.

36. Jain, S.K., Wild plant foods of the tribals of Bastar (Madhya Pradesh), in: *Proc. Nat. Inst. Sci. India B*, vol. 30, pp. 56–80, 1964.
37. Chandra, T.R., Suresh, C., Sanghamitra, D., Kumar, G.R., Kanchnara (Bauhinia variegatalinn): A critical review. *Int. J. Ayurveda Pharma Res.*, 3, 7, 39–46, 2015.
38. Kumari, A., Angmo, K., Bhalla, T.C., Traditional pickles of Himachal Pradesh, *Ind. J. Trad. Know.*, 15, 2, 330–336, 2016.
39. Singh, R.S. and Pandey, H.S., Two new long chain compounds from *Bauhinia variegata* Linn. *Indian J. Chem. B*, 45, 9, 2151–2153, 2006.
40. Zhao, Y.Y., Cui, C.B., Cai, B., Han, B., Sun, Q.S., A new phenanthraquinone from the stems of Bauhinia variegata L. *J. Asian Nat. Prod. Res.*, 7, 6, 835–838, 2005.
41. Naeem, M.Y. and Ugur, S., Nutritional and health consequences of Bauhinia variegata. *TURJAF*, 7, 27–30, 2019.
42. AlSharari, S.D., Salim., S., Al., R., Hatem, M.A., Mohamedand, M.A., Mohamed, M.H., Rutin attenuates Hepatotoxicity in high-cholesterol-diet-fed rats. *Oxid. Med. Cell. Longev.*, 2016, 1–11, 2016.
43. Crespo, I., San-Miguel, B., Mauriz, J.L., Ortiz de Urbina, J., Almar, M., Tuñón, M.J., González-Gallego, J., Protective effect of protocatechuic acid on TNBS-induced colitis in mice is associated with modulation of the SphK/S1P signaling pathway. *Nutrients*, 9, 288, 2017.
44. Kumari, I., Kaurav., H., Choudhary, G., Bauhinia variegate (kanchnara), an ornamental plant with significant value in ayurvedic and folk medicinal system. *Himalayan J. H. Sci.*, 6, 1–16, 2021.
45. Singh, K.L., Singh., D.K., Singh, V.K., Caracterização da atividade moluscicida dos extratos das plantas bauhinia variegata e Mimusops elengi contra o vetor da Fasciola, Lymnaea acuminata. *Rev. Inst. Med. Trop. São Paulo*, 54, 135–140, 2012.
46. Safaeiana, L., Emamia, R., Hajhashemia, V., Haghighatian, Z., Antihypertensive and antioxidant effects of protocatechuic acid in deoxycorticosterone acetate-salt hypertensive rats. *Biomed. Pharmacother.*, 100, 147–155, 2018.
47. Arain, S., Najma, M., Muhammad, T.R., Syed, T.H.S., Muhammad, I.B., Sarfaraz, A.M., Physico-chemical characteristics of oil and seed residues of Bauhinia variegata and bauhinia linnaei. *Pak. J. Anal. Environ. Chem.*, 13, 1, 6, 2012.
48. Hakkim, F.L., Shankar, C.G., Girija, S., Chemical composition and antioxidant property of holy basil (Ocimum sanctum L.) leaves, stems, and inflorescence and their *in vitro* callus cultures. *J. Agric. Food Chem.*, 55, 22, 9109–9117, 2007.
49. Cheng, Y.T., Lin, J.A., Jhang, J.J., Yen, G.C., Protocatechuic acid-mediated DJ-1/PARK7 activation followed by PI3K/mTOR signaling pathway activation as a novel mechanism for protection against ketoprofen-induced oxidative damage in the gastrointestinal mucosa. *Free Radic. Biol. Med.*, 130, 35–47, 2019.

50. Moraisa, C.A., de Rossoa, V.V., Estadellaa, D., Pisani, L.P., Anthocyanins as inflammatory modulators and the role of the gut microbiota. *J. Nutr. Biochem.*, 33, 1–7, 2016.
51. Sharma, K., Kumar, V., Kumar, S., Sharma, R., Mehta, C.M., Bauhinia variegata: A comprehensive review on bioactive compounds, health benefits and utilization. *Adv. Trad. Med.*, 21, 1–9, 2020.
52. Orwa, C., Mutua, A., Kindt, R., Jamnadass, R., Simons, A., *Agroforestree Database: A Tree Reference and Selection Guide. Version 4*, World Agroforestry Centre, Kenya, 2009.
53. Sachin, Sharma, O., Garg, N.K., A review article on: Kanchnara (bauhinia variegate linn). *World J. Pharm. Med. Res.*, 7, 6, 143–147, 2021.
54. Uddin, G., Sattar, S., Rauf, A., Preliminary phytochemical, *in vitro* pharmacological study of Bauhinia alba and Bauhinia variegata flowers. *MEJMPR*, 4, 75–79, 2012.
55. Frankish, N., de Sousa Menezes., F., Mills., C., Sheridan, H., Enhancement of insulin release from the β-cell line INS-1 by an ethanolic extract of Bauhinia variegata and its major constituent roseoside. *Planta Med.*, 76, 995–997, 2010.
56. Higgins, A.J. and Lees, P., The acute inflammatory process, arachidonic acid metabolism and the mode of action of anti-inflammatory drugs. *Equine Vet. J.*, 16, 3, 163–175, 1984.
57. hao, T., Mao, L., Yu, Z., Hui, Y., Feng, H., Wang, X., Sun, C.,Therapeutic potential of bicyclol in liver diseases: Lessons from a synthetic drug based on herbal derivative in traditional Chinese medicine. *Int. Immuno.*, 91, 107308, 2021.
58. Tripathi, A.K., Gupta, P.S., Singh, S.K., Antidiabetic, anti-hyperlipidemic and antioxidant activities of. *Bauhinia variegata flower. Biocatal. Agric. Biotechnol.*, 19, 8, 2019.
59. Yadava, R.N. and Reddy, V.M.S., A new flavone glycoside, 5-hydroxy 7, 3′, 4′, 5′-tetra-methoxyflavone 5-O-β-D-xylopyranosyl-(1→ 2)-α-L-rhamnopyranoside from Bauhinia variegata Linn. *J. Asian Nat. Prod. Res.*, 3, 4, 341–346, 2001.
60. de Souza, R.F.S.O., de Albuquerque, U.P., Monteiro, J.M., de Amorim, E.L.C., Jurema-Preta (mimosa tenuiflora [willd.] poir.): A review of its traditional use, phytochemistry and pharmacology. *Braz. Arch. Biol. Technol.*, 51, 5, 937–947, 2008.
61. Shirkande, A.A. and Shirkande, A.S., Kovidar-A holistic ayurvedic aroach. *Int. J. AYUSH (Ayurveda, Yoga, Unani, Siddha Homeopathy)*, 5, 310–321, 2016.
62. Sharma, S. and Kumar, A., Tribal uses of medicinal plants of Rajashthan: Kachnar. *Int. J. Life Sci. Pharma Res.*, 2, 4, 69–76, 2012.
63. Kumar, S., Baniwal, P., Kaur, J., Kumar, H., Kachnar (bauhinia variegata), in: *Antioxidants in Fruits: Properties and Health Benefits*, pp. 365–377, Springer, Singapore, 2020.

64. Gayathri, G., Saraswathy, A., Krishnamurthy, V., Antimicrobial activity of medicinal plant Bauhinia variegata Linn. *Int. J. Pharm. Biol. Sci.*, 1, 4, 400–408, 2011.
65. Kumar, A., Anand, V., Dubey, R.C., Goel, K.K., Evaluation of antioxidant potential of alcoholic stem bark extracts of Bauhinia variegata Linn. *J. Appl. Nat. Sci.*, 11, 1, 235–239, 2019.
66. Yadava, R.N. and Reddy, V.M.S., Anti-inflammatory activity of a novel flavonol glycoside from the Bauhinia variegata Linn. *Nat. Prod. Res.*, 17, 3, 165–169, 2003.
67. Pant, H.M. and Sharma, N., Inventory of some exotic cultivated tree species of Doon valley and their ethnobotanical uses. *J. Med. Plants Res.*, 4, 20, 2144–2147, 2010.
68. Singh, K.L., Singh, D.K., Singh, V.K., Toxicity of Bauhinia variegata and Mimusops elengi. *J. Biol.*, 2, 2, B76–B82, 2012.
69. Negi, A., Sharma, N., Singh, M.F., Spectrum of pharmacological activities from Bauhinia variegata: A review. *J. Pharm. Res.*, 5, 2, 792–797, 2012.
70. Patil, J.K., Jalalpure, S.S., Hamid, S., Ahirrao, R.A., *In-vitro* immunomodulatory activity of extracts of bauhinia vareigata linn bark on human neutrophils, *Iranian J. of Pharma and Ther.*, 9, 41–46, 2010.
71. Maldonado, P.D., Barrera, D., Rivero, I., Mata, R., Medina-Campos, O.N., Hernández-Pando, R., Pedraza-Chaverrí, J., Antioxidant s-allylcysteine prevents gentamicin-induced oxidative stress and renal damage. *Free Radic. Biol. Med.*, 35, 3, 317–324, 2003.
72. Khare, P., Kishore, K., Sharma, D.K., Historical aspects, medicinal uses, phytochemistry and pharmacological review of bauhinia variegate. *Asian J. Pharm. Pharmacol.*, 4, 5, 546–562, 2018.
73. Rao, Y.K., Fang, S.H., Tzeng, Y.M., Antiinflammatory activities of flavonoids and a triterpene caffeate isolated from bauhinia variegata. *Phytother. Res.*, 22, 7, 957–962, 2008.
74. Pokhrel, N.R., Adhikari, R.P., Baral, M.P., *In-vitro* evaluation of the antimicrobial activity of Bauhinia variegata, locally known as koiralo. *World J. Microbiol. Biotechnol.*, 18, 1, 69–71, 2002.
75. Pandey, S. and Agrawal, R.C., Effect of bauhinia variegate bark extract on DMBA-induced mouse skin carcinogenesis: A preliminary study. *Glob. J. Pharmacol.*, 3, 3, 158–162, 2009.
76. Bodakhe, S.H. and Ram., A., Hepatoprotective properties of Bauhinia variegata bark extract. *Yakugaku Zasshi*, 127, 9, 1503–1507, 2007.
77. Syeda, S. and Nikalje, A.P.G., A brief review on Bauhinia variegata: Phytochemistry, antidiabetic and antioxidant potential. *Am. J. Pharm. Tech Res.*, 7, 186–197, 2017.
78. Kumar, Y.R. and Rajani, G.P., Analgesic and anti-ulcer activities of ethanol and aqueous extracts of root of bauhinia variegata linn. *Int. J. Pharmacol.*, 7, 5, 616–622, 2011.

79. Singh, N., Singh, A., Pabla, D., A review on medicinal uses of bauhinia variegata linn. *PharmaTutor*, 7, 6, 12–17, 2019.
80. Pani, S.R., Mishra, S., Sahoo, S., Panda, P.K., Protective effect of herbal drug in cisplatin induced nephrotoxicity. *Indian J. Pharmacol.*, 43, 2, 200–202, 2011.
81. Verma, R., Awasthi, M., Modgil, R., Dhaliwal, Y.S., Effect of maturity on the physico-chemical and nutritional characteristics of kachnar (bauhinia variegata linn) green buds and flowers. *Indian J. Nat. Prod. Resour.*, 3, 2, 242–245, 2012.
82. Kumar, P., Baraiya, S., Gaidhani, S.N., Gupta, M.D., Wanjari, M.M., Antidiabetic activity of stem bark of Bauhinia variegata in alloxan-induced hyperglycemic rats. *J. Pharmacol. Pharmacother.*, 3, 1, 64, 2012.
83. Mishra, A., Sharma, A.K., Kumar, S., Saxena, A.K., Pandey, A.K., Bauhinia variegata leaf extracts exhibit considerable antibacterial, antioxidant, and anticancer activities. *BioMed. Res. Int.*, 2013, 1–10, 2013.
84. Reddy, M.V.B., Muntha, K.R., Gunasekar, D., Caux, C., Bodo, B., A flavanone and a dihydrodibenzoxepin from Bauhinia variegata. *Phytochemistry*, 64, 4, 879–882, 2003.
85. dos, S.K.M., Gonçalves, P.S., de Paiva, M.J.N., Lacerda, G.A., Acetylcholinesterase inhibition starting from extracts of bauhinia variegata l., Bauhinia var. candida (Aiton) buch.-ham., and Bauhinia ungulata l. *Rev. Soc. Bras. Med. Trop.*, 44, 6, 781–783, 2011.
86. Gupta, N., Bhattacharya, S., Sharma, R.K., Narang, E., *In vitro* comparative antioxidant activity of ethanolic extracts of glycosmis pentaphylla, bauhinia variegata and Leucas aspera. *J. Pharm. Res.*, 10, 1, 42–45, 2011.
87. Rajkapoor, B., Jayakar, B., Anandan, R., Kavimani, S., Anti-ulcer effect of bauhinia variegata linn. in rats. *J. Nat. Remedies*, 3, 2, 215–217, 2003.
88. Neil, F., de Sousa Menezes, F., Mills, C., Sheridan, H., Enhancement of insulin release from the β-cell line INS-1 by an ethanolic extract of Bauhinia variegata and its major constituent roseoside. *Planta Med.*, 76, 10, 995–997, 2010.
89. Dhonde, S.M., Siraskar, B.D., Kulkarni, A.V., Kullarni, A.S., Bingi, S.S., Haematinic activity of ethanolic extract of stem bark of bauhinia variegate linn. *Int. J. Green Pharm.*, 1, 3-4, 28–33, 2007.
90. Rajani, G.P. and Ashok, P., *In vitro* antioxidant and antihyperlipidemic activities of Bauhinia variegata Linn. *Indian J. Pharmacol.*, 41, 5, 227, 2009.
91. Rocha, A.C., Maciel, F.M., Silva, L.B., Ferreira, A.T.S., Da Cunha, M., Machado, O.L.T., Fernandes, K.V.S., Oliveira, A.E.A., Xavier-Filho, J., Isolation and intracellular localization of insulin-like proteins from leaves of bauhinia variegata. *Braz. J. Med. Biol. Res.*, 39, 11, 1435–1444, 2006.
92. Mali, R.G. and Dhake, A.S., Bauhinia variegata linn. (mountain ebony): A review on ethnobotany, phytochemistry and pharmacology. *Orient. Pharm. Exp. Med.*, 9, 3, 207–216, 2009.

93. Thiruvenkatasubramaniam, R. and Jayakar., B., Anti-hyperglycemic and anti-hyperlipidaemic activities of bauhinia variegata l on streptozotocin induced diabetic rats. *Der Pharm. Lett.*, 2, 5, 330–334, 2010.
94. Marimuthu, G., Rajeswary, M., Veerakumar, K., Muthukumaran, U., Hoti, S.L., Mehlhorn, H., Barnard, D.R., Benelli., G., Novel synthesis of silver nanoparticles using Bauhinia variegata: A recent eco-friendly approach for mosquito control. *Parasitol. Res.*, 115, 2, 723–733, 2016.
95. Gautam, S., Bauhinia variegata linn: All purpose utility and medicinal tree. *Forestry Bull.*, 12, 2, 61–64, 2012.

8

Hibiscus cannabinus

Deep Shikha[1*], Piyush Kashyap[2], Abhimanyu Thakur[3] and Madhusudan Sharma[1]

[1]Department of Food Engineering and Technology, Sant Longowal Institute of Engineering and Technology, Longowal, Punjab, India
[2]Department of Food Technology and Nutrition, School of Agriculture, Lovely Professional University, Phagwara, Punjab, India
[3]Department of Food Science and Technology, Dr YS Parmar University of Horticulture and Forestry, Nauni, Solan, Himachal Pradesh, India

Abstract
Hibiscus cannabinus (kenaf) is part of the family Malcaveae, a tall yearly herbaceous woody plant with high fiber, energy, and feedstock. Traditionally used as an herbal medicine, as flavouring agent in food industries and in various herbal drinks. Consumers generally show keen interest in natural plant-based food products having health properties. Kenaf and its parts particularly seeds are the rich source of vital nutrients and outstanding source for value added plant-based foods. Along with this, it is also good source of phytocompounds, antioxidant vitamins and amino acids, which are accountable for its antioxidant properties and pharmacological merits. The plant exhibits anti-oxidant, anti-microbial, anti-diabetic, hepatoprotective, antihypertensive, anti-cholesterol, and other therapeutic properties. Possessing high nutritional content and an important source of natural antioxidants, *Hibiscus cannabinus* can be utilized as a nutraceutical source and also as functional foods. The fibers of kenaf are not only used for pulping and cordage of paper but also are a promising bio-energy lingo-cellulosic feedstock. Oil from its seeds might be utilized for cooking and various industrial uses. The current chapter delves into history, present production trends, botany and nutritional profile *Hibiscus cannabinus*. The bioactive profile and health benefits interests of *Hibiscus cannabinus* will also be highlighted on the conditions with the researches in support with the systematic scientific approaches. The chapter concludes with a discussion of kenaf's potential industrial applications. The industries and scientists

Corresponding author: bansaldeepshikha789@gmail.com

Prerna Gupta, Navnidhi Chhikara and Anil Panghal (eds.) Harvesting Food from Weeds, (267–326)

can utilize this natural plant source as value added food source which might be of great significance for industries as well as human kind.

***Keywords*:** Kenaf (*Hibiscus cannabinus*), bioactive compounds, pharmacology, antioxidant activity, anti-cancerous activity

8.1 Origin and History

Hibiscus cannabinus (Kenaf) is tropical and subtropical wild plant, which is native to east-central Africa and Asia. It is found that diversified form of kenaf, with different Hibiscus species like roselle species (*Hibiscus sabdariffa* L.) are grown mainly in eastern part of Africa [1, 2] and also in India as well. Domestication of kenaf was found to be in 4000 BC in Sudan region which was justified by scientist through the surveys and investigations done on the field [3]. Cheng *et al.* [2] also suggested Africa as the origin country based on the amplified fragment length polymorphism (AFLP) analysis that reported that kenaf germ plasm was having 23 accessions and supported its dispersion from Africa to Asia and America. Kenaf have been utilized from the ancient times, for approximately more than 6000 years as basic material for making of cords and ropes. Apart from cordage, it was used as the feed for the livestock and was good source of textile fiber for the rags, ropes, twine, etc. For more than thousands of years it had been produced and was used as the source for rope making until it was domestically grown and used in northern America. It was also introduced to India back approx. 200 years ago, to China in 1935 and in Russia in 1902 [1]. Kenaf was commercially grown in Asia as the fiber crop in 1900s beginning. It spread from Taiwan to mainland of China [1, 4, 5] and then passed on to the USSR in the 1930. Kenaf was cultivated in the United States during World War II in the 1940s, and it was used to make cordage for use in war zones. With the increased demands of fiber in the USA during 1950s, more than 500 species of plants were investigated for the same purpose, in which kenaf was found to be most promising crop for the replacement of jute. Demand increase for fiber in United States in late 1970s was due to the production of various types of paper (corrugated board, newspaper, bond paper) for which kenaf was found to be efficient source of cellulose fiber to fulfill the emerged demands [6].

In Malaysia, kenaf was introduced in 1970 but came into consideration later as the economic and alternative basic material for textiles, fiberboard as well as fuel in late 1990s [6]. Due to the wide range of application of kenaf, many countries are researching for its cultivation and utilization. This crop is also considered as the "future crop" as it is wide spread ecologically and can be cultivated as varying condition of environment [7, 8].

8.2 Botanical Description

Kenaf, belongs to *Malvaceae* family, a tropical plant with tall herbaceous woody stem. It has attracted many of the grower's attention due to its versatility in its exploitation for different usage as fiber source, animal feed, raw material for paper production, etc. Height of these plants can go up to 20 feet under basic feasible condition of maturation otherwise under mild environmental condition it grows up to 8 to 14 feet in its growing months. Leaves on the plant stem are alternatively arranged, which are 8 to 15 cm in length and 3 to 5 deeply lobed. The leaves grow sidewise on stalk and are of two different shape (palmatifid and simple) (Figure 8.1h and i). Different variety of kenaf also varies in color of petioles and shape of leaves as in petioles color ranges from red, green to purple [6]. With the maturation of plant, newer small leaves grow additionally with the characteristic shape for that species of plant. In case of palmatifid leaf cultivars, it can grow up to 3 to 10 entire small leaves before producing the palmatifid leaf [9]. Flowers of the plant are produced in colour range of white to light yellow, which are of crimson colour at the base of it (Figure 8.1c and d). Flowers are usually of 8 to 10 cm diameter and have fleshy calyx at the base, which is of 1 to 2 cm. These are particularly borne at the apex of plant on short stalks at leaf axils and the main crowning meristem kept itself for the vegetative growth and plant mature in almost 6 months. Flowering period of these plants are longer and maturation of plant keep on increasing upon initiation of flowering of plant. Flowers on each plant blooms for the 1 day and it can last for 3 to 4 weeks. These flowers have reddish base with the cream or whitish colour of the petals and are produced annually. The flowering pattern of this flower starts from morning when it opens and started closing form mid-day and closes after noon and after that never opens again. Kenaf is characterized occasionally as the self-pollinating plant, which pollinates in short period of opening of flower for a day and also cross-pollinating crop. When flower opens, the anther releases the pollen of bright orange colour and thereafter its style emerges for pollination. Without touching anther (Figure 8.1e), five-part stigma opens and the lobe become rigid and corolla then closes in helical way so as to compress the anther to touch stigma for the self-pollination. Such way of self-pollination adopted when cross-pollination does not occur. Cross-pollination in kenaf varies from 2% to 24% and have moderate level of cross-pollination in crop which can be identified by seed of flower. Domesticated honey bees are responsible for cross-pollination whereas self-pollination occurs when the flower petals close in a twisting motion [6].

Kenaf plant is said to be drought tolerant as it has wide range lateral root system, which is deep, and have efficient tap root system [10]. The depth of root system expands up to 1 m deep down the soil and plants are much sensitive to compactions and plow pan. The color of stem varies from green to dark burgundy color and it have tiny thorns which spike out around the stem varying in size from tiny to large (Figure 8.1b). The presence of thorns depends on variety of kenaf such as blackberry bush have thorns on its stem. Stem of plant generally have thickness of bast 1 to 2 mm which is outer part of fiber and ranges in breadth from 1 to 2 cm. Plant crop of kenaf when grown in high density, are tends to grow straight and branched stem. High density of crop has range of plants 170,000 to 220,000 plants/ha. Growth rate of plant is high and it took about 6 to 8 months to grow up to height of 4 to 7 m. It is able to produce dry weight of 6500 to 10,000 kg/ha per year [11]. Bast represents outer portion of stem and accounts for 30 % dry weight of stalk whereas 70 % portion of dry matter is contributed by inside part of fiber, which has a white color. Inner core fiber

Figure 8.1 Botanical description of *Hibiscus cannabinus* (kenaf) plant components (a) kenaf whole plant (b) stem (c, d) flowers (e) internal morphology of flower (f) fruit (g) seeds (h, i) simple and palmated leaf.

accounting about 60% to 70% gives low quality pulp and high-quality pulp is produce by bast, i.e., the outer portion of fiber [12]. Pith of stem consists of parenchymatous cells which are extended in polygonal shape [13]. Seeds of fruit (Figure 8.1f) are brown in color, wedge shaped, glabrous, and present in seed capsule which contains about 20 to 26 seeds (Figure 8.1g). These seeds are present in five segments in the seed coat, which is hairy from outer surface. Seeds have 6 mm length and 4 mm breadth and have weight about 34,000 to 40,000 seeds/kg correlating to 25 to 29 g thousand grain weight [9].

8.3 Production

More than 20 countries have been involved in production of kenaf commercially in which 95% of total production is contributed by Thailand, China, and India. Native countries of kenaf plant are Vietnam, Malaysia, Thailand, Japan, India, Indonesia, and Pakistan [14]. *Hibiscus sabdariffa* and *Hibiscus cannabinus* are cultivated in bulk due to their easy handling and adaptability more than other fibre crop. Its cultivation is wide spread and had attained great attention in China, India and Thailand commercially. These two species of kenaf are much adaptive to harsh climatic conditions and can be grown with less care. These can even grow in such areas where cultivation of jute is difficult. In such way, these species of kenaf attained attention and grown in more than 26 lakhs hectares producing ample amount of bale accounting for approx. 12 lakhs. With the increased production, kenaf globally attained the production of 2.8 million tonnes in 1985. But after that there was decrease trend in its production and got stable at 0.2 million tons. Developing country are much dependent on the kenaf as it is important cash crop. In 1985, kenaf production in China accounts for mean cultivated area 150,000 to 400,000 hectare and afterwards decline in production was faced due to synthetic materials upcoming. From the data of 2014 to 2015, India accounts for 46% of total world kenaf production which is approx. 100 tones whereas China contributes 26 % of overall production in the world, whereas, Pakistan accounts 0.46% and Indonesia for 1.6 % as well as Africa for 6.8 % [15]. In Malaysia, kenaf gained attention in 2010 to replace tobacco production [16]. In India, kenaf is also used for bordering the garden areas so that it can create proper drainage as the border soil are loamy. It is substantial crop for fibres in south India and is grown as rain fed crop in Madhya Pradesh in huge hectare of land and also in Andhra Pradesh, Tamil Nadu. This crop can be cultivated with cereals in dry lands or can be as a lone crop [17]. The production statistics of kenaf

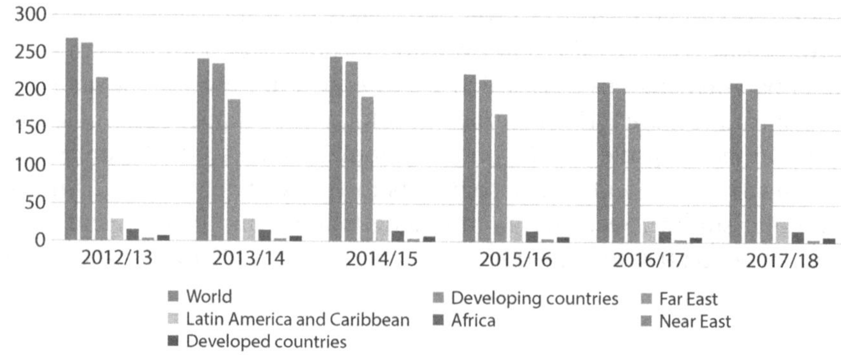

Figure 8.2 World production (100 tons) of kenaf and allied fiber crops (2012–2018) [18].

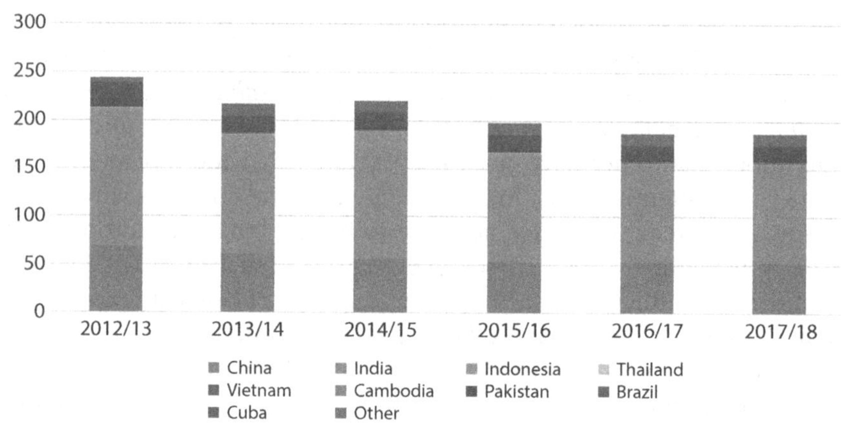

Figure 8.3 Kenaf and allied fiber crops production (100 tons) in major cultivation countries (2012–2018) [18].

and other allied crops around the world from 2012 to 2018 was shown in Figure 8.2 and Figure 8.3 showed the production statistics of major kenaf and other allied crops producing countries from 2012 to 2018.

8.4 Development and Maturation

Till now, kenaf phenology is divided into 4 different phases: "time to emergence, a basic vegetative phase, a photoperiod-induced phase and a flower development phase" [19]. Availability of water is very important for the germination and development of kenaf seeds [20, 21]. Kenaf mainly depends on the day length, especially for the floral budding [22], and stays abundant till the day length reach around 12.5 hours and then it switches to the reproductive growth [19]. Hence, seeds are normally viable only after 35 to 40 days of flowering. It is not produced at the places with high latitudes, like in the northern parts of USA.

Propagation of the kenaf plant occurred by seeds. In usual storage conditions, the seeds stay in abundant state for about 8 months. Deep (at least 20 cm) and thorough soil preparation is required. The plantation of seeds was done at the start of the rainy season, with about 6–30 kg seed/ha drilled 15 × 15 cm on dried soils and 12.5 × 12.5 cm on wet soils. Generally, two seeds are bored and one seedling detached if the germination is good to make sure a common stand for the generation of even stalks. The depth of planting is around 0.5–3.2 cm. Kenaf possess a deep and large root system, while its high-speed growing portion which is above the ground only has a single green stem and its various branches holds different shaped leaves. Kenaf stalks can achieve a length of 4 to 6 m and can have the yield up to 24 Mg ha-1 in 5 to 7 months [23]. It only needs low fertilizer (N-P-K) and can usually be grown in non-muddy soil of 5.7 – 8.2 pH [24].

It is crucial to weed for the first month. Planting takes place in May or June to allow as much vegetative growth as possible before flowering. Manure or green manure can be used to treat soils. The application of 585 kg/ha ammonium sulphate and *Mimosa envisa* green manure increases kenaf yield by 50%. Only recommendation of the fertilizer by the experts is ca 35–70 kg/ha N, 40–60 kg/ha P_2O_5, and 45–65 kg/ha K_2O. Kenaf crop of 50 MT/ha extract from the soil approx 175 kg N, 15 kg P, 75 kg K, 105 kg Ca, and 30 kg Mn [25].

8.5 Nutritional Profile

The nutritional composition of kenaf has been presented in Table 8.1. The moisture in the kenaf leaves is in the range of 5.09 to 11.82 %. Whereas in kenaf seed, seed oil, seed meal, defatted seed meal it is 8.3% to 9.6 %, 0.5% to 3.9 %, 4.65 to 10.34 g/100 g (raw and processed) and 9.34 % respectively. The carbohydrate content in the kenaf leaves is 37.67 % while it is low in seed/seed oil (18.7–24.4%) and very high in defatted kenaf seed meal (57.09 %). The fat content in the kenaf leaves has been reported as 2.33 % whereas crude fat content was 9.03 %. The seeds are rich source of fats due to the high oil content and 22% to 25% and 9 to 18.19 g/100 g DW fat content has been observed in seed and seed meal respectively [26]. Kenaf seeds are rich in oil content (16–22%) and because of high monounsaturated

Table 8.1 Proximate composition of kenaf [26, 29].

Nutrients	Leaves	Seed	Defatted seed meal	Stem	Flower
Moisture (%)	11.82	8.10	9.34	25.80	26.00
Protein (%)	18.0–30.0	30.88	57.09	4.70	4.90
Fat (%)	4.20	22–25	0.73	0.60	3.50
Crude fiber (%)	13.00	10.40	16.95	37.40	3.40
Carbohydrate (%)	37.67	18.7–24.4	57.09	-	-
Ash (%)	6.60	4.5–6.2	6.65	5.50	6.20
Minerals					
Calcium (mg/100 g)	214.20	27.0	-	89.0	43.0
Iron (mg/100 g)	34.0	7.0	-	13.0	42.0
Zinc (mg/100 g)	16.2	30.0	-	16.8	24.4
Potassium (mg/100 g)	157.0	153.0	-	315.0	260.0
Phosphorus (mg/100 g)	9.0	88.0	-	28.0	44.0
Tocopherol (mg/100 g)	-	65.7	-	-	-
Magnesium (mg/100 g)	30.0	61.0	-	48.0	34.0

fatty acids (MUFA) and polyunsaturated fatty acids (PUFA) content they can be considered edible for human consumption [27, 28]. The economic importance of seed production of kenaf is because of the high oil content in its seed [22].

The nutritional importance of kenaf is mainly due to its high protein content in its various plant parts. Its leaves, which are high in crude protein (14–34% in leaves and 2–12% in stalk), are eaten as a vegetable in human diets like spinach and consumed as food in India and some parts of Africa. [30]. A very wide range of protein content as 18% to 30% in the kenaf leaves has been reported [31]. In kenaf seeds the crude protein content

Table 8.2 Amino acid content of kenaf plant parts [26, 33, 34].

Amino acids	Leaf (mg/100 g)	Seed (mg/100 g)	Defatted seed (mg/100 g)	Stem (mg/100 g)	Flower (mg/100 g)
Aspartic acid	7.02	114.39	-	1.76	305.17
Serine	1.45	87.97	-	3.31	812.54
Glutamic acid	11.11	124.37	-	-	250.58
Arginine	5.02	178.72	-	1.31	65.29
Glycine	0.72	62.21	-	0.64	492.15
Alanine	1.65	46.61	-	2.95	1028.58
Proline	2.50	35.69	-	31.39	541.23
Tyrosine	3.06	34.15	-	-	281.65
Histidine	2.41	39.67	5.24–12.50	0.83	117.40
Threonine	3.25	54.65	1.21–3.68	1.67	621.65
Lysine	3.96	59.87	3.83–4.58	-	84.48
Valine	3.85	64.11	0.80–1.36	7.49	1017.76
Methionine	0.91	28.09	0.11–0.43	-	115.06

(Continued)

Table 8.2 Amino acid content of kenaf plant parts [26, 33, 34]. *(Continued)*

Amino acids	Leaf (mg/100 g)	Seed (mg/100 g)	Defatted seed (mg/100 g)	Stem (mg/100 g)	Flower (mg/100 g)
Isoleucine	2.81	54.07	0.26–2.44	2.50	520.10
Leucine	7.05	91.32	1.72–2.57	1.84	546.07
Phenylalanine	4.55	71.01	0.53–1.45	2.86	607.79
Cysteine	0.90	-	-	-	-
Essential amino acids	264.10	462.79	-	18.50	36.96.60

ranges from 24% to 35% [31, 32]. Dried leaves are also a potential source of crude proteins (30%) and used as livestock feed. The plant proteins (33.0%) present in extracts of kenaf seedcake might be responsible for its high antioxidant activity [6]. Other various researchers observed the protein content in leaf, seed/seed oil, seed meal and defatted seed meal as 9.4% to 30.0%, 22.31% to 30.88 %. 25.12% to 33.20 % (raw and processed) and 26.19 % respectively. Kenaf is a good source of essential and non-essential amino acids and the amino acid content in the kenaf leaves and other plant parts have been mentioned in Table 8.2.

Kenaf seed is having high oil content ranging from 19.84% to 25 % and a good source of fatty acids like oleic, linoleic and palmitic acid. Its oil is edible, and there are numerous health benefits of kenaf seed oil in human health due to the presence of MUFA and PUFA with functional compounds [31, 35]. The constitution of fatty acids in kenaf oil along with comparison with other oil crops has been presented in Figure 8.4. The kenaf seed oil manufactured from heated material is reddish brown in color with mild odor and the one prepared by unheated process is clear yellow in color without any odor [36].

The mineral content is represented by the ash content and the various plant parts of kenaf are rich in minerals. The ash content of kenaf leaves is in the range of 5.11 to 6.60 % whereas, in kenaf seed, seed meal and defatted seed meal it is 4.5% to 6.4 %, 4.40% to 10.60 % (raw and processed) and 6.65%, respectively. More calcium, iron and phosphorus content have been reported in kenaf plants [37]. The improvement of calcium content with the addition of powder from the dry kenaf leaves in various

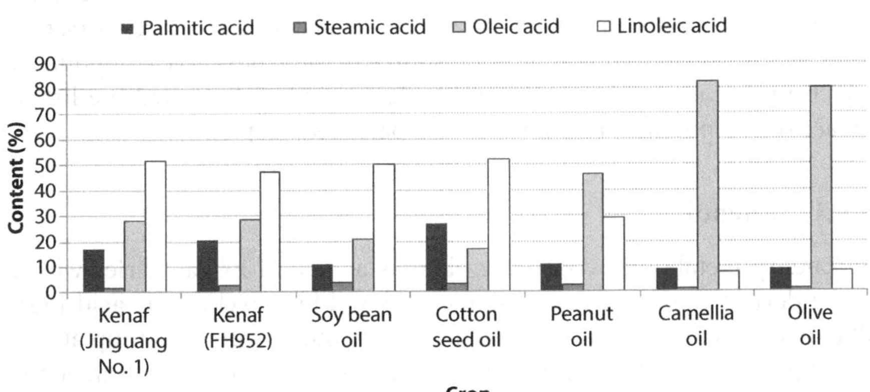

Figure 8.4 Oil composition. Comparison of kenaf and other plants seed oil composition [31].

types of food shows the high calcium content in its leaves [6, 38]. The other major minerals include magnesium, potassium, and phosphorus, which have been reported as >1 g/100 g DKSM (dried kenaf seed meal) through the energy dispersive X-ray spectrometric analysis technique [29]. The other various researchers observed the iron content as 41.50 mg/100 g and 2.69 to 3.39 mg/100 g (raw and processed) and phosphorus content as 171 mg/100 g. The calcium content as 27 mg/100 g has been reported in kenaf seed. The kenaf plant parts are also rich in vitamin C (120.0–180.0 65.7 mg/100 ml) and vitamin E (65.7 mg/100 g) content along with other antioxidants and phytochemicals [26, 37]. The other compounds in kenaf include total phenols, flavonoids, tannins, phytates, etc., which are the cause for its high antioxidant activity [39]. The total phenols, flavonoids and antioxidant activity of kenaf plant extracts reported by various researchers have been presented in Table 8.3.

8.6 Bioactive Compounds

Plant contains various secondary metabolites, which are also called bioactive compounds like flavonoids, polyphenols, alkaloids, essential oils, saponins, tocopherol, tannins, terpenes etc. which impart various benefits to body but are not count in nutritional profiling of food [50, 51]. These compounds show therapeutic effect and are produced in plant as the defense for physiological stress cause like UV radiation and microbial attack [52]. For quantification of their extent present in plant material, its extract is prepared and the chemical experiments are performed on extract. These bioactive compounds show therapeutic effect due to the antioxidant and antimicrobial activities which helps in maintaining blood sugar, blood pressure, ROS, free radicals, etc. Kenaf found to be good source of these bioactive compounds, which impart many pharmacological effects. Table 8.4 showed the list of bioactive compounds present in various part of kenaf plant.

8.6.1 Phenols

Secondary metabolites constitute phenols as one of its categories which is the derivative of hydroxycinnamic acid and hydroxybenzoic acid [53]. Phenol contains aromatic ring having one or more hydroxyl group at different positions on the ring which makes the derivative of phenols depending on the structural variation. Presence of these hydroxyl group makes the phenolic compounds antioxidant in nature as these have high scavenging activity [54].

Table 8.3 Total phenol, flavonoids and antioxidant activity of kenaf plant.

Plant part	TPC	TFC	Antioxidant activity	References
Leaf	178.1 mg/100 g (ME) 555.9 mg/100 g (WE) 119.1 mg/100 g (EE) 28.1 mg/100 g (CE) Old leaves: 67.87 mg/100 g Young leaves: 56.68 mg/100 g	126.7 mg/100 g (ME) 563.7 mg/100 g (WE) 40.2 mg/100 g (EE) 4.8 mg/100 g (CE) Old leaves: 117.42 mgQE/g Young leaves: 102.55 mgQE/g	62.1 % (ME) 73.2 % (WE) 37.7 % (EE) 4.0 % (CE) Old leaves: 52.54-81.13 µg/ml Young leaves: 49.83- 80.47 µg/ml	[40–42]
Flower	298.6 mg/100 g (ME) 308.5 mg/100 g (WE) 183.4 mg/100 g (EE) 2.9 mg/100 g (CE)	620.0 mg/100 g (ME) 755.2 mg/100 g (WE) 126.4 mg/100 g (EE)	71.84 % (ME) 80.6 % (WE) 2.0 % (CE)	[40]
Bark	23.2 mg/100 g (ME) 84.2 mg/100 g (WE) 24.8 mg/100 g (EE) 5.6 mg/100 g (CE)	70.2 mg/100 g (ME) 299.3 mg/100 g (WE) 38.0 mg/100 g (EE) 8.7 mg/100 g (CE)	15.2 % (ME) 29.1 % (WE) 15.4 % (EE) 1.0 % (CE)	[40]

(*Continued*)

Table 8.3 Total phenol, flavonoids and antioxidant activity of kenaf plant. (*Continued*)

Plant part	TPC	TFC	Antioxidant activity	References
Seed	52.6 mg/100 g; 536 mgGAE/100 g (ME)	17.5 mg/100 g; 177 mgGAE/100 g (ME)	5.1 %; 0.62αTE/g (ME)	[40, 43–47]
	162.7 mg/100 g; 1878 mgGAE/100 g (WE)	331.1 mg/100 g; 249 mgGAE/100 g (WE)	70.4 %; 2.30 αTE/g (WE)	
	83.7 mg/100 g (EE)	98.9 mg/100 g (EE)	21.1 %; 52.49 αTE/g (EE)	
	107.0 mg/100 g; 1226 mgGAE/100 g (CE)	154.2 mg/100 g; 294 mgGAE/100 g (CE)	0.56 αTE/g (CE)	
	216 mgGAE/100 g (HE)	164 mgGAE/100 g (HE)	0.40 αTE/g (HE)	
			54.45 αTE/g (EAE)	
			28.09 αTE/g (nHE)	
Seed oil	30.96 mgGAE/100 g, 4.7-24.9 mgGAE/100 g	52.94 mgCE/g	21.64 mg Teq/100 g, 9.4-20 mg Teq/100 g	[44, 48]
Defatted seed meal	3399.37 μgGAE/100 g	251.0 μg RE/g	3313.98 μg Teq./g	[49]

Table 8.4 Bioactive compounds present in Kenaf.

Class	Compounds	Structure	Solvent used	Plant part	References
Phenolic acids	Gallic acid	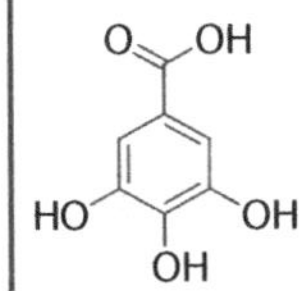	Water, ethanol, methanol	Seed, stem	[26, 40]
	Caffeic acid	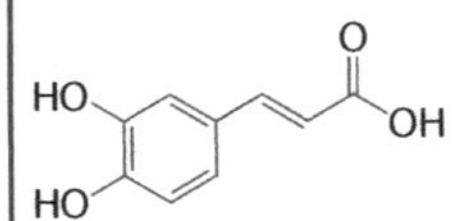	Water, ethanol, methanol,	Stem, flowers, leaves	[40, 55]
	Tannic acid	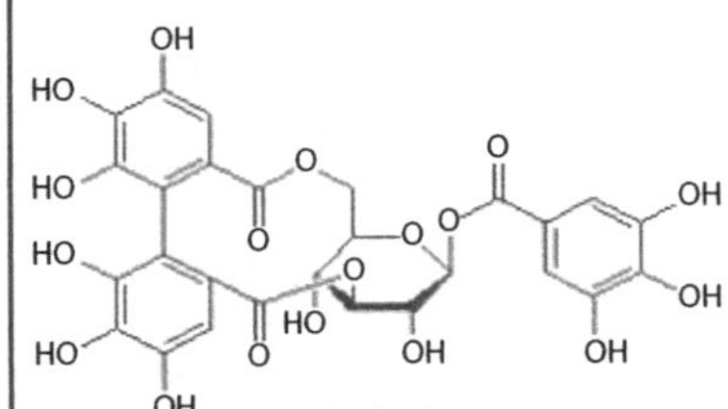 Corilagin	Water, ethanol, methanol, chloroform	Seed, leaves	[40, 55]
	Chlorogenic acid	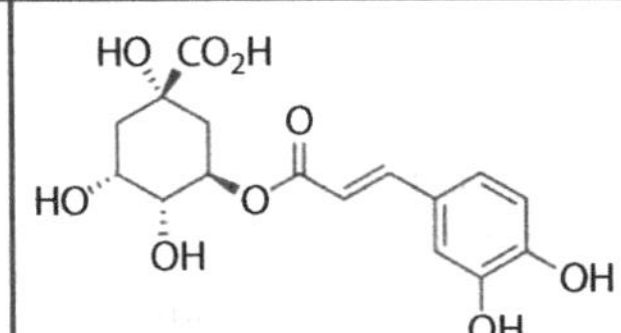	water	leaves	[55]

(*Continued*)

Table 8.4 Bioactive compounds present in Kenaf. (*Continued*)

Class	Compounds	Structure	Solvent used	Plant part	References
	Neochlorogenic acid	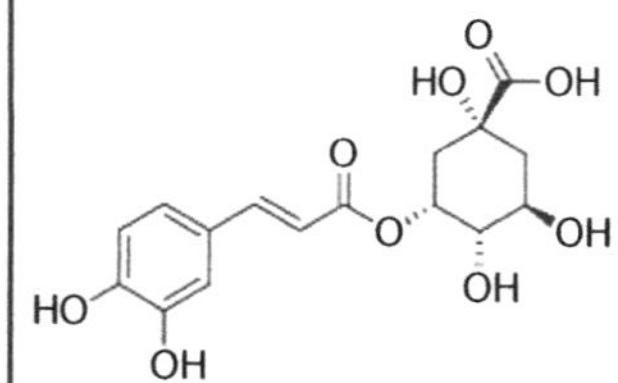	Methanol	leaves	[76]
	Syringic acid	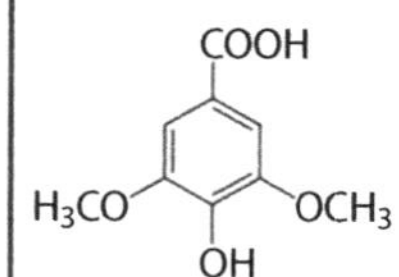	Water, methanol	Seed	[40]
	p-coumaric acid	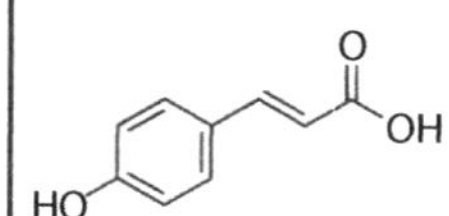	water	Seed	[40]
	p-hydroxybenzoic acid	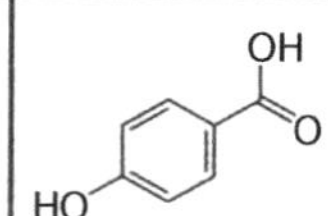	Water, methanol	Seed, Stem, flowers	[40]

(*Continued*)

Table 8.4 Bioactive compounds present in Kenaf. (*Continued*)

Class	Compounds	Structure	Solvent used	Plant part	References
	Vanillin	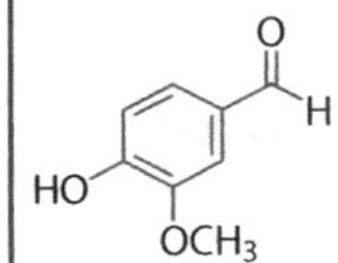	Water, methanol	Seed, Stem	[40]
	Feruloylquinic acid	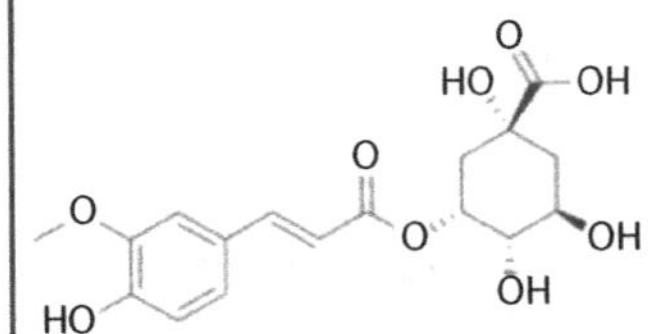	methanol	leaves	[77]
	Protocatechuic acid	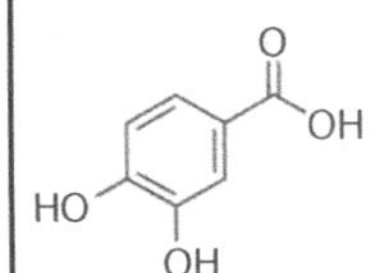	Petroleum ether	Seed	[43]
Flavonoids	kaempferitrin	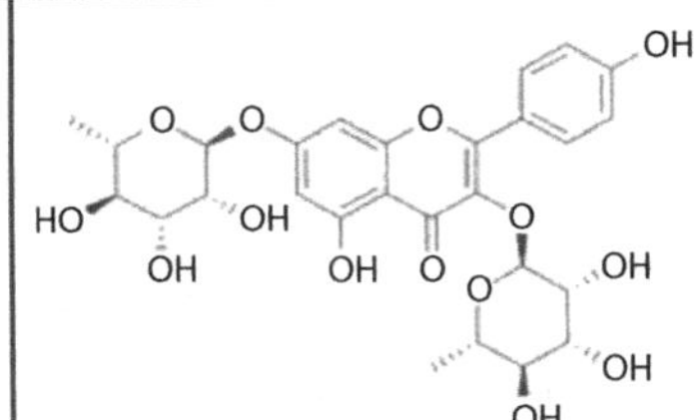	Water, ethanol, methanol, chloroform	Leaves, stem	[40]

(*Continued*)

Table 8.4 Bioactive compounds present in Kenaf. (*Continued*)

Class	Compounds	Structure	Solvent used	Plant part	References
	kaempferol	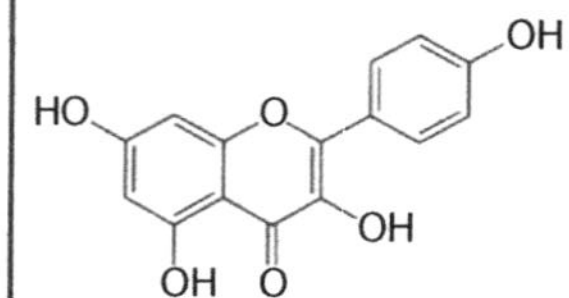	Water, ethanol, methanol, chloroform	leaves	[55]
	Isoquercetin	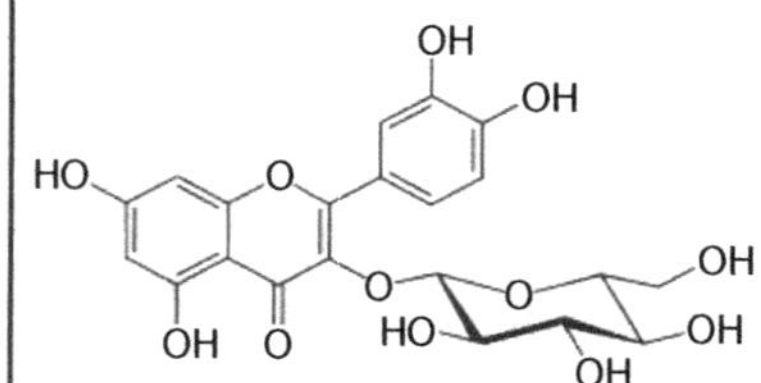	Water, ethanol, methanol,	Leaves, stem	[40]
	Myricetin glycoside	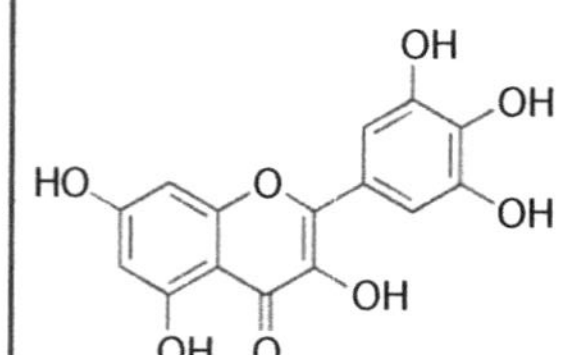	Water, ethanol, methanol,	Flowers	[40]

(*Continued*)

Table 8.4 Bioactive compounds present in Kenaf. (*Continued*)

Class	Compounds	Structure	Solvent used	Plant part	References
	Naringin	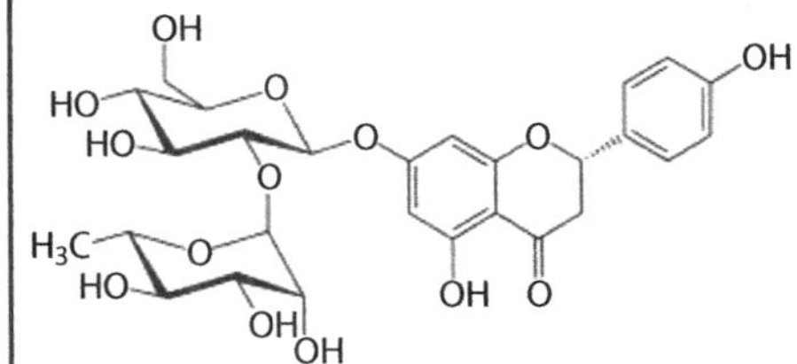	Ethanol, chloroform	Seed, leaves	[40]
	Kaempferol gylcoside	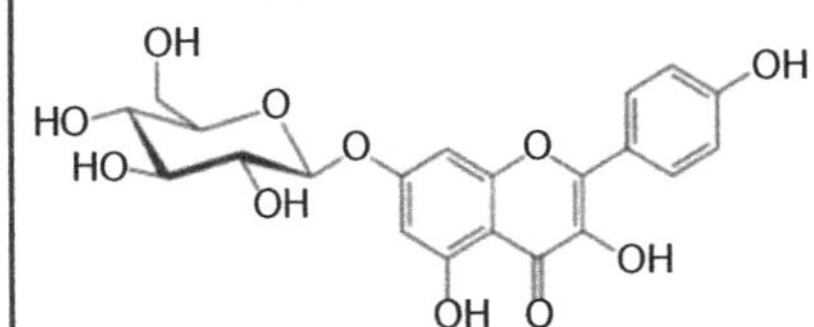	Water, ethanol, methanol,	Stem, leaves	[40]
	Catechin hydrate	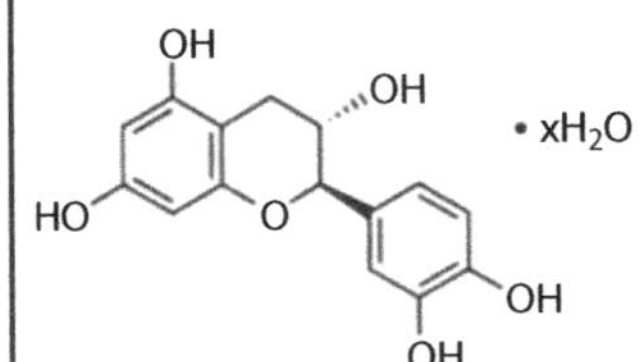	Water	Leaves	[55]

(*Continued*)

Table 8.4 Bioactive compounds present in Kenaf. (*Continued*)

Class	Compounds	Structure	Solvent used	Plant part	References
	Afzelin	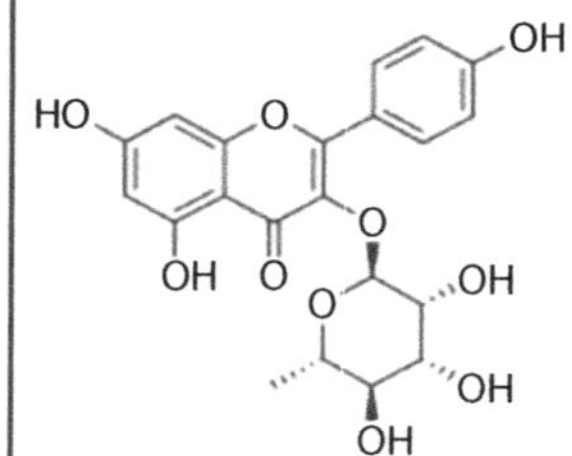	Water, ethanol, methanol	Stem, leaves	[40]
Carotenoids	Neoxanthin	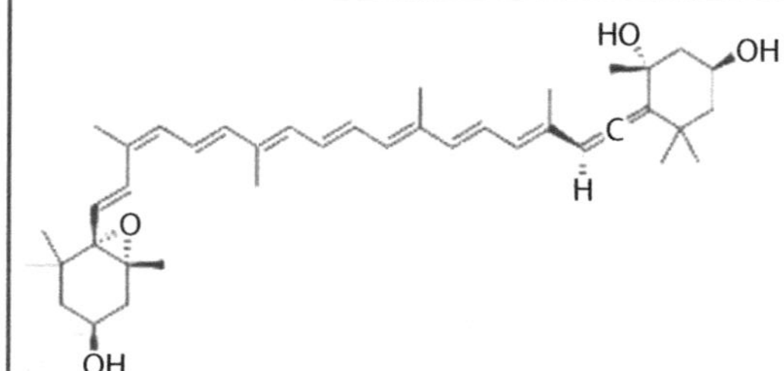	Acetone	Leaves	[40]
	leutein	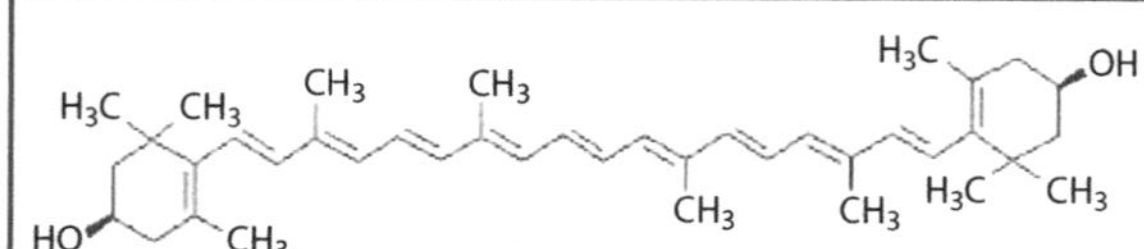	Acetone	Leaves	[40]

(*Continued*)

Table 8.4 Bioactive compounds present in Kenaf. (*Continued*)

Class	Compounds	Structure	Solvent used	Plant part	References
	Zeaxanthin	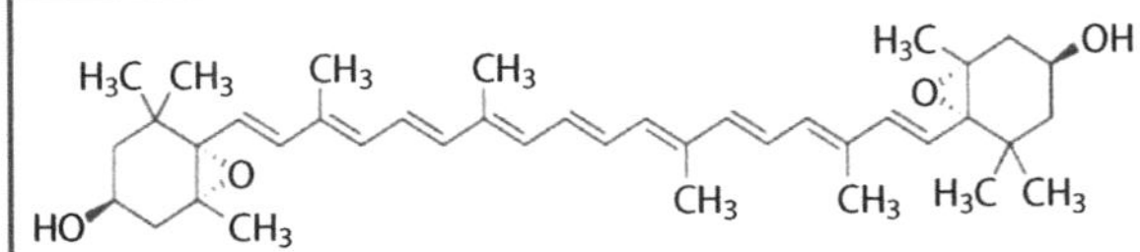	Acetone	Leaves	[40]
	β-carotene	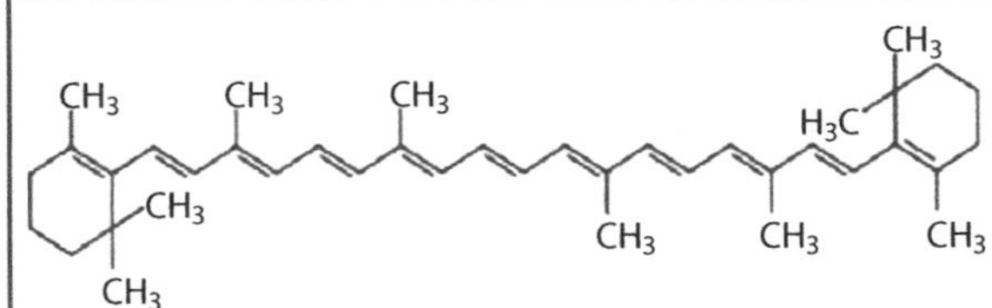	Acetone	Leaves	[40]
Tocopherol	α-tocopherol	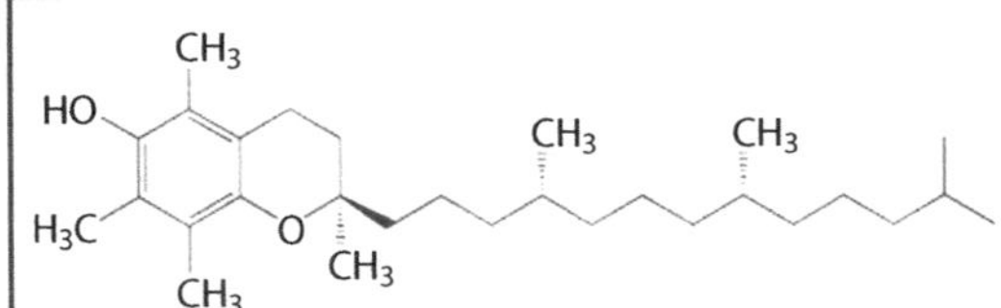	Petroleum ether	Seed	[43]

(*Continued*)

Table 8.4 Bioactive compounds present in Kenaf. (*Continued*)

Class	Compounds	Structure	Solvent used	Plant part	References
	γ -tocopherol	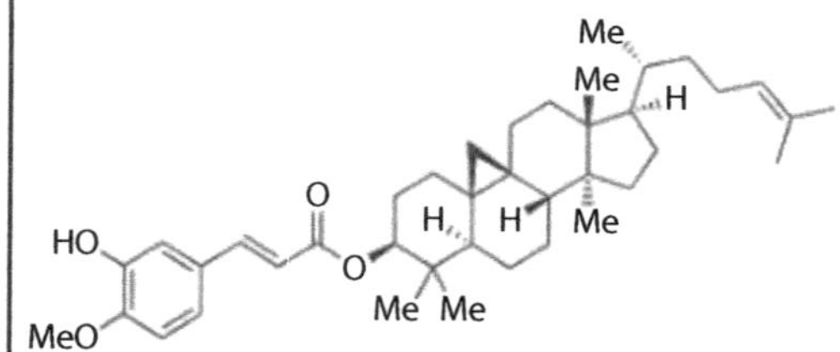	Petroleum ether	Seed	[43]
	δ-tocopherol	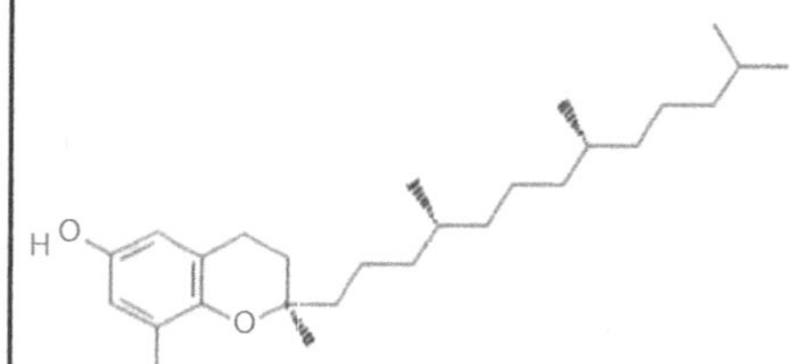	Petroleum ether	Seed	[43]
Tocotrienol	α-tocotrienol	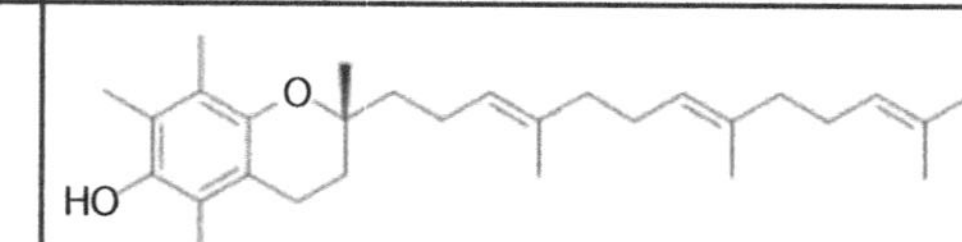	Hexane	Seed	[78]

(*Continued*)

Table 8.4 Bioactive compounds present in Kenaf. (*Continued*)

Class	Compounds	Structure	Solvent used	Plant part	References
	γ-tocotrienol	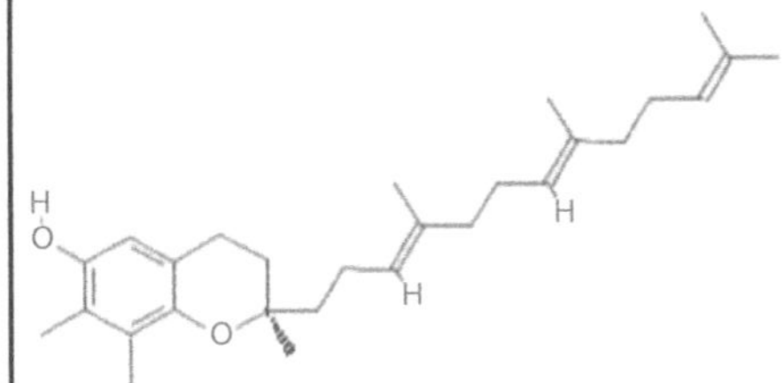	Hexane	Seed	[78]
	δ-tocotrienol	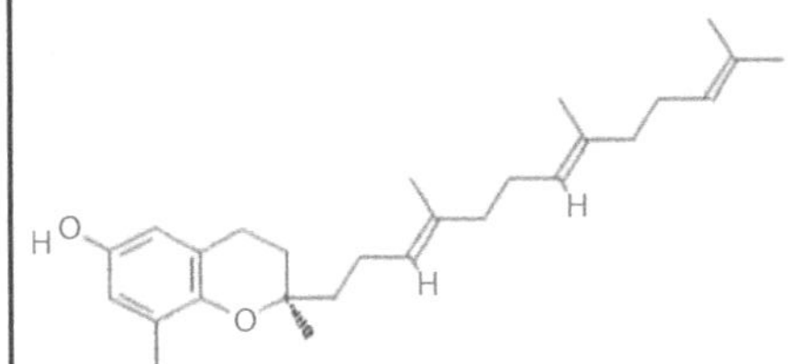	Hexane	Seed	[78]
Fatty acids	Linolenic acid	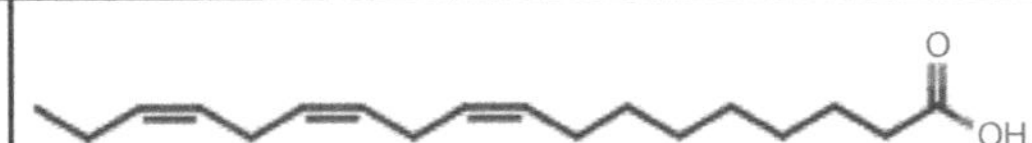	Seed	Hexane-isopropanol	[79]
	Oleic acid	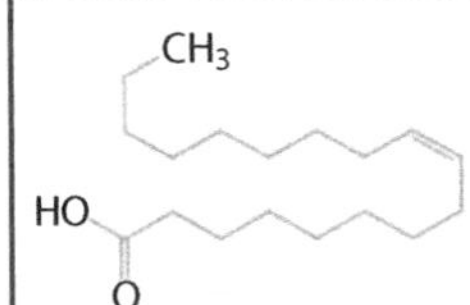	Seed	Hexane-isopropanol	[79]

(*Continued*)

Table 8.4 Bioactive compounds present in Kenaf. (*Continued*)

Class	Compounds	Structure	Solvent used	Plant part	References
	9-octadecanoic acid	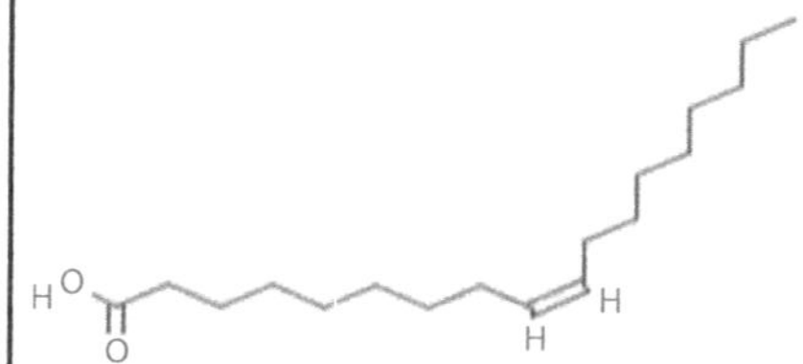	Leaves, stem, Flowers, seed	Hexane	[40, 70]
	9,12-octadecadienoic acid	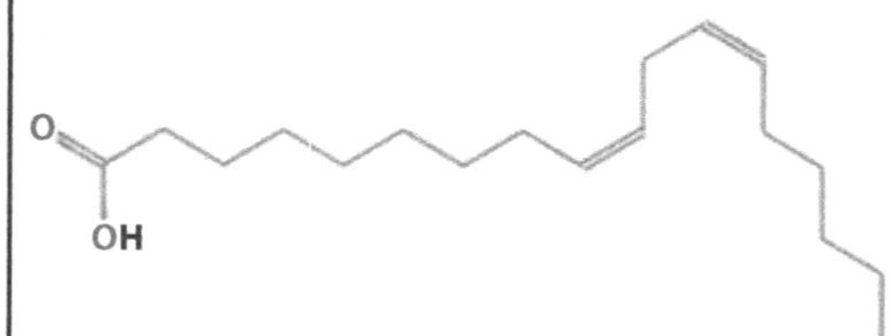	Leaves, stem. Flowers, seed	Hexane	[40]
	9,12,15-octadecatrienoic acid	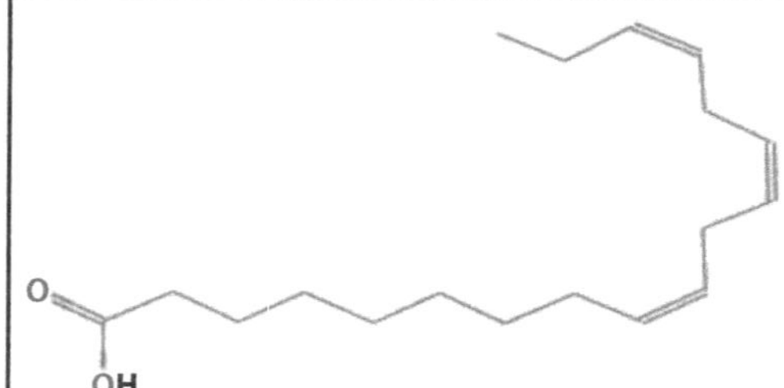	Leaves, stem. Flowers, seed	Hexane	[40]

(*Continued*)

Table 8.4 Bioactive compounds present in Kenaf. (*Continued*)

Class	Compounds	Structure	Solvent used	Plant part	References
	Hexadecanoic acid		Leaves, stem. Flowers, seed	Hexane	[40]
	9-Hexadecanoic acid	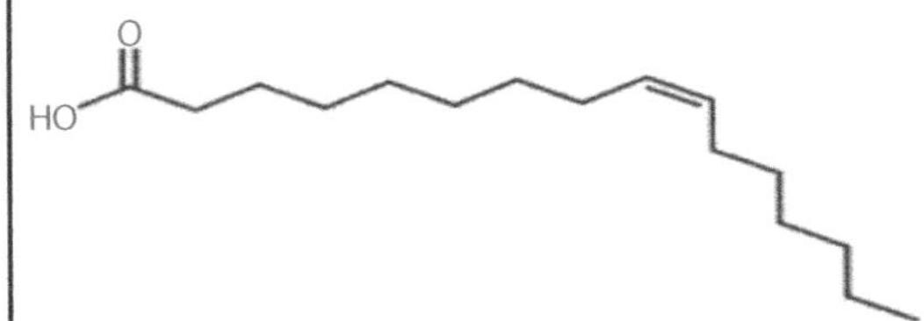	Seed, stem	Hexane	[40]
	Tetradecanoic acid	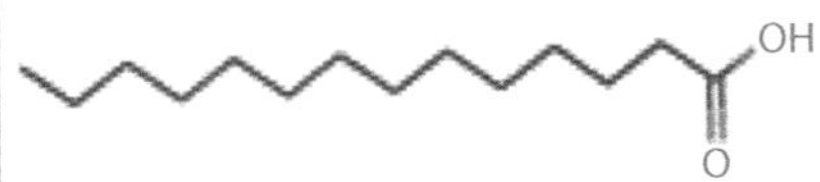	Seed, leaves	Hexane	[40]
	Nonadecadenoic acid	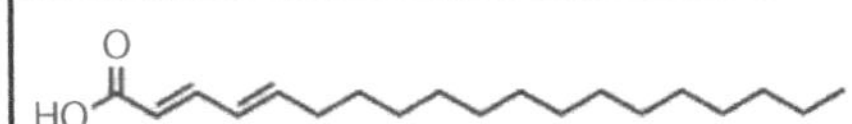	Seed	Hexane	[40]
Terpenes	Phytol	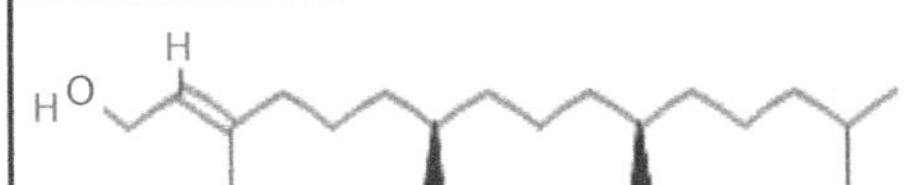	Leaves, stem	Hexane	[40]

(*Continued*)

Table 8.4 Bioactive compounds present in Kenaf. (*Continued*)

Class	Compounds	Structure	Solvent used	Plant part	References
	2-Pentadecanone, 6,10,14-trimethyl	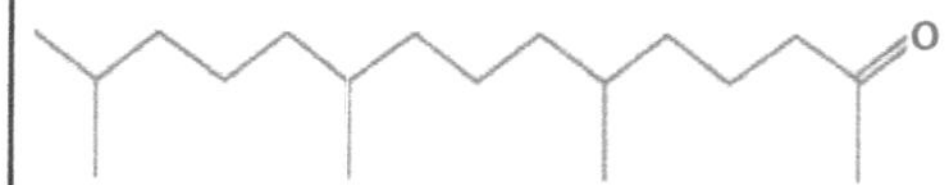	Leaves	Hexane	[40]
	3,7,11,15 tetramethyl-2 hexadecen-1-ol (Zphytol)	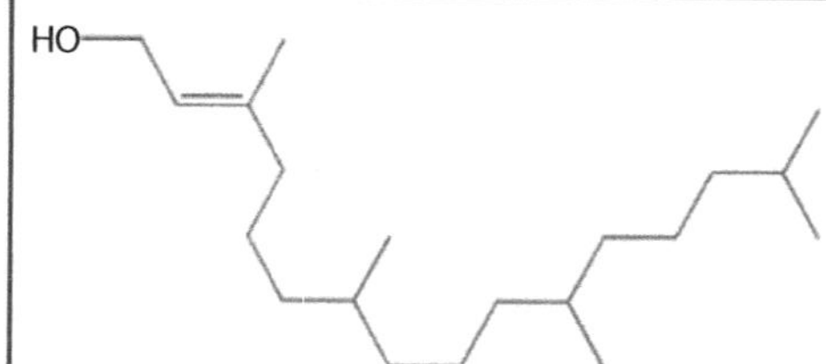	Leaves	Hexane	[40]
	2,6,10,14,18,22-Tetracosahexaene		Leaves	Hexane	[40]
Citric acid derivative	Hydroxycitric acid	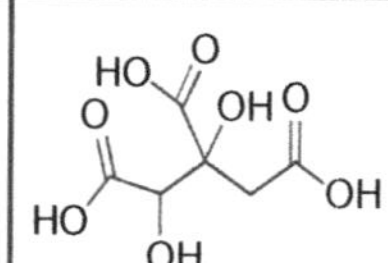	Leaves	Methanol	[77]
	Methoxycitric acid	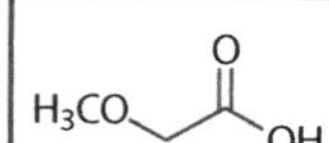	Leaves	Methanol	[77]

(*Continued*)

Table 8.4 Bioactive compounds present in Kenaf. (*Continued*)

Class	Compounds	Structure	Solvent used	Plant part	References
Siloxane	Octasiloxane		Flowers	Hexane	[40]
	Hexasiloxane	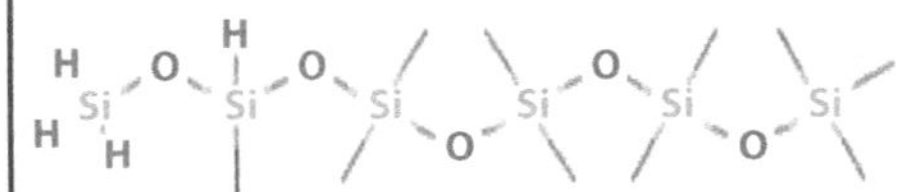	Flowers	Hexane	[40]
	Heptasiloxane	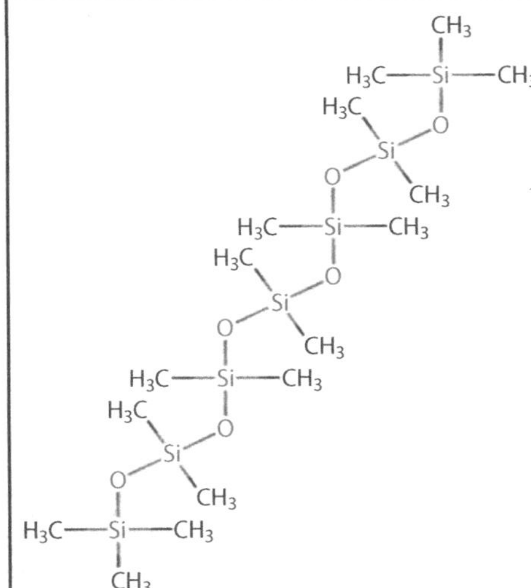	Flowers	Hexane	[40]

(*Continued*)

Table 8.4 Bioactive compounds present in Kenaf. (*Continued*)

Class	Compounds	Structure	Solvent used	Plant part	References
	Trisiloxane	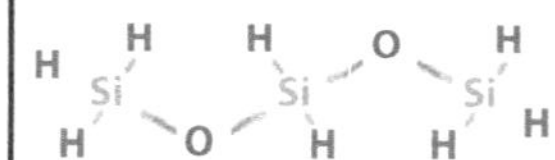	Flowers	Hexane	[40]
	3-Isopropoxy-1,1,1,7,7,7-hexamethyl3,5,5-tris(trimethylsiloxy) tetrasiloxane	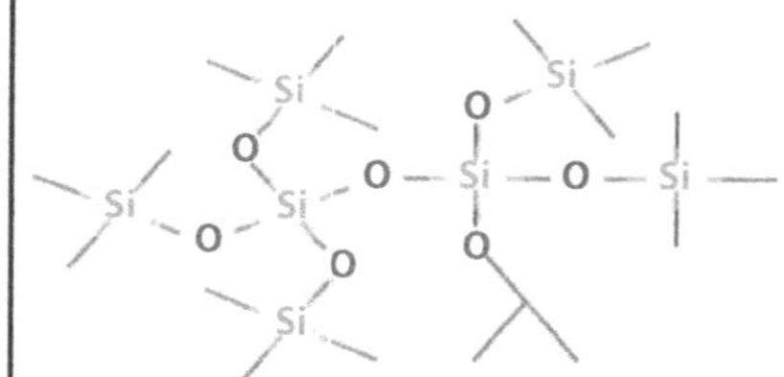	Flowers	Hexane	[40]
Lignans	Boehmenan H	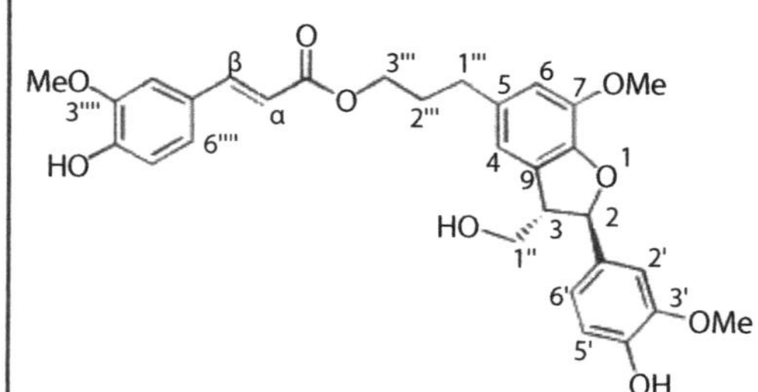	Stem	Light Petroleum and acetone	[80]

(*Continued*)

Table 8.4 Bioactive compounds present in Kenaf. (*Continued*)

Class	Compounds	Structure	Solvent used	Plant part	References
	Boehmenan K	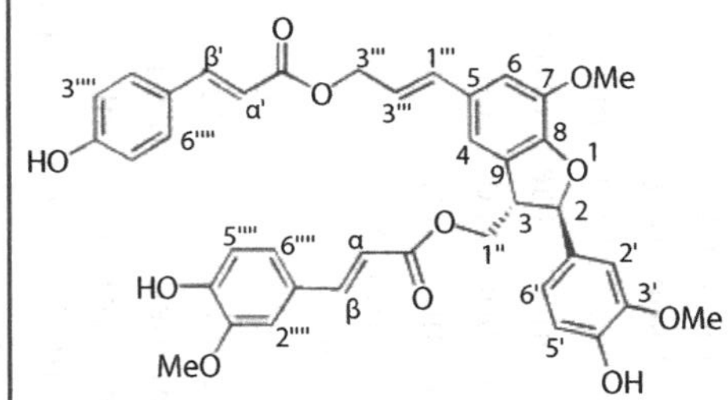	Stem	Light Petroleum and acetone	[80]
	Threo-carolignan H	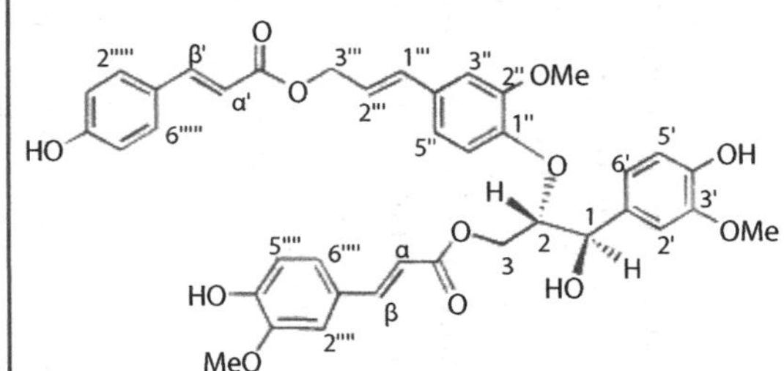	Stem	Light Petroleum and acetone	[80]
	Threo-carolignan K	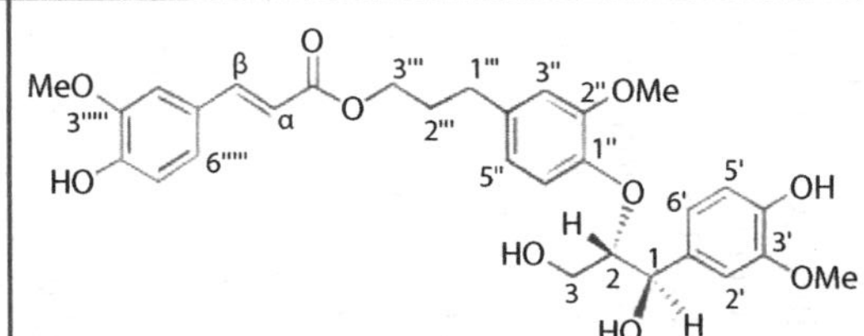	Stem	Light Petroleum and acetone	[80]

Kenaf being rich in phenolic compounds shows the antioxidant activities [40, 49] and can be utilized as source of antioxidants for the food preservation and to enhance food quality. Ryu *et al.* [40] reported that the kenaf leaves containing highest amount of total phenolic content found to be 555.90 mgGAE/100 g. Flowers were having less amount than leaves which accounts for 308.50 mgGAE/100 g, which was followed by seed having 162.70 GAEmg/100 g, and then stem accounts for 84.20 mgGAE/100 g. Most abundant phenolic compound among 12 phenols present in various parts of kenaf was caffeic acid, which accounted for 76.40 mg/100 g in leaves, 51.30 mg/100 g in stem, and 17.70 mg/100 g in flowers. Apart from caffeic acid, tannic acid, and catechin hydrate were also found in kenaf leaves with an amount of 2915.20 mg/100 g and 143.50 mg/100 g, respectively [55]. Jin *et al.* [56]; Abd Gafar [57]; Esmail *et al.* [58] provided that the phenolic and flavonoids in kenaf seed were having potential to act as inhibitor for angiotensin I-converting enzyme and helps in reducing peroxidation of lipids.

Removal of phenolic components from kenaf seeds were done using different solvents and total phenolic content value varied largely, *i.e.*, 2.16 to 18.78 mg GAE/g according to the polarity of solvent used [45]. In this study, aqueous extracts were having highest amount of total phenolic content, i.e., 18.78 mg GAE/g extract in comparison to other solvents used. Significant difference was found in the extracts obtained from water and other solvents ($p < 0.05$). Polar solvents were much appreciated for the extraction purpose as kenaf seed were found to consist of polar compounds. Thus, selecting polar solution increase the efficiency of extraction process. Because of the presence of hydrophilic groups on the phenolic ring in the structure makes the phenolic compounds more soluble in water as water is polar in nature as solvent. Adnan *et al.* [47] studied about the removal of phenolic components from kenaf seed using ethanol as solvent which gave highest yield in comparison to water as solvent. Ethanolic extracts were having highest amount of extracted bioactive compounds, whereas water extract was more suitable for recovery of compounds such as flavonoids, phenols, and maintaining their antioxidant activity. Ethanolic extraction of oil with ultrasound technique attained highest total phenolic content, i.e., 71.02 mg GAE/100 g, which was high in comparison to extraction of oil obtained from ethanol without ultrasound technique, i.e., 66.78 GAE/100 g. Ethanol being able to solubilize and extract both hydrophilic and lipophilic polyphenols make it potent solvent for extraction purpose [59, 60]. In general, enzyme extracted oil was having more polyphenolic content than in hexane/solvent extracted oil [61, 62].

8.6.2 Flavonoids

Flavonoids are considered to be UV-B absorbing compounds, which come under phenolic category with the structural variation having C6-C3-C6 chain. Such activity of flavonoids makes them useful in restricting growth of many bacterial species and viral enzymes such as protease and transcriptase and also has ability to kill some pathogenic protozoans [63]. Kenaf is a source of flavonoids which contains flavonols and flavanols in both, simple and polymeric form [64]. Flavonoid content for kenaf have been observed and till the date only 10 flavonoid compounds have been found in it. Pharmacological value of kenaf was found mainly due to presence of kaempferol which is a potent antioxidant, anti-inflammatory agent, have anti-microbial, anti-cancer and anti- diabetic activity [65].

Ryu *et al.* [40] suggested in a study carried out on leaves, stem and flowers of kenaf that the leaves and stem are rich source of kaempferitrin and found their content 178.2 mg/100 g and 25 mg/100 g, respectively whereas, flowers were found to have myricetin glycoside which is approx. 142.5 mg/100 g. The least flavonoid content was present in stem (299.3 mg/100 g and highest in flower (755.2 mg/100 g). The extract of kenaf was investigated for the presence of phytochemical and observed the presence of flavonoids, phytosterols, and glycosides [66]. Yusri *et al.* [45] analyzed the effect of solvents on the removal of total flavonoid content. The comparative study was done among methanol, water, chloroform, and hexane, in which hexane was found to have minimum value 1.64 mg/g and chloroform was having maximum value 2.94 mg/g. Water extract were also having comparable results with chloroform as 2.49 mg/g and followed by methanolic extract with 1.77 mg/g value. Kenaf flower also consists of coloring/ pigment flavonoids such as anthocyanins or natural pigments which gives color to flower and are used in industries as the source of natural color such as red, blue, purple [67].

8.6.3 Carotenoids

Carotenoids are the natural pigment compounds which are lipophilic in nature and are produced by photosynthetic plants and helps in preventing damages which are mediated by the excess energy on photosynthetic apparatus [68]. With the pigmentation effect these compounds also possesses the antioxidant capacity which helps in protecting cells from stress mediated damage to them, also prevent aging process and other chronic diseases. Prime application of carotenoids is as the natural colorant in food products.

Raju *et al.* [69] study about the proximate analysis of carotenoids in *Hibiscus cannabinus* and reported that kenaf constitutes xanthophyll, which are present as neoxanthin 5.95 mg/100 g, zeaxanthin 0.14 mg/100 g, leutin 33.97 g/100 g, and total xanthophyll 40.06 mg/100 g, provitamin A (β-carotene) 26.02 mg/100 g. β-carotene content was estimated in kenaf seed oil obtained from ultrasonication using different solvents. Highest content was found in ethanolic extracted oil which was 7.75 mg/kg oil, followed by enzymatic extracted oil with water as the solvent having β-carotene content as 2.73 mg/kg oil.

8.6.4 Tocopherols and Tocotrienols

Tocopherols and tocotrienols are the naturally occurring antioxidant present in oil from plant source. α-tocopherol also called vitamin E is a natural antioxidant which reduce the occurrence of chronic diseases by preventing the oxidation of fatty acid and lipids, producing free radical which in turns creates free radial stress in the body. This helps in preventing heart disease, cancer and delay in occurrence of Alzheimer's disease [43].

Kenaf seed oil was observed to have natural antioxidant, i.e., δ-tocopherol with the highest value of 63.86 mg/100 g. Kenaf was found to have α-tocopherol (20.01 mg/100 g) and γ-tocopherol (0.79 mg/100 g) [43]. In a study, tocopherol content was compared with the roselle variety and found that γ-tocopherol was low 5.6 mg/100 g, and bigger values of δ-tocopherol 0.16 mg/100 g and α-tocopherol 3.36 mg/100 than that of roselle seed extract [70, 71]. Adnan *et al.* [47] studied gas chromatography of kenaf leaves extract which shows the presence of vitamin E covering highest peak area 4.45%, below which alpha-amyrin and clionasterol with 3.93%, followed by 2-linoleoylglycerol 3.40%.

Zhang *et al.* [59] provided the data for the presence of tocopherols and tocotrienols in kenaf seed oil. In this investigation, 7 components were reported in which vitamin E (92.38–105.01 mg/100 g) was found highest and this analysis was consistent to the results gained by Mariod *et al.* [72], who provided the data as the highest amount of vitamin E, i.e., 88.20 mg/100 g in kenaf seed oil extracted using super critical fluid extraction. The extraction of seed oil was solvent dependent and resulted in similar composition when extracted using hexane and ethanol. The slight difference was found in the profile and vitamin E content of kenaf seed oil as, ethanolic extract was having higher amount of γ-tocopherol [43, 44, 59]. In the storage study of kenaf seed oil exhibits high amount of γ-Tocopherol, followed by α-tocopherol, β-tocopherol, and δ-tocotrienol when comparison of two category of oil was compared in which one sample was of bulk refined kenaf seed oil and other was microencapsulated oil. δ-Tocopherol, γ-tocotrienol, and α-tocopherol were present in small amounts in seed oil.

8.6.5 Fatty Acids

Kenaf seeds contains ample amount of fatty acid which can be extracted through removal of seed oil. Seed oil of kenaf consists of linoleic acid, oleic acid and palmitic acid as the main fatty acids [26]. Linoleic acid found in highest amount below, which comes the oleic, palmitic, stearic, and palmitoleic acid respectively which as a whole accounted for 96% of total fatty acid profile [59]. Both saturated and unsaturated fatty acids are present in kenaf seed oil which are 21.6–28.2% and 71.8–78.3% respectively. Unsaturated fatty acid is further subdividing as monounsaturated (MUFA) and polyunsaturated fatty acids (PUFA). Seed oil of Kenaf consists of 28.0–44.5% MUFA and 28.8–50.3% PUFA of total unsaturated fatty acid [73]. Seed oil varies with the variation in species of kenaf which in turns variate the amount of fatty acid composition of different variety of kenaf seeds [27]. Kenaf seed oil fatty acid profile consists of 43.32% linoleic acid, 30.35% oleic acid, 22.80% palmitic acid, 2.80% stearic acid, and 0.74% palmitoleic acid. Storage conditions of kenaf seed oil studied to justify the variation in amount of fatty acid with the duration of storage. Decrease in linolenic acid up to 9.57% was found when oil was set at accelerated storage conditions for 24 days. Apart from reduction in linolenic acid, increase in palmitic acid 4.44%, oleic acid 4.25%, stearic acid 0.69%, and palmitoleic acid by 0.18%. The decrease in polyunsaturated fatty acid can be attributed to oxidation of fatty acid, which resulted increased amount of saturated fatty acid. Same results were also detected by [16, 74].

Adnan *et al.* [47] uses chromatographic analysis for the confirmation of fatty acid in kenaf seed oil which showed the presence of 77.46% 9-octadecanoic acid (z), 10.25% Hexadecanoic acid, 9, 12-octadecadienoic acid and other many fatty acids. 15-methylhexadecanoic acid was found only in kenaf flowers extract.

8.6.6 Other Bioactive Compounds

Apart from polyphenols, flavonoids, fatty acids, etc. many more bioactive compounds are present in kenaf in minor percentage. Citric acid derivative, siloxane and terpenes are also present in kenaf where terpene is found in stem and leaves of kenaf. Derivative of terpene, phytol present in kenaf which is utilized as for the manufacturing of synthetic form of Vitamin K1 and vitamin E [75]. Siloxane and hydroxycitric acid were present in kenaf flowers and silicons which are siloxane polymers have potential application as food contact materials [26].

8.7 Pharmacology

Being a promising source of pharmacological components, kenaf shows therapeutic properties and activity to cure various chronic diseases. Epidemiological and clinical research work have identified the retardation in the propagation of chronic illnesses with regular consumption of kenaf. Table 8.5 gives the overview of pharmacological activities of kenaf.

8.7.1 Antioxidant Activity

Free radical stress in the body arises because of the development of reactive oxygen species which oxidizes biological molecules. When antioxidant defense system weakens, free radicals thus developed in the body induces damage to the cell membrane, and molecules such as protein, carbohydrate, DNA. Chronic illnesses such as hypertension, diabetes, heart failure and many pathological situations occur due to oxidative stress [81]. For reducing their effect, artificial antioxidant compounds, such as butylated hydroxy tyrosine (BHT) and butylated hydroxy anisole (BHA), are used which are cheap and highly stable in food system. However, their use in food is limited as these compounds are precursor of cancerous diseases as well as consumers are also demanding for natural additives such as antioxidant, bioactive compounds. Natural antioxidants are better than synthetic ones and have become the first choice of consumers [54]. Kenaf leaves ethanolic extract was found to have *in vitro* antioxidant activity. In different researches, it has been noticed that kenaf leaves effectively reduced the promotion levels of superoxide dismutase and could increase scavenging activity and ferric reducing ability [40]. In another study, a kenaf seed extract was found to have antioxidant activity, demonstrating strong radical scavenging activity through beta-carotene bleaching and a reduction in the formation of hydroperoxides and thiobarbituric acid reactive components in the total antioxidant activity test [45]. Extracts from Hibiscus cannabinus flowers [HCF] obtained using parameters investigated for *in vitro* scavenging activity of free radicals. Investigations regarding activity to prevent oxidative damage of DNA and inhibiting gelatinolytic activity of collagenase types I and II were also performed. Different extract showed 440 to700 μg/ml DPPH free radical scavenging activity and visible similarity found in reducing power activity. With the inference of these two

Table 8.5 Pharmacological properties of kenaf.

Pharmacological properties	Sample type	Model type	Result summary	References
Antioxidant	Ethanolic and water extract of leaves	TPC, DPPH and SOD activity	TPC: 555.9 mg/100 g (water extract), 119.1 mg/100 g (ethanolic extract) DPPH: 73.2% (water extract); 37.7% (ethanolic extract) SOD: 80% (water extract); 58% (ethanolic extract)	[40]
	Different extract of seed, flower, stem	TPC, DPPH and SOD activity	TPC (seed): 52.6-216 mg/100 g TPC (flower): 2.9- 308.5 mg/100 g TPC (stem): 5.6-84.2 mg/100 g DPPH (seed): 5.1-70.4% DPPH (flower): 2-80.6% DPPH (stem): 8.7-299.2% SOD (seed): 40-81% SOD (flower): 3-82% SOD (stem): 2-58%	

(Continued)

Table 8.5 Pharmacological properties of kenaf. (*Continued*)

Pharmacological properties	Sample type	Model type	Result summary	References
Antimicrobial	Ethanolic extract of leaves	MIC and MBC	39.06-10000 μg/mL	
	Water extract of leaves	Disc diffusion method, MIC, MBC	12-12000 μg/mL; 0.98-250 mg/mL	[82, 96]
	Ethanol extract of leaves and seed	Disc diffusion method	**Zone of inhibition against *E. coli*** Leaves: 9.8 mm (ethyl acetate extract); 11.3 mm (water extract) Seed: 13.1 mm (ethyl acetate extract); 13.7 mm (water extract); 15.2 mm (ethanol extract) **Zone of inhibition against *B. cereus*** Leaves: 9.7 mm (water extract) Seed: 11.2 mm (ethyl acetate extract); 12.5 mm (water extract); 12.8 mm (ethanol extract) **Zone of inhibition against *B. subtilis*** Seed: 10.1 mm (ethyl acetate extract); 9.3 mm (water extract); 11.6 mm (ethanol extract)	[47]

(*Continued*)

Table 8.5 Pharmacological properties of kenaf. (*Continued*)

Pharmacological properties	Sample type	Model type	Result summary	References
Anticancerous	80% ethanol, Hexane	MTT; SRB (HeLa, MCF-7, HCT-116, SK-LU1),	Seed extract showed lower IC_{50} than seed oil. Experimental dose of 15.63-1000 μg/mL.	[86] [50]
	Hexane	MTT, Trypan blue dye exclusion method (CaOV3)	All seed oil showed cytotoxicity against ovarian cancer and colon cancer cell line. Oil from sonication was most toxic towards ovarian cancer. Experimental dose of 50-300 μg/mL.	
	CO_2	*In vivo*-mice Immunofluorescence staining; Histopathological evaluation	Population of T cells increased and monocytes qand granulocytes population decreased in peripheral blood of mice with kenaf seed oil treatment. Weight of spleen and liver also decreased	[87]
	CO_2	*In vivo* rats	Experimental dose of 1-5 g/kg Aberrant crypt foci reductions compared with the untreated group were 45.3, 51.4 and 53.1% in rats fed with 500, 1000 and 1500 mg/kg body weight, respectively. No significant difference was seen in weight of rats	[57]

(*Continued*)

Table 8.5 Pharmacological properties of kenaf. (*Continued*)

Pharmacological properties	Sample type	Model type	Result summary	References
Anti-inflammatory	Methanolic and water leaf extract	Rats (*in vivo*)	Both extract showed inhibition of raw edema in dose-dependent manner. Maximum inhibition (52%) was seen in methanolic extract at dose of 400 mg/kg Lipid peroxidation level was 31.10 nmol/g (Methanolic extract) and 35.23 nmol/g (water extract)	[89]
	80% ethanol, Hexane	Rats (*in vivo*)	Seed extract showed anti-inflammatory effect on edema induced rats at a dose of 500 mg/kg	[97]
Hepatoprotective	Aqueous leaf extract	*In vivo*- rats AST; ALT; bilirubin level; lipid peroxidation; Histopathological evaluation	The concentration of plasma transaminases and bilirubin was significantly reduced. Absence of necrosis in pre-treated liver cells Experimental dose: 1.6 g/kg	[92]

(*Continued*)

Table 8.5 Pharmacological properties of kenaf. (*Continued*)

Pharmacological properties	Sample type	Model type	Result summary	References
Antihypolipidemic activity	50% alcoholic extract	*In vivo*-rats	Significant decrease in total cholesterol, triglycerides, low density lipoprotein (LDC-C), very low density lipoprotein (VLDC-C), TBARS	[95]
	Seed (80% ethanol)	*In vivo*-rats Body and liver weight; Serum lipid profile; AI; CRI; MDA; Histopathological evaluation	Experimental dose: 400 mg/kg Seed extract showed higher efficiency of cholesterol lowering activity than seed oil at a dose of 400 mg/kg.	[94]
	Seed (Hexane)	*In vivo*-ratsBody and liver weight; Serum lipid profile; AI; CRI; GSH; MDA; TBA; Histopathological evaluation	Kenaf seed oil-in-water nanoemulsions showed highest effect on cholesterol lowering effect, weight control and decrease in liver fat.	[98]

activities, HCF can be considered as potential antioxidant to combat radical stress in body. Extracts with concentration 100 μg/ml, were able to provide protection against oxidative damage of DNA and showed inhibition of gelatinolytic activity of collagenase type I and type II up to 87% and 65% respectively [82]. Presence of high total phenolic content and total flavonoid content give kenaf strong antioxidant activity, implying that it could be used as a source of health-promoting compounds in the future.

8.7.2 Antimicrobial Activity

Resistance strain of pathogenic bacteria is increasing and causing increased rate of mortality throughout the world. For which there is requirement for the production of improved and novel antimicrobial agents [83, 84]. For such microbial development, medicinal herbs are coming in demand with better health outlooks and free from the adverse effects due to synthetic chemicals. Ethanolic and aqueous extracts of *Hibiscus cannabinus* leaves [120,000–12 μg/10 ml] were investigated for antibacterial effects against *Salmonella typhimurium*. The extracts obtained gave distinct activity; the growth inhibition areas ranged between 12.67±1.52 and 6.67±1.15 mm for the aqueous extract and 12.33±2.08 to 6.33±0.58 mm for the ethanol extract [82]. In another study, kenaf leaves extract exhibit effective antimicrobial activity against *Bacillus subtilis, Staphylococcus aureus, Bacillus cereus, Staphylococcus epidermis, Salmonella,* and *Pseudomonas* with MIC value156.25 μg/mL, 625 μg/mL, 10000 μg/mL, 10000 μg/mL, 5000 μg/mL, 5000 μg/mL respectively. Antibacterial activity of chloroform extract against E. coli [10, 8 and 10 mm at concentration of 10, 20 and 30 μl], against Klebsiella spp. [12 mm at concentration of 10 and 30 μl], against Pseudomonas Sp. [14 and 12 mm at concentration of 20 and 30 μl] and against Staphylococcus spp. [11 mm at concentration of 30 μl] [85]. The studies are not sufficient now on the antimicrobial activity of kenaf seeds.

8.7.3 Anticancer Activity

Cancer is the one of the leading chronic diseases which had been a cause of mortality worldwide. Surgery, chemotherapy and radiation, etc. are used for the treatment of cancer apart from conventional drugs. Being expensive, these treatments also induce many side effects to the body. Hence, to reduce the adverse effects of drugs, medicinal natural sources like plants, their leaves, fruits etc. are focused by scientific community because of the occurrence of highly effective phytochemicals. Bioactive components present in plant are potential antioxidant, which could be associated with their ability to inhibit different cancerous diseases. In a study, Wong *et al.* [86] showed MCF-7 (breast cancer), CCL-2 (HeLa), SK-LU1 (lung cancer) and HCT-116 (colon cancer), after kenaf seed extract (KSE) and kenaf seed oil (KSO) treatment at 72 h, exhibited signs of apoptosis under an inverted light microscope in different ways as membrane blabbing, cellular shrinkage and apoptotic body formation. KSE had lower IC50 value than KSO in cancer cell tissue culture. Supercritical CO2 extraction aqueous substance was utilized for the removal of kenaf seed oil and was analyzed for the cytotoxicity using MTS assay with regards to human colorectal cancer cell lines [HT29] and mouse embryonic fibroblast [NIH/3T3] cell lines. KSO with IC50 of 200 μg/ml exhibited strongest cytotoxicity towards HT29. A substantial rise in the accumulation of KSO-SFE-treated cells at sub-G1 phase was observed using cell cycle analysis, which shows initiation of apoptosis, by kenaf seed oil [57]. In another report done by Foo *et al.* [87] provided that treatment with KSO reduced the development of immature monocytes, lowering spleen and liver weights, the harshness of leukemia in WEHI-3B cells by increasing the growth of cytotoxic T cells to kill the leukemia cells. Not only oil, lignan extracted from kenaf too found to have cytotoxic activity against Hep-2, HeLa, and A-549 cell lines. When cellular division was advanced stage, one compound showed moderate activity on HeLa cells [88].

8.7.4 Anti-Inflammatory Activity

Morbidity and reduction in work force is highly due to the Diseases caused by inflammation. This inflammation problem can be cured by the intake of steroidal and non-steroidal drugs, but consumption of these drugs has toxic effect on the human organs. In many studies, kenaf have been studied for anti-inflammatory activity as a natural source for the cure apart from commercial drug. Kenaf leaf extract significantly ($p < 0.05$) showed inhibition of rat paw edema in dose-dependent manner, but not in much amount to be compared to standard control- indomethacin [89]. KLE has potential anti-inflammatory activity which can be tested by reduction in lipopolysaccharides-induced nitric oxide manufacturing in murine macrophage RAW 264.7 cell which is much dependent on the dose of extract [90]. Kenaf seed extract also exhibited anti-inflammatory activity by inhibiting 83.33% in the arachidonic acid induced paw edema, 73.18% histamine induced paw edema, 74.72% of inhibition in the carrageenan-initiated paw edema and after 5 h injection [26]. Ethanolic removal of Hibiscus cannabinus fresh leaves notably subdued TNF-alpha production and the mRNA expression of interleukin [IL]-3 and IL-12 in the RAW264.7

cells, stimulated by lipopolysaccharide [LPS, 2.5 microg/ml]. Ethanolic extract also reduced the secretion of inflammatory mediators like nitric oxide, reactive oxygen species and prostaglandin E2. Expression of heme oxygenase-1 [HO-1] mRNA which is a potent cytoprotective molecule was induced by extract. Expression of co-stimulatory molecules [CD80 and CD86] of LPS-activated RAW264.7 cells and the phagocytic uptake were suppressed by extract. The extract also down-regulated both the functional activation of beta1-integrin [CD29] and the LPS-induced up-regulation of the surface CD29 level [91].

8.7.5 Hepatoprotective Activity

When the functioning of liver in the body alters, hepatic damage is considered as the prime reason behind it. Liver functioning includes regulation of secretion, detoxifying, metabolism and storage. Hepatic diseases are a global problem and among the main dangers to public health. Agbor *et al.* [92] investigated about the safe effects of aqueous kenaf leaves extract which can protect liver from the oxidative damage initiated by paracetamol and carbon tetrachloride (CCl4). KLE treatment exhibited significant decrease in the plasma alanine aminotransferase (ALT), aspartate aminotransferase (AST), bilirubin level and lipid peroxidation in rodents with induced liver damage. In comparative histopathological study between the control and the KLE treated rats, less inflammation was found in KLE treated rats. Kenaf leaves showing hepatoprotective effect could be associated with inhibitory action against cytochrome P450 enzyme, encouragement of glucoronidation activity and natural antioxidant components in KLE that can scavenge the CCl4 and *N*-acetyl-p-benzoquinone-imine electrophilic metabolite [93].

8.7.6 Hypolipidemic Effect

Elevated serum lipid in terms of total cholesterol (TC), low density (LDL-C), very low-density lipoprotein cholesterol (VLDL-C) and decreased in high-density lipoprotein cholesterol (HDL-C) is the chronic disease named as Hyperlipidemia. Hyperlipidemia is also the precursor of many more diseases such as arteriosclerosis, hypertension and heart disease. Study on protective effect of kenaf against hyperlipidemia, showed a higher antihyperlipidemic effect at low concentration it demonstrated effect similar to the control. 400 mg/kg KLE can markedly stop the liver micro vesicular steatosis and showed comparable antihyperlipidemia effects with the standard drug-atorvastatin. Kenaf seed oil nano emulsions also exhibit the cholesterol-lowering qualities (decreased serum TC, TG, LDL-C, atherogenic index (AI), and coronary risk index (CRI)), weight control and reduced liver fat [98]. Alike effects on hepatic lipid peroxidation (MDA) and hepatic decreased glutathione (GSH) in liver tissues were found in KSON and simvastatin treated groups. Apart from seed oil, seed extract also shows similar effects with the commercial hypocholesterolemia drug [94]. 50% hydro alcoholic extract of Hibiscus cannabinus leaves were investigated for the hypolipidemic effect in rat model giving them high fat feed. The extract showed a strong dose dependent anti-hyperlipidemic activity and at dose level of 400 mg/kg po, there was decrease in serum level (TC, TG, LDL-C, VLDL-C and TBARS) also observed. The extract also stopped the liver micro vesicular steatosis in hyperlipidemic rats [95].

8.8 Health Benefits

The biological dysfunctions or by products of various cellular metabolisms in the living cells generates harmful pro-oxidants and reactive oxygen species (ROS) which are the reason for oxidative damage to cellular constituents and cause various chronic diseases like cancer, neurodegenerative disease, diabetes mellitus and various chronic diseases [99–101]. The various anti-oxidants acts as bacteriostatic and bactericidal agents and its therapeutic applications can decrease or reverse various symptoms related with neurodegenerative diseases, like Alzheimer's disease (AD), Parkinson's disease (PD) and spongiform encephalopathies [102–104]. Therefore, plants that possess high anti-oxidant compounds like phenolics, flavonoids and other bioactive compounds must be checked for various anti-oxidative properties. Kenaf (*Hibiscus cannabinus*) is a good source of various bioactive compounds and other natural oxidants [47]. Other than high anti-oxidant compounds kenaf plant extracts have more quantity of monounsaturated and polyunsaturated fatty acids [27, 105]. Thereof bioactive phytochemicals in kenaf seeds have a great nutritional, phyto-pharmaceutical and therapeutic values [35]. Kenaf fruits have significant effects in lowering blood pressure by reducing oxidized lipids in the blood. Kenaf is having very high vitamin C, antioxidants and phytochemicals content which is useful in treating various diseases like atherosclerosis, liver disease, cancer and diabetes [4]. Kenaf seed oil also possesses different biological properties like anti-hypercholesterolemic, anti-oxidation, anti-cancer, anti-inflammatory and others because of the presence of phytochemicals [36]. The MUFA and PUFA content in kenaf seed oil is very high which provides health benefits in cardiovascular health [106]. The various compounds like phytosterols, flavonoids, polyphenols, tannins, steroids, alkaloids, saponins, lignans, essential oils, glucosides like cannabiscitrin, cannabiscetin, and anthocyanin glycoside has been identified by phytochemical analysis of *Hibiscus cannabinus* and further various health effects like cytotoxic, anthelmintic, antibacterial, antiulcer, antidiabetic, hypolipidemic, antioxidant, immunological, hematinic and hepatoprotective effects has been reported [58]. The details of various health effects in different plant parts of kenaf reported by various researchers has been given in Table 8.6.

8.9 Industrial Use

The first use of kenaf fiber in history is as rope, burlap cloth and fish net. The resistance of kenaf fibers towards rot and mildew made its best use in the above applications [111]. In modern era, the industrial uses of kenaf includes as food use for humans, feed use for animals, fuel wood, fiber use in pulp and paper industry, textile industry, bio-energy production, phytoremediation, medicinal and cosmetic industry (Table 8.7). The fiber extracted from kenaf has multiple applications in various fields like textile and packing substances, pulp and paper industry, composite media, animal feed and in potting, building, filtration materials and as a source of biomass energy [112]. The various industrial uses of kenaf plant had been described in detail in following sub heads.

8.9.1 Food Use

Oily kenaf seeds are utilized for the production of edible oil (16-22 % oil content) which has healthful benefits because of high amount of MUFA and PUFA. The other uses include margarine manufacturing and can also be utilized for cooking purpose in form of flour [22, 27, 105]. The leaves are used as a vegetable in some parts of the world, and adding dried kenaf leaf powder to various foods has been shown to increase fiber and calcium content [6, 92]. Kenaf seed have vast potential in food applications because of its various functional properties of its protein concentrate [33]. Defatted kenaf seed meal (DKSM) with high nutritional content (57.09% carbohydrate; 26.19% protein and 6.65% ash) can be utilized as edible flour and which is further used in various food preparations [49]. Seeds of Kenaf have also been used as an ingredient in making vegetable-based protein substitute like tempeh and texturized vegetable protein [113]. The defatted seed meal has been used in the production of noodles, pasta and as an

Table 8.6 Health benefits of different parts of kenaf plant.

Plant part	Medicinal property	Treatment of disorders	References
Kenaf whole plant	• Anti-tumor compounds • Antioxidant, cardio-protective, anti-inflammatory, anti-hypertensive and anti-proliferative activities • Anti-anemic, antihyperlipidemic, antidiabetic, anthelmintic, hepatoprotective, haematinic, antiulcer, anti–hypercholesterolemic, antibacterial, glycosylation inhibition	• Anodyne, aperitif, aphrodisiac, as well as fattening, purgative and stomachic • Blood and throat disorders, bilious conditions, fever and puerperium • As atherosclerosis, liver disease, cancer and diabetes	[92, 107, 108]
Leaves	• Anticancer, antioxidants, analgesic, anti-inflammatory, aphrodisiacs, and hepatoprotective activities • Aperitif, aphrodisiac, fattening, purgative • Antimicrobial activity • Ayurvedic medicine	• Dysentery • Throat disorders. • Artherosclerosis • Prevention and treatment of scurvy, • Treatment of jaundice • Applied to sores and boils • Treatment of cough • Antimicrobial activity against *Escherichia coli*, *Proteus mirabilis* and *Shigella flexneri*	[26, 31, 37, 91, 109]

(*Continued*)

Table 8.6 Health benefits of different parts of kenaf plant. *(Continued)*

Plant part	Medicinal property	Treatment of disorders	References
Seeds	• Anticancer, antioxidative, analgesic, anti-inflammatory, aphrodisiacs, and hepatoprotective activities	• Biliousness and bruises	[31, 86]
Flower or juice	• Hepatoprotective	• Biliousness and bruises • Emollient and infusion of the petals used as a demulcent. • Decoction given in bronchial catarrh	[6]
Stem	• Hematinic agent • Cytotoxic effect	• Treat fatigue and anaemia • Anaemia treatment in Africa	[26]
Defatted kenaf seed meal	• Antioxidant and anti-inflammatory properties, cardio protective functional food	• Hypercholesterolemia-associated oxidative stress and systemic inflammation.	[29, 110]

Table 8.7 Various uses of Kenaf.

Application	Functional use	References
Food use	• Edible oil (cooking oil) and margarine production. • Dry kenaf leaves powder: ingredient in food preparations • Leaves and flowers: vegetables and food supplements • Calyx: Health drinks	[6, 22, 31, 35, 38, 117]
Feed use	• For ruminant livestock on pasture (24% to 35%) • Dried kenaf leaves: livestock feed (30% crude proteins) • Kenaf seeds: protein supplement	[31, 32, 117, 134]
Fiber use	• Used as wood substitute • Good source of cellulosic fiber and used for manufacturing of various paper products • Used in making specialty papers: security paper, tea bags, currency notes, newsprint, bond, coating raw stock and surfaced sized paper. • Long fibers: manufacturing high grade pulps in which are used for packing for horticultural commodities • Short fiber/core fiber: manufacturing animal bedding • Plastic compounds from kenaf: Replacement of glass-reinforced plastics in automotive industry, food packaging and construction/housing.	[6, 31, 35, 117, 135, 136]
Bioenergy production	• Production of bioenergy: bioethanol, biohydrogen • Used as biofuel in the USA and China • Production of bio-ethanol of second generation • Used as lubricants and biofuel: alternative towards mineral oils	[35, 36, 120, 121]

(*Continued*)

Table 8.7 Various uses of Kenaf. (*Continued*)

Application	Functional use	References
Textile	Kenaf textiles: absorbent and fire-retardant used in manufacture of outerwear or shoes Woven into fabrics Southeast Asia and other tropical countries: used as textile fibre	[31]
Phyto-remediation	Phytoremediation of contaminated soil. Improvement of ecological and economic values of marginal lands Removal of excess amount of Zinc and Copper from the soils Economic and environmental benefits in the areas of soil remediation Reduced soil erosion due to wind and water	[119, 125, 126]
Oil and chemical absorbents	Used to clean-up of toxic waste, oil spills and chemicals	[119]
Medium for mushroom cultivation	Japan and China: commercially used in mushroom cultivation in Kenaf potting soil: as a substitute for peat moss	[8, 132]
Environment cleaning	Absorb CO_2 and NO_2 from the atmosphere (3 to 5 times faster than forests) Highest carbon monoxide (CO) absorption capacity, (1.5 tons of atmospheric CO absorption/ton of kenaf)	[137]
Cosmetic industry	Kenaf Seed Oil (KSO): As sunscreen formulation Cosmetic formulation development (vitamin E gel-cream prepared by 9% KSO)	[36, 133, 138]

ingredient in cookery [114]. Dried seeds crushed into powder form are used to prepare a beverage which can be a substitute to the coffee as the preparation is carried out in similar way [115]. Kenaf leaf aqueous extract can be used for the development of RTS beverage (standard °Brix: acid ratio 15 °B and 0.15% acidity) which has functional food properties [116].

8.9.2 Feed Use and Bio-Energy Production

Kenaf leaves can also be utilized as a feed for the livestock because of its high crude protein content (up to 30 %) [36, 117]. The seeds are also rich in protein (24% to 35%) and has been used as protein supplement for livestock. In addition to that it also acts as a rich source of roughage which helps in keeping normal functions of rumen along with other health benefits [32]. The top leafy portion can be used as forage for the livestock feed and an also be used in ensiled form [118, 119]. The high cellulose content and low lignin content makes the Kenaf a promising eco-friendly and economically feasible feedstock for producing bio-fuels like bioethanol, biodiesel, biogas, bio hydrogen and average biomass yield from kenaf is approximately 12.35 to 18.53 t/ha [120]. In term of production kenaf can produce 61.4 gal of biofuels/ton biomass which has a calorific value of 18,000 kJ/kg of kenaf [120]. Kenaf seed oil has also wide applications in industry as lubricants and biofuel which is an alternative to mineral oils due to its good lubricating properties [121]. The properties and effects of seed oils are comparable to the conventional lubricants and due to its biodegradable nature seed oil is also more environmentally friendly than mineral oil [122].

8.9.3 Pulp, Paper, and Textile Industry

Hibiscus spp., is one of the most important commercial crops for the extraction of natural fiber after cotton and jute [35]. Two types of fibers are present in kenaf plant: long fibers and short fibers which are removed from bast and core respectively. Pulp extracted from kenaf can be used for manufacturing the paper of low (core fiber) and high quality (bast fiber) which are further used for printing, writing as well as for wrapping and packaging purposes [123, 124]. Kenaf fiber's application in the manufacturing of a wide range of paper products is increasing gradually due to various benefits such as less energy and chemical inputs for processing. The use of kenaf fiber has gotten a lot of attention around the world because of environmental concerns (artificial fiber causes long-term pollution) and enhanced paper consumption. The short fibers are used to manufacture animal bedding and horticultural mixes whereas long bast fibers are used for manufacturing high-grade pulps which is further used as protective packing for horticultural produce [6]. The kenaf pulp can be used for making high quality kenaf fiber and the paper manufactured from this pulp is stronger, whiter, with better ink adherence than wood paper [13, 117]. Kenaf fibers blended with cotton is made into yarn, woven into fabrics, and used for cordage, woven, and non-woven textile products. The fire retarding properties of kenaf textiles makes it ideal for manufacturing shoes.

8.9.4 Phytoremediation

Kenaf is used in phytoremediation process of contaminated soil or substrates and the ecological and economical values of marginal lands can be improved with the cultivation practices of Kenaf [125]. The excess amount of zinc and copper in soils can be removed by growing Kenaf) and later on the harvested plant biomass is decomposed [126]. The kenaf plants act as bio remedial tool for removal of selenium and boron from contaminated soil as these minerals are also accumulated by Kenaf plant. The other properties like rapid growth, large biomass, stress resistance, and adaptability makes it a suitable plant for bioremediation [127]. Like papyrus and sorghum, a very high rate of nitrogen and phosphorus removal as 1.3-1.7 g $m^{-2} d^{-1}$ and 0.24-0.26 g $m^{-2} d^{-1}$ has been reported [128]. Kenaf is also used as oil and chemical absorbents due to the high absorption capacity of its fiber and used in removing oil spills in Gulf of Mexico in 2010 [129]. The porous nature of kenaf core makes it a suitable material for absorbing urine and odors which are used for wipes and hygienic products, filtration system and bedding for livestock and poultry [130, 131].

8.9.5 Other Uses

The kenaf has been considered as important medicinal crop and used as a folk medicine for treating metabolic diseases like atherosclerosis, liver disease, cancer and diabetes [4]. Kenaf potting soil acts as substitute for peat moss and kenaf core with wood powder is commercially used as plant medium in mushroom cultivation [8, 132]. Kenaf seed oil has been used in cosmetic industry as sunscreen formulation and cosmetic formulation development [133].

8.10 Conclusion and Future Prospectus

Kenaf (*Hibiscus cannabinus*) plant, leaves, seeds, and its derivative are known to be rich source of phytochemicals as various studies showed it by various analyses. The extracts of this plant and its various parts have a good amount of polyphenolic compounds, flavonoids, lignins, tocopherols, tocotrienol, anthocyanins, and other bioactive compounds and having good nutritional quality in terms of carbohydrates, proteins and seed oil with high amount of MUFA and PUFA added more value to nutrition. Having these phytochemical compounds lead to provide not only antioxidant activity but also give pharmacological benefits including antiproliferative, analgesic, anti-cancerous and anti-inflammatory effect.

Future studies must be focused to increase value of plant in terms of its uses to not only in food uses as to increase product value with enhanced flavors as well as to cure chronic diseases but also in fiber production as fibers produced from this plant could be used to replace jute and used in textile, pulp, and paper industry, medicinal, bio-energy production, phyto-remediation, and cosmetic industry.

References

1. Dempsey, J.M., *Fiber Crops*, pp. 203–304, The University Presses of Florida, Gainesville, 1975.
2. Cheng, Z., Lu, B.R., Sameshima, K., Fu, D.X., Chen, J.K., Identification and genetic relationships of kenaf (*Hibiscus cannabinus* L.) germplasm revealed by AFLP analysis. *Genet. Resour. Crop Evol.*, 51, 4, 393–401, 2004.
3. Wilson, F.D. and Menzel, M.Y., Kenaf (*hibiscus cannabinus*), roselle (*hibiscus sabdariffa*). *Econ. Bot.*, 18, 80–91, 1964.
4. Webber III, C.L. and Bledsoe, V.K., Kenaf yield components and plant composition, in: *Trends in New Crops and New Uses*, pp. 348–357, 2002.
5. Li, D., Kenaf production, research and development in China. *International Kenaf Symposium*, TN, USA, 2002.
6. Ayadi, R., Hanana, M., Mzid, R., Hamrouni, L., Khouja, M.I., Hanachi, A.S., *Hibiscus cannabinus* L. E kenaf: A review paper. *J. Nat. Fibers*, 14, 4, 466–484, 2017.
7. Mazumder, B.B., Ohtani, Y., Cheng, Z., Sameshima, K., Combination treatment of kenaf bast fiber for high viscosity pulp. *J. Wood Sci.*, 46, 5, 364–370, 2000.
8. Cheng, Z., Kenaf research, products and applications in Japan (in Chinese). *Plant Fibers Prod.*, 23, 3, 16–24, 2001.
9. Webber, C.L., Bhardwaj, H.L., Bledsoe, V.K., Kenaf production: Fiber, feed and seed, in: *Trends in New Crops and New Uses*, J. Janick and A. Whipkey (Eds.), pp. 327–339, ASHS Press, Alexandria, VA, 2002a.
10. Stricker, J.A., Prine, G.M., Riddle, T.C., *Kenaf–A Possible New Crop for Central Florida*, University of Florida Cooperative Extension Service. Institute of

Food and Agricultural Sciences, Gainesville, Florida, 2001, On the Internet: http://www.edis.ifas.ufl.edu/BODY_AA220.
11. Summerscales, J., Dissanayake, N.P., Virk, A.S., Hall, W., A review of bast fibres and their composites. Part 1–Fibres as reinforcements. *Compos. Part A Appl. Sci. Manuf.*, 41, 10, 1329–1335, 2010.
12. Abdul Khalil, H.P.S., Ireana Yusra, A.F., Bhat, A.H., Jawaid, M., Cell wall ultrastructure, anatomy, lignin distribution, and chemical composition of Malaysian cultivated kenaf fiber. *Ind. Crops Prod.*, 31, 113–121, 2010.
13. Tahir, P.M., Ahmed, B.A., Saiful Azry, S.O.A., Ahmed, Z., Retting process of some bast plant fibers and its effect on fiber quality: A review. *Bio Res.*, 6, 4, 5260–5281, 2011.
14. Ashori, A., Harun, J., Raverty, W., Yusoff, M., Chemical and morphological characteristics of Malaysian cultivated kenaf (*Hibiscus cannabinus*) fibre. *Polym - Plast. Technol. Eng.*, 45, 1, 131–134, 2006.
15. Alexopoulou, E., Papatheohari, Y., Christou, M., Monti, A., Origin, description, importance, and cultivation area of kenaf, in: *Kenaf: A Multi-Purpose Crop For Several Industrial Applications*, pp. 1–15, Springer, London, 2013.
16. Ng, S.H., Tahir, P.M., Mohamad, R., Abdullah, L.C., Choo, A.C.Y., Liong, Y.Y., Effect of pretreatment process on bioconversion of kenaf (*Hibiscus cannabinus* L.) core to glucose. *Bioresources*, 8, 2, 2010–2017, 2013.
17. http://www.jatrophaworld.org/kenaf_94.html#:~:text=a%20pure%20crop., Kenaf%20is%20an%20important%20fibre%20crop%20in%20South%20India.,small%20patches%20in%20garden%20lands.&text=Kenaf%20cultivation%20requires%20specific%20climate%20and%20land.
18. FAO, *Jute, Kenaf, Sisal, Abaca, Coir and Allied Fibred Statistics December 2018*, Food and Agriculture Organization of the United Nations, Rome, 2019.
19. Carberry, P., Muchow, R., Williams, R., Sturtz, J., McCown, R., A simulation model of kenaf for assisting fibre industry planning in northern Australia. I. General introduction and phenological model. *Crop Pasture Sci.*, 43, 1501–1513, 1992.
20. Angelini, L., Macchia, M., Ceccarini, L., Bonari, E., Screening of kenaf (*Hibiscus cannabinus* L.) genotypes for low temperature requirements during germination and evaluation of feasibility of seed production in Italy. *Field Crop Res.*, 59, 73–79, 1998.
21. Mosley, J. and Baldwin, B.S., The effects of tillage treatment on the growth and development of kenaf, in: *Proceedings of the 2nd Annual Conference of the American Kenaf Society*, pp. 41–47, 1999.
22. Patanè, C. and Sortino, O., Seed yield in kenaf (*hibiscus cannabinus* L.) as affected by sowing time in South Italy. *Ind. Crop Prod.*, 32, 381–388, 2010.
23. Brown, R.C. and Brown, T.R., *Biorenewable Resources: Engineering New Products From Agriculture*, John Wiley & Sons, New York, NY, 2014.
24. Cook, C.G., Effect of soil salinity on kenaf performance in California, in: *Annual Conference Proceedings*, vol. 1, The Society, p. 21, 1998.

25. Whitely, E.L., Kenaf, hibiscus cannabinus, in: *CRC Handbook of Biosolar Resources, vol. II. Resource Materials*, T.A. McClure, and E.S. Lipinsky, (Eds.), pp. 259–268, CRC Press, Inc., Boca Raton, FL, 1981.
26. Sim, Y.Y. and Nyam, K.L., *Hibiscus cannabinus* l. (kenaf) studies: Nutritional composition, phytochemistry, pharmacology, and potential applications. *Food Chem.*, 344, 128582, 2021.
27. Coetzee, R., Labuschagne, M.T., Hugo, A., Fatty acid and oil variation in seed from kenaf (*Hibiscus cannabinus* L.). *Ind. Crops Prod.*, 27, 1, 104–109, 2008.
28. Bakhtiari, M.R., Ahmad, D., Othman, J., Ismail. Physical, N., and mechanical properties of kenaf seed. *Appl. Eng. Agric.*, 27, 2, 263–268, 2011.
29. Chan, K.M., Ismail, M., Esa, N.M., Alitheen, N.B.M., Imam, M.U., Ooi, D.J., Khong, N.M.H., Defatted kenaf (*hibiscus cannabinus* L.) seed meal and its phenolic-saponin-rich extract protect hypercholesterolemic rats against oxidative stress and systemic inflammation via transcriptional modulation of hepatic antioxidant genes. *Oxid. Med. Cell. Longev.*, 2018, 1–11, 2018.
30. Webber, C.L. and Bledsoe, R.E., Kenaf: Production, harvesting, processing and products. *Second National Symposium on New Crops; Exploration, Research and Commercialization,* John Wiley and Sons, Inc., Indianapolis, IN, New York, 1993.
31. Xu, J., Tao, A., Qi, J., Wang, Y., Bast fibres: Kenaf, in: *Handbook of Natural Fibres*, pp. 71–92, 2020.
32. Oh, S., Mbiriri, D.T., Ryu, C., Lee, K., Cho, S., Choi, N.J., *In vitro* and *in vivo* evaluation of kenaf (*hibiscus cannabinus* L.) as a roughage source for beef cattle. *Asian Australas. J. Anim. Sci.*, 31, 10, 1598–1603, 2018.
33. Ibrahim, S.G., Karim, R., Saari, N., Abdullah, W.Z.W., Zawawi, N., Razak, A.F.A., Umar, R.A., Kenaf (*hibiscus cannabinus* L.) seed and its potential food applications: A review. *J. Food Sci.*, 84, 8, 2015–2023, 2019.
34. Mariod, A.A., Fathy, S.F., Ismail, M., Preparation and characterization of protein concentrates from defatted kenaf seed. *Food Chem.*, 123, 3, 747–752, 2010.
35. Dhar, P., Kar, C.S., Ojha, D., Pandey, S.K., Mitra, J., Chemistry, phytotechnology, pharmacology and nutraceutical functions of kenaf (*Hibiscus cannabinus* L.) and roselle (*Hibiscus sabdariffa* L.) seed oil: an overview. *Ind. Crops Prod.*, 77, 323–332, 2015.
36. Cheng, W.Y., Akanda, J.M.H., Nyam, K.L., Kenaf seed oil: A potential new source of edible oil. *Trends Food Sci. Technol.*, 52, 57–65, 2016.
37. Bindhu, M.R. and Umadevi, M., Synthesis of monodispersed silver nanoparticles using *Hibiscus cannabinus* leaf extract and its antimicrobial activity. *Spectrochim. Acta A Mol. Biomol. Spectrosc.*, 101, 184–190, 2013.
38. Alexopoulou., E., Li, D., Papatheohari, Y., Siqi, H., Scordia, D., Testa, G., How kenaf (*Hibiscus cannabinus* L.) can achieve high yields in Europe and China. *Ind. Crops Prod.*, 68, 131–140, 2015.

39. Abu-Tarboush, H.M., Ahmed, S.A.B., Alkahtani, H.A., Some nutritional properties of karkade (*Hibiscus sabdariffa*) seed products. *Cereal Chem.*, 74, 3, 352–355, 1997.
40. Ryu, J., Kwon, S.J., Ahn, J.W., Jo, Y.D., Kim, S.H., Jeong, S.W., Lee, M.K., Kim, J.B., Kang, S.Y., Phytochemicals and antioxidant activity in the kenaf plant (*Hibiscus cannabinus* L.). *J. Plant Biotechnol.*, 44, 2, 191–202, 2017.
41. Nor, M.M., Noor Amri, N.F.A., Rahman, M.M., Liew, J.Y., Omar, M.N., Basri, S., Kinetic study of phytochemical extraction and antioxidant potential of *Hibiscus cannabinus* L. *Int. J. Eng. Res. Technol.*, 12, 2502–2509, 2019.
42. James, S.A., Ladan, M.J., Goje, D.J., Antioxidant potential of *hibiscus cannabinus* methanolic leaf extract. *Sci. World J.*, 8, 1, 8–12, 2013.
43. Nyam, K.L., Tan, C.P., Lai, O.M., Long, K., Man, Y.C., Physicochemical properties and bioactive compounds of selected seed oils. *LWT-Food Sci. Technol.*, 42, 8, 1396–1403, 2009.
44. Chew, S.C., Tan, C.P., Long, K., Nyam, K.L., Effect of chemical refining on the quality of kenaf (*hibiscus cannabinus*) seed oil. *Ind. Crops Prod.*, 89, 59–65, 2016.
45. Yusri, N.M., Chan, K.W., Iqbal, S., Ismail, M., Phenolic content and antioxidant activity of *hibiscus cannabinus* l. seed extracts after sequential solvent extraction. *Molecules*, 17, 11, 12612–12621, 2012.
46. Leong, M.H., Tan, C.P., Nyam, K.L., Effects of accelerated storage on the quality of kenaf seed oil in chitosan-coated high methoxyl pectin-alginate microcapsules. *J. Food Sci.*, 81, 10, 2367–2372, 2016.
47. Adnan, M., Oh, K.K., Azad, M.O.K., Shin, M.H., Wang, M.H., Cho, D.H., Kenaf (*hibiscus cannabinus* L.) leaves and seed as a potential source of the bioactive compounds: Effects of various extraction solvents on biological properties. *Life*, 10, 10, 223, 2020.
48. Chew, S.C. and Nyam, K.L., Oxidative stability of microencapsulated kenaf seed oil using co-extrusion technology. *J. Am. Oil Chem. Soc.*, 93, 4, 607–615, 2016a.
49. Chan, K.W., Khong, N.M.H., Iqbal, S., Mansor, S.M., Ismail, M., Defatted kenaf seed meal (DKSM): Prospective edible flour for agricultural wastes with high antioxidant activity. *LWT-Food Sci. Technol.*, 53, 308–313, 2013.
50. Yazan, L.S., Foo, J.B., Ghafar, S.A., Chan, K.W., Tahir, P.M., Ismail, M., Effect of kenaf seed oil from different ways of extraction towards ovarian cancer cells. *Food Bioprod. Process.*, 8, 9, 328–332, 2011.
51. Mariod, A.A., Fathy, S.F., Ismail, M., Preparation and characterization of protein concentrates from defatted kenaf seed. *Food Chem.*, 123, 3, 747–752, 2010.
52. Khoddami, A., Wilkes, M.A., Roberts, T.H., Techniques for analysis of plant phenolic compounds. *Molecules*, 18, 2, 2328–2375, 2013.
53. Singh, P.P. and Saldaña, M.D., Subcritical water extraction of phenolic compounds from potato peel. *Food Res. Int.*, 44, 8, 2452–2458, 2011.

54. Kashyap, P. and Anand, S., Phytochemical and GC-MS analysis of *Rhododendron arboreum* flowers. *Int. J. Farm Sci.*, 6, 4, 145–151, 2016.
55. Sim, Y.Y., Ong, W.T.J., Nyam, K.L., Effect of various solvents on the pulsed ultrasonic assisted extraction of phenolic compounds from *Hibiscus cannabinus* L. leaves. *Ind. Crops Prod.*, 140, 1–8, 2019b.
56. Jin, C.W., Ghimeray, A.K., Wang, L., Xu, M.L., Piao, J.P., Cho, D.H., Far infrared assisted kenaf leaf tea preparation and its effect on phenolic compounds, antioxidant and ACE inhibitory activity. *J. Med. Plants Res.*, 7, 1121–1128, 2013.
57. Ghafar, A.S.A., Ismail, M., Yazan, L., Saiful, L., Fakurazi, S., Ismail, N., Tahir, P.M., Cytotoxic activity of kenaf seed oils from supercritical carbon dioxide fluid extraction towards human colorectal cancer (HT29) cell lines. *Evid. Based Complement. Altern. Med.*, 2013, 1–8, 2013.
58. Al-Snafi, A.E., Pharmacological effects and therapeutic properties of *hibiscus cannabinus*-a review. *Indo Am. J. Pharm. Sci.*, 05, 04, 2176–2182, 2018.
59. Zhang, Z., Xie, Q., Che, L., Synergistic effects of ultrasound and extraction solvent on the bioactive compound in kenaf seed oil. *J. Food Sci. Technol.*, 57, 2118–2128, 2020.
60. Lee, S.Y., Fu, S.Y., Chong, G.H., Ultrasound-assisted extraction kinetics, fatty acid profile, total phenolic content and antioxidant activity of green solvents' extracted passion fruit oil. *Int. J. Food Sci. Technol.*, 50, 8, 1831–1838, 2015.
61. Latif, S. and Anwar, F., Aqueous enzymatic sesame oil and protein extraction. *Food Chem.*, 125, 2, 679–684, 2011.
62. Li, H., Zhang, Z., He, D., Xia, Y., Liu, Q., Li, X., Ultrasound-assisted aqueous enzymatic extraction of oil from perilla seeds and determination of its physicochemical properties, fatty acid composition and antioxidant activity. *Food Sci. Technol.*, 37, 71–77, 2017.
63. Havsteen, B.H., The biochemistry and medical significance of the flavonoids. *Pharmacol. Ther.*, 96, 2-3, 67–202, 2002.
64. Da-Costa-Rocha, I., Bonnlaender, B., Sievers, H., Pischel, I., Heinrich, M., *Hibiscus sabdariffa* l.–a phytochemical and pharmacological review. *Food Chem.*, 165, 424–443, 2014.
65. Badal, S. and Delgoda, R., *Pharmacognosy: Fundamentals, Applications and Strategy*, M. Shields (Ed.), pp. 295–313, Chemotherapeutics, Elsevier, Netherlands, 2017.
66. Salib, J.Y., Polyphenolic compounds from flowers of hibiscus, in: *Polyphenols in Plants*, pp. 231–239, 2014.
67. Wang, H., Li, J., Tao, W., Zhang, X., Gao, X., Yong, J., Duan, J.A., *Lycium ruthenicum* studies: Molecular biology, phytochemistry and pharmacology. *Food Chem.*, 240, 759–766, 2018.
68. Ahmed, F., Fanning, K., Netzel, M., Turner, W., Li, Y., Schenk, P.M., Profiling of carotenoids and antioxidant capacity of microalgae from subtropical coastal and brackish waters. *Food Chem.*, 165, 300–306, 2014.

69. Raju, M., Varakumar, S., Lakshminarayana, R., Baskaran, V., Carotenoid composition and vitamin A activity of medicinally important green leafy vegetables. *Food Chem.*, 101, 4, 1598–1605, 2007.
70. Mohamed, R., Fernández, J., Pineda, M., Aguilar, M., Roselle (*Hibiscus sabdariffa*) seed oil is a rich source of γ-tocopherol. *J. Food Sci.*, 72, 3, S207–S211, 2007.
71. Peng, W.L., Mohd-Nasir, H., Setapar, S.H.M., Ahmad, A., Lokhat, D., Optimization of process variables using response surface methodology for tocopherol extraction from Roselle seed oil by supercritical carbon dioxide. *Ind. Crops Prod.*, 143, 1–11, 2020.
72. Mariod, A.A., Matthaus, B., Ismail, M., Comparison of supercritical fluid and hexane extraction methods in extracting kenaf (*hibiscus cannabinus*) seed oil lipids. *J. Am. Oil Chem. Soc.*, 887, 931–935, 2011.
73. Chew, S.C. and Nyam, K.L., Kenaf (hibiscus cannabinus L.) seed oil, in: *Fruit Oils: Chemistry and Functionality*, M. Ramadan, (Ed.), Springer, Cham, 2019.
74. Zhang, Z., Zhang, L., Xie, Q., Che, L., Effect of accelerated storage on fatty acids, thermal properties and bioactive compounds of kenaf seed oil. *J. Food Sci.*, 84, 8, 2121–2127, 2019.
75. Byju, K., Vasundhara, G., Anuradha, V., Nair, S.M., Kumar, N.C., Presence of phytol, a precursor of vitamin e in *chaetomorph antinnina*. Mapana. *J. Sci.*, 12, 2, 57–65, 2013.
76. Kapepula, P.M., Kabamba Ngombe, N., Tshisekedi Tshibangu, P., Tsumbu, C., Franck, T., Mouithys-Mickalad, A., Mumba, D., Tshala-Katumbay, D., Tits, M., Frédérich, M., Comparison of metabolic profiles and bioactivities of the leaves of three edible congolese hibiscus species. *Nat. Prod. Res.*, 31, 24, 2885–2892, 2017.
77. Pascoal, A., Quirantes-Pine, R., Fernando, A.L., Alexopoulou, E., Segura-Carretero, A., Phenolic composition and antioxidant activity of kenaf leaves. *Ind. Crops Prod.*, 78, 116–123, 2015.
78. Chew, S.C., Tan, C.P., Nyam, K.L., Application of response surface methodology for optimizing the deodorization parameters in chemical refining of kenaf seed oil. *Sep. Purif. Technol.*, 184, 144–151, 2017b.
79. Mohamed, A., Bhardwaj, H., Hamama, A., Webber III, C., Chemical composition of kenaf (hibiscus cannabinus l.) seed oil. *Ind. Crops Prod.*, 4, 3, 157–165, 1995.
80. Seca, A.M., Silva, A.M., Silvestre, A.J., Cavaleiro, J.A., Domingues, F.M., Pascoal-Neto, C., Phenolic constituents from the core of kenaf (*Hibiscus cannabinus*). *Phytochemistry*, 56, 7, 759–767, 2001.
81. Gonenc, A., Hacısevki, A., Tavil, Y., Cengel, A., Torun, M., Oxidative stress in patients with essential hypertension: A comparison of dippers and nondippers. *Eur. J. Intern. Med.*, 24, 139–144, 2013.
82. Nilugal, K.C., Fattepur, S.R., Kazi, H.A., Dela Cruz Bacayo, M.F., Asmani, M.F.M., Chittur, M.A., II, *In-vitro* screening of *hibiscus cannabinus* L. leaves extracts for antibacterial activity. *World J. Pharm. Res.*, 5, 6, 369–378, 2016.

83. Kraiczy, P. and Wurzner, R., Complement escape of human pathogenic bacteria by acquisition of complement regulators. *Mol. Immunol.*, 43, 31–44, 2014, 1–7, 2006.
84. Oskay, M., Oskay, D., Kalyoncu, F., Activity of some plant extracts against multi-drug resistant human pathogens. *Int. J. Pharm. Sci. Res.*, 8, 293–300, 2010.
85. Karthik, M. and Gayathri, C., Effect of ethanolic extract of *hibiscus cannabinus* leaf on high cholesterol diet induced obesity in female albino rats. *Asian J. Pharm. Clin. Res.*, 6, 4, 65–67, 2013.
86. Wong, Y.H., Lau, H.W., Tan, C.P., Long, K., Nyam, K.L., Binary solvent extraction system and extraction time effects on phenolic antioxidants from kenaf seeds (*hibiscus cannabinus* L.) extracted by a pulsed ultrasonic assisted extraction. *Sci. World J.*, 2014, 1–7, 2014.
87. Foo, J.B., Yazan, L.S., Mansor, S.M., Ismail, N., Tahir, P.M., Ismail, M., Kenaf seed oil from supercritical carbon dioxide fluid extraction inhibits the proliferation of WEHI-3B leukemia cells *in vivo*. *J. Med. Plants Res.*, 6, 1429–1436, 2012.
88. Moujir, L., Seca, A.M., Silva, A.M., López, M.R., Padilla, N., Cavaleiro, J.A., Neto, C.P., Cytotoxic activity of lignans from *Hibiscus cannabinus*. *Fitoterapia*, 78, 5, 385–387, 2007.
89. Shaikh, S., Joshi, Y.M., Kadam, V., Comparative study of anti-inflammatory activity of aqueous and methanolic extracts of *hibiscus cannabinus* leaf (malvaceae). *Int. J. Pharm. Pharm. Sci.*, 8, 4, 64–68, 2016.
90. Zhen, J., Villani, T.S., Guo, Y., Qi, Y., Chin, K., Pan, M.H., Wu, Q., Phytochemistry, antioxidant capacity, total phenolic content and anti-inflammatory activity of *Hibiscus sabdariffa* leaves. *Food Chem.*, 190, 673–680, 2016.
91. Lee, Y.G., Byeon, S.E., Kim, J.Y., Lee, J.Y., Rhee, M.H., Hong, S., S., J.C., Lee, H.S., Kim, M.J., Cho, D.H., Cho, J.Y., Immunomodulatory effect of Hibiscus cannabinus extract on macrophage functions. *J. Ethnopharmacol.*, 113, 1, 62–71, 2007.
92. Agbor, G.A., Oben, J.E., Nkegoum, B., Takala, J.P., Ngogang, J.Y., Hepatoprotective activity of *hibiscus cannabinus* (linn.) against carbon tetrachloride and paracetamol induced liver damage in rats. *Pak. J. Biol. Sci.*, 8, 10, 1397–1401, 2005.
93. Porchezhian, E. and Ansari, S.H., Hepatoprotective activity of *Abutilon indicum on* experimental liver damage in rats. *Phytomedicine*, 12, 1–2, 62–64, 2005.
94. Kai, N.S., Nee, T.A., Ling, E.L.C., Ping, T.C., Kamariah, L., Lin, N.K., Antihypercholesterolemic effect of kenaf (*hibiscus cannabinus* L.) seed on high-fat diet sprague dawley rats. *Asian Pac. J. Trop. Med.*, 8, 1, 6–13, 2015.
95. Pradeep, K., Gagandeep, K., Nanjaian, M., Antihyperlipidemic effect of hydroalcoholic extract of Kenaf (*hibiscus cannabinus* L.) leaves in high fat diet fed rats. *Ann. Biol. Res.*, 1, 3, 174–181, 2010.

96. Zakaria, Z.A., Zakaria, M.L., Amom, Z., Desa, M.N.M., Antimicrobial activity of the aqueous extract of selected Malaysian herbs. *Afr. J. Microbiol. Res.*, 5, 30, 5379–5383, 2011.
97. Nyam, K.L., Sin, L.N., Kamariah, L., Phytochemical analysis and anti-inflammatory effect of kenaf and roselle seeds. *Malays. J. Nutr.*, 21, 2, 245–254, 2015.
98. Cheong, A.M., Koh, J.X.J., Okechukwu, P.N., Tan, C.P., Nyam, K.L., Hypocholesterolemic effects of kenaf seed oil, macroemulsion, and nanoemulsion in high-cholesterol diet induced rats. *J. Food Sci.*, 83, 3, 854–863, 2018a.
99. Young, I.S. and Woodside, J.V., Antioxidants in health and disease. *J. Clin. Pathol.*, 54, 3, 176–186, 2001.
100. Harman, D., Aging: Phenomena and theories. *Ann. N. Y. Acad. Sci.*, 854, 1–7, 1998.
101. Adnan, M., Mohammad, K.I., Manik, M.E.H., Anticancer agents in combination with statins. *J. Bioequivalence Bioavailab.*, 9, 4, 463–466, 2017.
102. Brambilla, D., Mancuso, C., Scuderi, M.R., Bosco, P., Cantarella, G., Lempereur, L., Benedetto, G.D., Salvatore, P., Renato, B., The role of antioxidant supplement in immune system, neoplastic, and neurodegenerative disorders: A point of view for an assessment of the risk/benefit profile. *Nutr. J.*, 7, 29, 1–8, 2008.
103. Kashyap, P., Anand, S., Thakur, A., Evaluation of antioxidant and antimicrobial activity of *rhododendron arboreum* flowers extract. *Int. J. Food Ferment. Technol.*, 7, 1, 123–128, 2017.
104. Thakur, A., Joshi, V.K., Thakur, N.S., Immunology and its relation with food components: an overview. *Int. J. Food. Ferment. Technol.*, 9, 1, 1–16, 2019.
105. Peças, P., Carvalho, H., Salman, H., Leite, M., Natural fibre composites and their applications: A review. *J. Compos. Sci.*, 2, 4, 66, 2018.
106. Whitaker, J.M., *Quick & Health: Recipes for Vibrant Living*, p. 19, 775 Silverback Books, USA, 2003.
107. Maganha, E.G., da Costa Halmenschlager, R., Rosa, R.M., Henriques, J.A.P., de Paula Ramos, A.L.L., Saffi, J., Pharmacological evidences for the extracts and secondary metabolites from plants of the genus. *hibiscus. Food Chem.*, 1, 1–10, 2010.
108. Kowti, R., Majumder, P., Harshitha, H.S., Joshi, V., Kumar, M.R., Ahmed, S.S., hibiscus herbs-a comprehensive botanical, chemical and biological overview. *World J. Advanced Res. Rev.*, 7, 3, 291–303, 2020.
109. Foyet, H.S., Hritcu, L., Ciobica, A., Stefan, M., Kamtchouing, P., Cojocaru, D., Methanolic extract of. *hibiscus asper leaves improves spatial Memory deficits 6-hydroxydopamine-lesion rodent Model Parkinson's disease, J. Ethnopharmacol.*, 133, 773–779, 2011.
110. Yazan, L.S., Abd Rahman, N., Chan, K.W., Abd Ghani, W.N.H.W., Tor, Y.S., Foo, J.B., Phenolics-saponins rich fraction of defatted kenaf seed meal

exhibits cytotoxicity towards cancer cell lines. *Asian Pac. J. Trop. Biomed.*, 6, 5, 404–409, 2016.
111. Cook, J.G., *Handbook of Textile Fiber*, Merrow Publishing, Watford, UK, 1960.
112. Ayadi, R., Hamrouni, L., Hanana, M., Bouzid, S., Trifi, M., Khouja, M.L., *In vitro* propagation and regeneration of an industrial plant kenaf (*hibiscus cannabinus* l.). *Ind. Crops Prod.*, 33, 2, 474–480, 2011.
113. Vital, R.J., Bassinello, P.Z., Cruz, Q.A., Carvalho, R.N., Paiva, J.C.M., Colombo, A.O., Production, quality, and acceptance of tempeh and white bean tempeh burgers. *Food*, 7, 9, 136, 2018.
114. Zawawi, N., Gangadharan, P., Ahma Zaini, R., Samsudin, M.G., Karim, R., Maznah, I., Nutritional values and cooking quality of defatted kenaf seeds yellow (DKSY) noodles. *Int. Food Res. J.*, 21, 2, 603–608, 2014.
115. Katende, A.B., Ssegawa, P., Ann, B., *Wild Food Plants and Mushrooms of Uganda*, p. 329, Sida RelmaTechnical Handbook, Regional Land Management Unit, 1999.
116. Sulochanamma, G., Rao, D.M., Rao, P.G.P., Balaswamy, K., Development of a low calorie ready-to-serve beverage from *hibiscus cannabinus* L. *Biomed. J. Sci. Technol. Res.*, 11, 2, 8418–843, 2018.
117. Zhang, T., *Improvement of Kenaf Yarn For Apparel Application*, Master Thesis of Louisiana State University, US, 2003.
118. Kipriotis, E., Heping, X., Vafeiadakis, T., Kiprioti, M., Alexopoulou, E., Ramie and kenaf as feed crops. *Ind. Crops Prod.*, 68, 126–130, 2015.
119. Islam, M., Kenaf (*hibiscus cannabinus* l., malvaceae) research and development advances in Bangladesh: A review. *J. Nutr. Food Process.*, 2, 1, 2019.
120. Saba, N., Jawaid, M., Hakeem, K.R., Paridah, M.T., Khalina, A., Alothman, O.Y., Potential of bioenergy production from industrial kenaf (*hibiscus cannabinus* L.) based on Malaysian perspective. *Renew. Sustain. Energy Rev.*, 42, 446–459, 2015.
121. Bassam, N.E., *Handbook of Bioenergy Crops: A Complete Reference to Species, Development and Applications*, 1st ed, p. 227, Earthscan, UK, 2010.
122. Black, M. and Bewley, J.D., *Seed Technology and Its Biological Basis*, 1st ed, CRC Press, UK, 2000.
123. Neto, C.P., Seca, A., Fradinho, D., Coimbra, M., Domingues, F., Evtuguin, D., Silvestre, A., Cavaleiro, J., Chemical composition and structural features of the macromolecular components of *hibiscus cannabinus* grown in Portugal. *Ind. Crop Prod.*, 5, 3, 189–196, 1996.
124. Saikia, C., Goswami, T., Ali, F., Evaluation of pulp and paper making characteristics of certain fast growing plants. *Wood Sci. Technol.*, 31, 467–475, 1997.
125. Arbaoui, S., Evlard, A., Mhamdi, M.E.W., Campanella, B., Paul, R., Bettaieb, T., Potential of kenaf (*hibiscus cannabinus* L.) and corn (*zea mays* L.) for phytoremediation of dredging sludge contaminated by trace metals. *Biodegradation*, 24, 4, 563–567, 2013.

126. Aishah, R.M., Shamshuddin, J., Fauziah, C.I., Arifin, A., Panhwar, Q.A., Phytoremediation of copper and zinc in sewage sludge amended soils using *Jatropha curcas* and *hibiscus cannabinus*. *J. Chem. Soc. Pak.*, 38, 6, 1230–1243, 2016.
127. Deng, Y., Li, D., Huang, Y., Huang, S., Physiological response to cadmium stress in kenaf (*hibiscus cannabinus* L.) seedlings. *Ind. Crops Prod.*, 107, 453–457, 2017.
128. Abe, K. and Ozaki, Y., Comparison of useful terrestrial and aquatic plant species for removal of nitrogen and phosphorus from domestic wastewater. *Soil Sci. Plant Nutr.*, 44, 4, 599–607, 1998.
129. Essiet, D., Oil spill clean-up: Kenaf to the rescue, in: *The Nation*, 2012.
130. Zaveri, M.D., *Absorbency Characteristics of Kenaf Core Particles*, Textiles and Apparel, Technology and Management, North Carolina State University, Raleigh, 2004.
131. Han, J.S., Miyashita, E.S., Lin, Y.Y., Roa, A., Stormwater filtration of a harvested, ensiled, and mature kenaf by sheep. *Prof. Anim. Sci.*, 12, 99–104, 1999.
132. Ramaswamy, G.N., Sellers, T., Tao, W., Crook, L.G., Kenaf nonwovens as substrates for laminations. *Ind. Crops Prod.*, 17, 1, 1–8, 2003.
133. Chu, C.C. and Nyam, K.L., Kenaf (*hibiscus cannabinus* l.) seed oil: Application as cosmetic product ingredients. *Ind. Crops Prod.*, 156, 112871, 2020.
134. Odetola, O.M. and Eruvbetine, D., Nutritional evaluation of whole kenaf (*hibiscus cannabinus* L.) seed meal in rats. *J. Adv. Lab. Res. Biol.*, 3, 3, 152–157, 2012.
135. Villar, J.C., Revilla, E., Gomez, N., Carbajo, J.M., Simon, J.L., Improving the use of kenaf for kraft pulping by using mixtures of bast and core fibers. *Ind. Crops Prod.*, 29, 2-3, 301–307, 2009.
136. Dutt, D., Upadhyay, J.S., Singh, B., Tyagi, C.H., Studies on *hibiscus cannabinus* and *hibiscus sabdariffa* as an alternative pulp blend for softwood: An optimization of kraft delignification process. *Ind. Crops Prod.*, 29, 16–26, 2009.
137. Othman, M.R. and Akil, H.M., The CO2 adsorptive and regenerative behaviors of Rhizopus oligosporus and carbonaceous *hibiscus cannabinus* exposed to thermal swings. *Microporous Mesoporous Mater.*, 110, 2-3, 363–369, 2008.
138. Lee, X.Y., Chu, C.C., Hasan, Z.A.A., Chua, S.K., Nyam, K.L., Novel nanostructured lipid carriers with photoprotective properties made from carnauba wax, beeswax, and kenaf seed oil. *J. Am. Oil Chem. Soc.*, 96, 201–211, 2019.

9

Dhatura: Nutritional, Phytochemical, and Pharmacological Properties

K.M. Manju[1], Ritu Sindhu[1*], Priyanka Rohilla[1] and Rohit Kumar[2]

[1]*Centre of Food Science and Technology, Chaudhary Charan Singh Haryana, Agricultural University, Hisar, Haryana, India*
[2]*Department of Microbiology, Central University of Haryana, Mahendergarh, India*

Abstract
In Ayurveda, several formulations are made by utilizing herbal resources to treat various diseases according to the nature of sickness. These herbals may also include some poisonous plants though they are generally used after multiple purification methods. Datura is one of these weedy plant which is applicable in several therapeutical applications. It is called as "devil's trumphet" and first reported by Linnaeus in 1753. The genus Datura belongs to family Solanaceae, popular for presence of various tropane alkaloids. Datura genus is popularly known for their medicinal and hallucinogen properties. There are about ten species of *Datura* used as drug plants but from these most important is *D. stromonium* and *Datura anoxia*. A large number of alkaloids found in datura species are as hyoscine, hyoscyamine, meteloidine, scopolamine, tigloidine, tropine, withametelline, datumetine, etc. and similarly whole plant has various therapeutic effects as, bronchodilator, painkiller, hypnotic, narcotic and anesthetic, cough suppressants, antiasthmatic, antispasmodic (suppress muscle spasm). This chapter will provide information about nutritional and phytochemical approaches, applications for curing diseases and toxicity of datura plant.

Keywords: Herbal, Datura, alkaloids, hallucinogen, therapeutic and toxicity

9.1 Introduction

In present scenario, herbal medicinal products or supplements has risen enormously with respect to health care. Medical plants had numerous

**Corresponding author*: ritu3400@gmail.com

Prerna Gupta, Navnidhi Chhikara and Anil Panghal (eds.) Harvesting Food from Weeds, (327–350)

bioactive substances, which are responsible for their pharmacological properties. Datura plant usually grows in average soils, can tolerate different soil environment but preferably requires nutrient rich and moist or alkaline soil. It is well known with different names, some are Kanaka, Matula, Kitava, Mahamohi, Dhurta etc., their seeds known as jimson seed, grown in temperate and tropical region in unwanted places and roadside throughout India, especially in Kashmir, Bihar, Orissa, Madhya Pradesh, Karnataka and Tamil Nadu [1, 2], grown in the fields along with the other grains like wheat, rye, corn. Datura has been reported to exhibit both the poisonous and medicinal uses like possess anti-inflammatory, relieve in pain, fever, and stimulation of central nervous system etc. [3] including some poisoning cases when consumed in large amount affect the central nervous system and cause uncertainty, bizarre behavior and subsequent amnesia but patient recover within sometime [2]. Alkaloids are majorly responsible for its deliriant and hallucinogen property, which is highly toxic in large amount. Various components isolated and identified from datura, such as saponin, tannins, steroids, alkaloids, phenols, and gylycosides, potentially exhibit therapeutic activities [4]. All together properties, this plant fall under Schedule E-1 of drugs and cosmetic Act. In India, the seed of Datura stramonium has been generally used during God Shiv puja in Hindu religion, which encourage the misuse of datura, such as Shivaratri and Swasthanipuja [5]. Datura plant showed high larvicidal effect against red flour beetle and act as mosquito repellent; plant may also exhibit antimicrobial, insecticidal, antidiabetic, antiasthmatic, antioxidant, anti-inflammatory, wound healing, and neural activities [6, 7]. Oral administration of leaf extract can cure asthma, ulcer, heal burn, swelling, and sinusitis [8]. In the light above benefits, this chapter pivots on phytochemistry, pharmacognocy, toxicology, and medical uses of datura.

9.2 Botanical Description

Datura species are either prostrate annuals or short-lived perennials having brown cylindrical roots. These are characterized by trumpet shaped fragrant flowers in different colors of white, yellow, pink, or purple. A large number of seeds are enclosed within spiny fruit. As shown in Figure 9.1, the plant *Datura metel* is plumb, tender, spreading annual herb with height meter or more. These have divaricate branches (Figure 9.1a). Leaves are triangular with elliptical from outline and unequal at base (Figure 9.1b).

Unaccompanied flowers are short with purple exterior and white interior part (Figure 9.1d). The fruits are ovoid, sheath covered all parts with thick pricks (Figure 9.1c), bend when fruit get mature, contains numerous seeds of yellow-brown color [1]. However, *Datura stramonium* is of about 3 to 4 feet in height, even 6 feet when grown on rich soil, large and coarse shrub. Its root spread with taproots system along with off many fibers, which are whitish in color.

The stem is green or purple, tubular, leafy and diverging branched from which flowers arise. Leaves are approx 8 cm long and 6 cm broad found on nodes of stem and on branches also, petiolated, sharp, unicosted, greyish green smooth surface without any projections and ordered in parallel venation. Flowers are developed from dry, spiny and ellipsoid-ovoid fruit form. Each fruit is pruned and residuum of calyx that steeply down, green when immature and turn brown after maturity; finally divides into four parts that give off seeds. The seeds are dark colored with pitted or somewhat reticulated surface [9, 10].

Figure 9.1 Parts of datura plant—(a) plant, (b) leaves, (c) fruit, (d) flower.

9.3 Nutritional Properties and Phytochemistry

Datura plant is divided in many parts as root, stem, bark, leaves, fruit, seed, and seed coat, and each part has poisonous and therapeutic potential because of presence of bioactive compounds. On the basis of proximate analysis, datura seed contain crude protein (13.90–20.73), carbohydrate (25.07–51.22), fat (14.72–16.60), ash (5.14–8.70), and crude fiber (6.55–17.35). Also, seeds are rich source of calcium 174 mg/100 g, phosphorus 690 mg/100 g, potassium 0.50 mg/100 g, sodium 0.085, iron 16.8 mg/100 g, zinc 2.60 mg/100 g and magnesium 390 mg/100 g. Moreover, seed oil is filled with an abundance of mono-unsaturated fatty acid (around 27.49%), poly-unsaturated fatty acid (54.25%), total saturated fatty acid (18.03%) and total unsaturated fatty acid (81.74%). Among fatty acids, palmetic acid is of around 15.31%, steric acid 2.72%, oleic acid 25.97%, linolenic acid 54.25% and palmitolic acid 1.52% [11]. The flower part of the plant contains ketone of 23.61% and ethyl palmitate of 15.84% are main component. Predominant component in leaves and petiole were ketone (18.84% and 39.45%) and phytol (18.71% and 31.32%) respectively, while seed and stem of datura contains palmitic acid of around 30.60% and 38.38% and ethyl linoleate of 21.56% and 17.38% [12]. Datura leaves extract have been reported to contain glycoside, phenols, lignins, saponins and tannins [12]. The seeds contain a high amount of antinutrients as phytate, tannin and oxalate while minerals, i.e., calcium, iron, potassium, sodium, and phosphorus, were found higher in seed coat [9]. Seed coat is highly rich in protein and ash than seed. However, seeds were higher in fat, carbohydrate and fiber than seed coat. Sterols and their derivatives of essential oil are major constituents along with 16.53% of 5-α-ergosta-7, 22- dien-3-β-ol, 14.97% of 3-Hydroxycholestan-5-yl acetate and 10.39% of 26, 26-Dimethyl-5 investigated in stramonium [13]. The major phytochemicals present in D. stramonium are tropane alkaloids, i.e., atropine and scopolamine [14] while daturine is major alkaloid isolated from seeds. These two major alkaloids gradually increase with all growing stages and most prominent at the end of its reproductive phase [15, 16] and maximum scopolamine accumulated in roots after 16 weeks. The concentration of atropine in *Datura metel* leaves and seeds were found to be 0.426%, while 0.43% in flower [17] while root contains higher concentration of atropine among all aerial parts of plant [18]. Whereas in *D. stramonium*, these two alkaloids reached maximum amount at or after ten week of seed germination and decrease with initiation of generative stage in plant [19]. All species of Datura contain major tropane alkaloids, such as scopolamine, hyoscyamine, and atropine. *D. stramonium,* in general contains approx. 64 tropane alkaloids. Apart from these, two tropane alkaloids; 3-phenylacetoxy-6,7-epoxynortropane and 7-hydroxyapoatropine are chemically identified. Some other alkaloids further identified as scopoline, 3-(hydroxyacetoxy) tropane, 3-hydroxy-6-(2-methybutryoxy) tropane, 3a-tigloyloxy-6-hydroxyl propane, 3,7-dihydroxy-6-tigoyloxytropane, 3-tigloyloxy-6-propionloxytropane, 3-phenylacetoxy-6,7-epoxytropane, 3-phenylacetoxy-6-hydroxytropane, aponorscopolamine, 3a,6a–ditigloyoxytropane and 7–hydroxyhyoscyamine [20].

Presently, datura is a well-known plant owing to its multipurpose potential values and its medicinal properties. The macro and micro content in different parts of different species of datura is presented in Table 9.1. The major alkaloid group is chemically characterized by ring structure attach to pyrrolidine and piperidine with single nitrogen and two carbons, whilst amino group present in the methylated form. Numerous studies have reported that datura leaves and flower contain abundance of withanaolides included with afastuosins A, B, C and E; withammetelin; dmetelins A-D and baimantuoluoline A, B and C from different species (83–86). Methanolic extract of datura metel have found 10 new chemical withanaloids includes; withametelins I-P. Daturafolisides A-I, (22R)-27-hydroxy-7α-methoxy-1-oxowitha-trienolide-27-O-β-d-gluco-pyranoside, daturataturin A, daturametelin J, daturataurin B, baimantuoluoside B, and 12-deoxywithastramonolide are isolated from the leaves. However, cis-2-nitro-4-t-butylcyclohexanone (36.54%), acetic acid, trifluoro-, 2,2-dimethylpropyl ester (31.97%), 4-trifluoroacetoxyoctane (19.28%) and 1,4-cyclohexadiene, 1-methyl- (12.22%) are another four compounds were isolated from the methanolic extract of datura flower [21] and including; five megastigmanesesquiterepenes from ethanolic extract [22]. The seed extract shows 268.6 μg of gallic acid equivalent and 8.84 μg of quercetin equivalent from datura fruit pulp [23]. Isofraxidin, scopatone, daturadiol, 1, 4-benzenediol, arenarine D, vanillin, N-trans-feruloyl-tyramine, scopoletin, G-sitosterol, and hyoscyamilactol were isolated from the seeds of *Datura metel* from Xinjiang Province [24].

Table 9.1 Nutritional composition of different Datura species.

Species	Phytoconstituent	Leaves	Seeds	Stem	References
D. metel	Fat	11.5	14.72		
	Carbohydrate	43.42	51.22		
	Protein	29.68	20.73		
	Ash content	15	5.14		
	Moisture		4.63		
	Crude fiber	1	17.35		[17]
D. stramonium	Fat	2.19	16.60	7.41	
	Carbohydrate	8.81	26.20	60.80	
	Protein	7.87	16.20	22.30	[25]
	Ash content	5.38	8.70		
	Moisture	60.5	8.50	74.20	
	Crude fiber	16.61	23.70		[9, 26]
D. innoxia	Fat	7.50	15.52	3.42	[27, 28]
	Carbohydrate	30.09	25.07	51.00	[29]
	Protein	17.21	13.90	0.74	
	Ash content	16.59	8.26	13.58	
	Moisture	7.50	10.00	6.47	
	Crude fiber	8.95	6.55	31.42	

9.4 Properties of Plant

9.4.1 Pharmacological Properties

Datura is well known for its pharmacological properties to exhibit analgesic, antioxidant, antimicrobial, anticancer as well as effective painkiller and potent analgesic activities (Table 9.2). *Datura stamonium* has been utilized to cure Parkinson's, anticholinergic syndrome and parasympathetic stimulation due to presence of biologically active substance known as atropine and scopolamine [30]. The extract of these valuable plant parts exhibited antibacterial activity against *Salmonella typhii, Pseudomonas aeruginosa, Proteus vulgaris, Styphylococcus aureus, E. coli,* and antifungal activity against *Fusarium mangiferae* and *Fusarium oxysporum* [31]. Atropine present in the plant has an anti-asthmatic compounds that prompt paralysis of pulmonary branches of lungs results removing of convulsion which is responsible for asthma attack [32]. Also, if pregnant women consume it for curing asthma then it would make nicotine receptors inactive and a continuous release of acetylcholine will lead to the harm to fetus [2]. There are various bioactive compounds as shown in Figure 9.2, whose structural units are responsible for different properties.

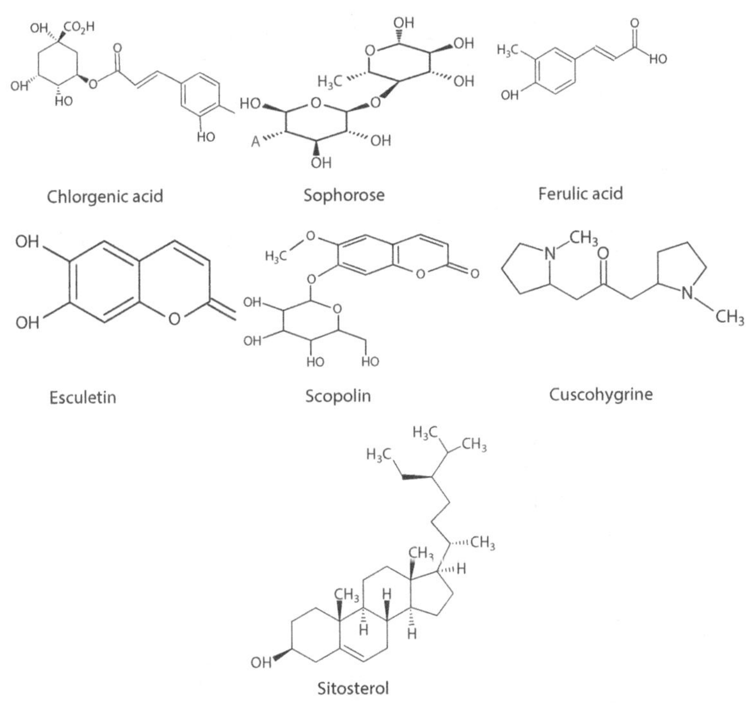

Figure 9.2 Phytochemical compounds of Datura species.

Table 9.2 Pharmacological studies of different Datura species.

Sr. no.	Plant part	Medicinal property	Related study	References
1.	*D. innoxia* leaf	Inhibited proliferation of leukemia cells	Study on human chronic myeloid leukemia cells (K562 cell-line) by aqueous extract of *D. innoxia*.	[33]
2.	*D. stramonium* root	Antidiabetic	The plant extract significantly lowers blood glucose level of diabetic mice or oral glucose injected mice.	[34]
3.	*D. fastuosa* seeds	Antimicrobial	Antibacterial activity of methanolic and hydroalcoholic extract of *D. fastuosa* seed against *E. coli, Bacillus subtilis* and *Staphylococcus* while antifungal activity against *Candida albicans* and *Aspergillus niger* was reported.	[35, 36]
4.	*D. fastuosa* root	anti-inflammatory	Anti-inflammatory activity in carrageenan induced rat paw edema at 200 mg/kg dose of ethanolic and ethyl acetate extract of root was reported effectively.	[17]
5.	*D. fastuosa* leaves and seeds	Wound healing	Ethanolic extract of 5% and 10% concentration along with Nitofurazone ointment as standard was given to Wistar albino rats. 10% conc. extract shows effective wound healing activity.	[37]
6.	*Datura metel* seeds and *D. fastuosa*	Antifertility effects	1% and 2% Acetone extracts of *Datura metal* was orally given to female albino mice causes 100% and 80% antifertility. Serum levels of testosterone, Follicle-stimulating hormone, Luteinizing hormone also declined.	[38]

(*Continued*)

Table 9.2 Pharmacological studies of different Datura species. (*Continued*)

Sr. no.	Plant part	Medicinal property	Related study	References
7.	*D. fastuosa* seeds	Antioxidant activity	Methanolic and hydroalcoholic extract of seeds has proanthocyanidines, flavonoid, flavonols, and phenolic compounds which attributed to radical scavenging activity.	[34]
8.	*D. stramonium* seed extract	Organophosphate (dichlorvos) poisoning	Seeds extracts increases the survival rate by 90% in male rat comparative to organophosphate poisoning which is of 10%.	[39]
9.	*D. stramonium* leaves	Mosquito repellent	Ethanolic extract had larvicidal and mosquito repellent activity against *Aedes aegypti*, *Anopheles stephensi*, and *Culex quinquefasciatus* by LD_{50} 86.25, 16.07, and 6.25.	[40]
10.	*D. metel* seeds	Insecticidal effect	1.5 and 2% of methanolic extract of seeds shows adverse effects on larval survival, weight and duration, pupal period, % of pupation and adult emergence of *Helicoverpaarmigera* (Hubner).	[41]
11.	Datura leaves	Anti-viral activity	200 µg/ml blocks of atropine blocked the glycosylation of Herpes virus. New virions formed under atropine presence were non-infectious.	[42, 43]

(*Continued*)

Table 9.2 Pharmacological studies of different Datura species. (*Continued*)

Sr. no.	Plant part	Medicinal property	Related study	References
12.	*D. fastuosa* leaves and seeds	Analgesic effect	400 and 800 mg/kg doses of extract given to acetic acid jolted and hot plate reaction mice was observed analgesic effects.	[44]
13.	*D. fastuosa* leaves	Antipyretic	Seed extract reduces the rectum temperature by increasing the water uptake.	[45]
14.	*D. metel* leaves and roots	Antispasmodic activity	5% leaves and 20% root extract was given to Wistar rats isolated uterus and antispasmodic factor was found.	[46]
15.	*D. metel* leaves	Xanthine oxidase inhibitory activity	Methanolic datura extract exhibit 50% xanthine oxidase inhibition activity compared to standard allopurinal antigout drug of 93.21% at 100 mg/ml.	[47]

9.4.2 Anti-Inflammatory and Analgesic Activities

A unique class of bioactive compounds present in plant parts of Datura is popular for its anti-inflammatory activity. The possible mechanism could be that to stop the release of chemical mediators which is responsible for trigger of nociceptors and causes inflammation [48, 49]. Datura has ability to inhibit the production of cyclooxygenase (COX1 and COX2) and then conquer of prostaglandin release and this inhibitory action may be responsible for anti-inflammatory and narcotic effect of Datura species [49, 50]. It was experimentally depicted that aqueous extract of Datura leaf has protective effect in inflammatory activity to develop active herbal pain-related drugs [50]. Owing to these inhibition properties, edema accelerating phase selectively stops, whereas ethanolic extract possess anti-inflammatory effect against carrageenan induced paw edema in rat. Oral administration at the rate of 200 mg/kg for 3 hours significantly (39.43%) relieve in edema, which results from the inhibition of mast cell mediator release of histamine and serotonin that is first phase of progression of edema [51].

9.4.3 Antioxidant Activities

The relieving action for the pharmacological attribute of datura is due to the abundance of phytochemicals, which are powerful free radical scavenger, aid to prevent cellular damage, thereby ability to cure many disease, such as cancer, in that cellular damage is a general pathway and other various disease [34]. The abundance of major polyphenolic content in datura leads 53.17% superoxide radical scavenging activity at 60 μg/ml, while hydroxyl radical scavenging activity was found 57.88% at concentration of 30 μg/ml [52].

9.4.4 Antimicrobial Potential of Datura

Datura plant parts provide noteworthy phytochemicals which can suppress the growth of certain microbes. The antibacterial activity against particular strains of bacteria is dose dependent. Different solvent extracts possessed antibacterial activities against *Enterobacter* sp., *Micrococcus luteus, P. aeruginosa, E. coli,* and *S. aureus.* It was observed that benzene extract at a dose of 3.12 mg/ml was effective against *P. aeruginosa* while same dose of chloroform extract inhibit *S. aureus, P. aeruginosa* and *M. luteus* by the method of MBC (Minimum bactericidal concentration) [53] along with antifungal activity against *A. fumigatus*, *A. niger* and *S. cerevisiae;* among them highest activity was shown for *S. cerevisiae* and lowest for *A. niger* [17]. However, methanolic extract showed MBC of 25 µg/ml against *E. coli* and hydro-alcoholic extract showed MBC and MIC (minimum inhibitory concentration) of 25 µg/ml against *B. subtilis,* while, on other hand, *A. niger* showed resistance and thus antifungal activity varied from 18.29% to 85.36% [54]. *D. metel* seed oil was very effective as antibacterial activity against various bacterial strains with maximum zone of inhibition for *L. delbrueckii lactis* (19 mm) and minimum was against *P. aeruginosa* (18 mm), indicated more vulnerability of bacterial strains to *D. metel* seed oil with increase in seed oil concentration [55]. Furthermore, hexane, chloroform, acetone, and methanolic extracts of *D. metel* have been reported antifungal properties against *Aspergillus* sp. The MIC of chloroform extract was found 625 µg/ml by micro broth dilution method and 12.5 µg/disc by disc diffusion assay. A concoction brewed from *D. stramonium*, *Calotropis gigantean*, *Azadirachta indica*, and cow manure was used against floral malformations caused by *Fusarium mangiferae* [56].

9.4.5 Antiasthmatic and Bronchodilating Effects

The presence of atropine and scopolamine in *D. stramonium* is strongly responsible for anticholinergic and bronchodilating activities, results from blocking of muscarinic acetylcholine receptors (M1, M2 and M3) which help in regulating airway [57, 58], in these M1 and M3 are responsible for bronchoconstriction (causes asthma) whereas M2 weakens the release of acetylcholine [59]. Acetylcholine released by cholinergic neurons comes in contact with cholinergic muscarinic acetylcholine receptors and cause contraction of smooth muscle which are less functional in asthmatic. There may be continuous release of neurotransmitter by inhibiting M2 receptor after consumption of datura (Figure 9.3). However, used by pregnant women for asthma lead to damaging the fetus by desensitizes nicotinic receptors, which slow down the reaction to repetitive booster modulatory effect on brain functioning [60].

9.4.6 Anticancer Potential of Datura

Datura fustuosa has been demonstrated to exhibit ten new withanaloids in its flower part, out of which BGC-823 (gastric), K562 (leukemia), and A549 (lung) possesses cytotoxic effect on different cancer cells [31]. These withanaloids also have cytotoxicity against human colorectal adenocarcinoma

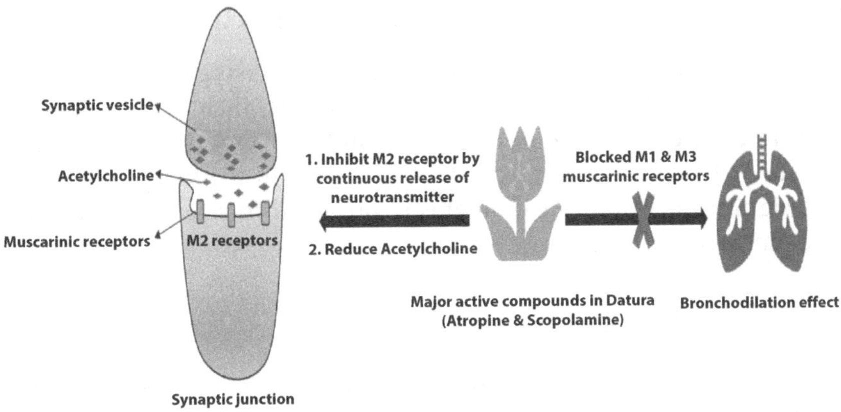

Figure 9.3 Antiasthmatic and broncho dilating effects of Datura plant.

cells (DLD-1) and human lung cancer cells (A549), with IC50 values of 2.0 and 7.0 M, respectively [61]. Furthermore, the root and stem extracts of datura plant possess cytotoxicity effect against human liver cell (HepG-2) with IC_{50} value of 613.88 and 341.12 μg/ml. while leaf and root extracts are strongly effective against cervical carcinoma cells (HeLa) with IC_{50} values of 267.76 and 348.35 μg/ml [17]. It have been demonstrated (51) that the cytotoxicity of datura methanolic seed extract on human breast adenocarcinoma (MCF-7) cell with IC_{50} dose at 113.05 μg/ml (Figure 9.4).

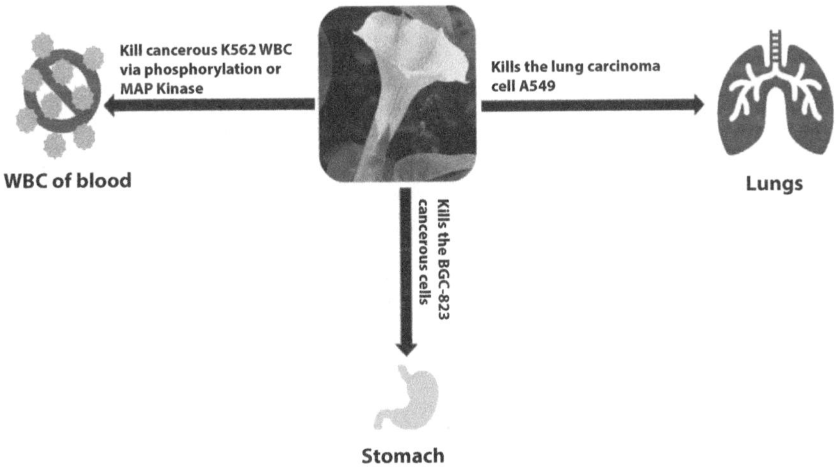

Figure 9.4 Anticancerous activity of Datura flower.

9.4.7 Cell Protection and Wound-Healing Activities

A positive effect showed on Wistar albino rats by using ethanolic extract of *D. fastuosa* on wound healing activities [62], including factors like meantime for epithelialization, hydroxyproline, DNA and protein level. A paste of *D. fastuosa* part extract can be used in the form of ointment have significant wound healing effect compared to reference nitrofurazone ointment [6].

9.4.8 Antiulcer Activity

Datura extract found to serve with analoids (20 mg/kg) leads to reduce the ulcer significantly along with decrease the amount of gastric secretion, acid, and peptic in rats but it does not affect the mucin release and total glycoprotein terms of total carbohydrate and cell replication, it also increases prostaglandins [63].

9.4.9 Hypoglycemic Effect

The administration of *datura metel* seed powder to normal and alloran-induced diabetic rat at the rate of 25, 50, and 75 mg/kg significantly reduced the glucose level after 8 hours [64]. Investigation on diabetic mice, orally loaded mice remarkably reduced the glucose level at relatively high doses (100, 200, and 400 mg/kg) might be attributed to datura powder damage the β-cells responsible for insulin secretion [34] (Figure 9.5).

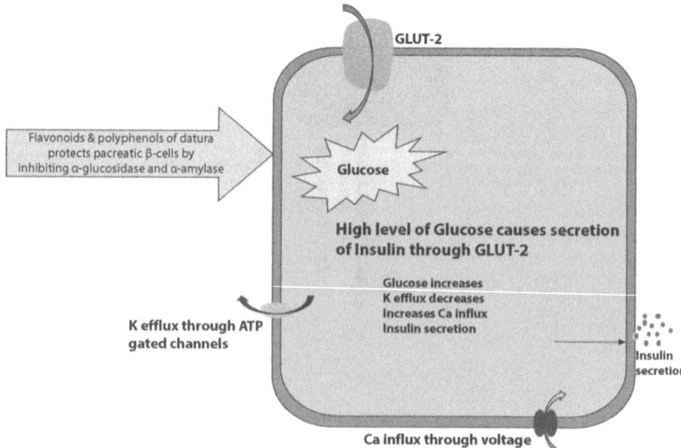

Figure 9.5 Hypoglycemic effect (by pancreatic β-cells) of Datura plant.

9.5 Applications

Utilization of all parts of datura confers its usefulness as traditional remedic plant. Its leaves are used for curing cough, asthma, rheumatism, gastric, ulcer, hemorrhoids, and seasickness. The crushed dried leaves and flower filled into cigarettes for antiasthamatic. Datura seeds, leaves, and roots are reported to be help in treating fever with catarrh (inflammation of mucous membrane), diarrhea, cerebral complications, eczema, and antiseptic. In addition, leaves extract and seed oil is effective in relieving pain, rheumatoid arthritis and tumors. Along with milk and butter, datura consumed to treat insanity [65]. The usage of burnt seed ash with clove and honey up to 7 days to treat malaria and asthma. In fact, leave juice is applied on effected parts externally help in to treat fistula of spinal cord. The unripe fruit is used on affected areas of cattle to treat swelling and skin rashes [66]. Dried leaves used in smoking for curing cough and sneezing. Leaves of this plant boiled with 50 ml of sesame oil used as paste to remove complete poison without any side effect. Fruit poultice applied on belly of infant to treat "Kodabat disease."

Datura powder is mixed with "zeera, laung, sonthi, elaichi" in equal quantity and used twice a day to relieve joint pain [67]. Also, root and flower decoctions used as soothing agent to relaxing patient during setting of fractures [68]. In Europe, datura used in the form of ointment made by leaf juice and young shoot for the application of running sores. For curing Gonorrhea, datura leaves juice is taken with curdled milk [69]. Owning to the medicinal properties of datura, it has long been used to treat falij (paralysis). The decoction of half kilogram of datura seed with 1 kg of khaskhas seed in 5 l of water till the oily layer appear and that oily layer used for massage body. In kashmir, leaves of datura useful in treatment of bronchitis and asthma [70]. Seeds were crushed in mustard oil and its paste is applied for rheumatism, tumors, parasitic skin disease, and decaying teeth [71]. The administration of datura seed powder to cattle helps in treatment of pneumonia for purulent discharge from ailments [72].

9.6 Toxic Effects of Datura Plant

Datura is used as hallucinogens and poisoning agent from decades due to the presence of various compounds in it especially in flowers and seeds, such as scopolamine, hyoscyamine, and atropine. Datura toxication depends on various factors like species type, age of plant, growing conditions, and local weather conditions. Anticholinergic compounds present in datura gives individual a fantasy drive, sometimes aggressive behavior, tachycardia, photophobia due to severe mydriasis, hyperthermia, and amnesia [73]. The overdose of datura can causes cardiopulmonary arrest due to tropane alkaloids, even death also. To control the damage of overdose activated charcoal was given to wipe out the GI tract of individual. In severe case, when only datura ingestion is confirmed, little doses of physostigmine given to revert back the symptoms. Benzodiazepines were also administered to patients having strong hallucinogen effects to calm down them. Anticholinergic toxicity individuals were admitted to ICU until symptoms went out through different therapies. Experimental demonstration showed the reason behind most occurring symptoms is blockage of peripheral muscarinic receptors that innervate exocrine glands, parasympathetic blockade of salivary secretion, smooth muscle and cardiac tissue, fever and erythems due to vasodilation and inhibition of seating. The high concentration of alkaloids consumption leads to "garden eye" and "cornpickers eye" by result of entering to papillary sphincter and ciliary muscle and nervous system affected by belladonna alkaloids which result in agitation, confusion and disoriented problem arise in patients [74]. Physostigmine is generally given to adults in dose of 0.5 to 2 mg, while children should take not more than 0.5 mg to intravenously effect of datura. Death studies are less but adverse effect is very common. Therefore, suggested physostigmine is better and supportive treatment in case of datura poisoning [75].

9.6.1 Toxicity Mechanism

At present there are 28 different belladonna alkaloids determine the toxicity due to seeds of datura plants; atropine and scopolamine are main engineers in the toxicity [76]. Presence of these alkaloids varies from species to species and composition also differs in same species specimens also. The toxicity level may differ in different parts in plants as well as difference in individual. Order of toxicity level due to belladonna alkaloids is flowers > stem > seeds > leaves > root. Muscarinic receptors, present in smooth muscles and exocrine glands are blocked by datura alkaloids [77].

9.7 Conclusion

Datura, the Jimson weed is one of most used plants by human. It contains a lot of bioactive compounds, such as steroids, alkaloids, flavonoids, phenols, atropine, phenols, glycosides and so many others. Datura shows a variety of pharmacological, medicinal, and therapeutical properties. Phytochemistry suggests that datura has ample amount of protein, ash, carbohydrate, and lipid content. These bioactive compounds show unique properties while ingested in by human as a form of drug, i.e., atropine and alkaloids give us protection of gut by antiulcer activities and diarrhea. Atropine shows antiasthmatic feature. Other bioactive compounds like saponin, tannins also give some protective measures from diabetes, cancer, and rheumatic arthritis. Antimicrobial as well as antifungal properties are also performed by bioactive compounds of various parts of plant. Datura plants also have insecticidal and antihelmintic significance. Datura plant has hallucinogen and analgesic effects are very useful in making of drugs in pharma industry. Consumption of various parts is tradition in various countries for treatment and abuse also. It is a good medicinal plant if used wisely for well being otherwise having lethal consequences for ingested in irregular amount.

References

1. Sharma, P.C., Yelne, M.B., Dennis, T.J., Joshi, A., *Database on medicinal plants used in Ayurveda*, vol. 3, pp. 292–312, Central Council for Research in Ayurveda and Siddha, New Delhi, India, 2001.
2. Gaire, B.P. and Subedi, L., A review on the pharmacological and toxicological aspects of Datura stramonium L. *J. Integr. Med.*, 11, 2, 73–79, 2013, https://doi.org/10.3736/jintegrmed2013016.
3. Norton, S., Toxic effects of plants, in: *Casarett and Doull's Toxicology, The Basic Science of Poisons*, Seventh Ed, pp. 1103–1115, McGraw Hill, New York, 2008.
4. Shagal, M.H., Modibbo, U.U., Liman, A.B., Pharmacological justification for the ethnomedical use of Datura stramonium stem-bark extract in treatment of diseases caused by some pathogenic bacteria. *IRJPP*, 2, 1, 16–19, 2012.
5. Gaire, B.P., *Monograph on Datura stramonium*, pp. 1–114, The School of Pharmaceutical and Biomedical Sciences Pokhara University, Dhungepatan-12, Lekhnath, Kaski, Nepal, 2008.
6. Alam, W., Khan, H., Khan, S.A., Nazir, S., Akkol, E.K., Datura metel: A review on chemical constituents, traditional uses and pharmacological activities. *Curr. Pharm. Des.*, 27, 13, 2545–2557, 2021, https://doi.org/10.2174/1381612826666200519113752.
7. Meselhy, K.M., Cytotoxic and insect-repellent activities of surface flavonoids from Datura stramonium L. grown in Egypt. *Life Sci. J.*, 9, 4, 3154–8, 2012.

8. Naik, V., Babu, K.S., Latha, J., Kolluru, B., A review on phytochemical and pharmacological activity of Datura stramonium (medicinal plant). *RJPP*, 10, 1, 77–80, 2018.
9. Oseni, O.A., Olarinoye, C.O., Amoo, I.A., Studies on chemical compositions and functional properties of thorn apple (Datura stramonium L) solanaceae. *Afr. J. Food Sci.*, 5, 2, 40–44, 2010, https://doi.org/10.5897/AJFS.9000274.
10. Das, S., Kumar, P., Basu, S.P., Phytoconstituents and therapeutic potentials of Datura stramonium linn. *J. Drug Deliv. Ther.*, 2, 3, 4–7, 2012, https://doi.org/10.22270/jddt.v2i3.141.
11. Rai, I., Bachheti, R.K., Joshi, A., Pandey, D.P., Physicochemical properties and elemental analysis of some non-cultivated seed oils collected from Garhwal region, Uttarakhand (India). *Int. J. ChemTech Res.*, 5, 1, 232–236, 2013.
12. Xue, J., Sun, Y., Wei, Q., Wang, C., Yang, B., Kuang, H., Wang, Q., Chemical composition and cytotoxicity of the essential oil from different parts of Datura metel L. *Nat. Prod. Res.*, 30, 17, 1938–1940, 2016, https://doi.org/10.1080/14786419.2015.1088541.
13. Wang, S. and You, L., Allelopathic potential in different polarity phases of Datura stramonium L. *Adv. Biomed. Eng.*, 6, 1–16, 2012.
14. Jakabová, S., Vincze, L., Farkas, Á., Kilár, F., Boros, B., Felinger, A., Determination of tropane alkaloids atropine and scopolamine by liquid chromatography–mass spectrometry in plant organs of Datura species. *J. Chromatogr. A*, 1232, 295–301, 2012, https://doi.org/10.1016/j.chroma.2012.02.036.
15. Lim, K.M.C., Dagalea, F.M.S., Vicencio, M.C.G., Antibacterial activity of Datura metel Linn. (TALONG-PUNAY) fruit extract. *J. Pharm. Res. Int.*, 32, 21, 96–101, 2020, DOI: 10.9734/jpri/2020/v32i2130758.
16. Nandakumar, A., MayilVaganan, M., Sundararaju, P., Udayakumar, R., Phytochemical analysis and nematicidal activity of ethanolic leaf extracts of Datura metel, Datura innoxia and Brugmansia suaveolens is against meloidogyne incognita. *Asian J. Cell Biol.*, 2, 1–11, 2017, http://krishi.icar.gov.in/jspui/handle/123456789/6141.
17. Al-Snafi, A.E., Medical importance of Datura fastuosa (syn: Datura metel) and Datura stramonium-A review. *IOSR J. Pharm.*, 7, 2, 43–58, 2017, DOI:10.9790/3013-0702014358.
18. Afsharypuor, S., Mostajeran, A., Mokhtary, R., Variation of scopolamine and atropine in different parts of Datura metel during development. *Planta Med.*, 61, 04, 383–384, 1995, DOI: 10.1055/s-2006-958114.
19. Al-Snafi, A.E., Medicinal plants alkaloids, as promising therapeutics-a review (part 1). *IOSR J. Pharm.*, 11, 51–67, 2021.

20. Berkov, S., Zayed, R., Doncheva, T., Alkaloid patterns in some varieties of Datura stramonium. *Fitoterapia*, 77, 3, 179–182, 2006, https://doi.org/10.1016/j. fitote.2006.01.002.
21. Kiruthika, K.A. and Sornaraj, R., Screening of bioactive components of the flower Datura metel using the GC-MS technology. *IJPRT*, 3, 2025–2028, 2011.
22. Kuang, H.X., Yang, B.Y., Xia, Y.G., Feng, W.S., Chemical constituents from the flower of Datura metel L. *Arch. Pharm. Res.*, 31, 9, 1094–1097, 2008, https://doi.org/10.1007/s12272-001-1274-6.
23. Roy, S., Pawar, S., Chowdhary, A., Evaluation of *in vitro* cytotoxic and antioxidant activity of Datura metel Linn. and cynodondactylon linn. Extracts. *Pharmacognosy Res.*, 8, 2, 123, 2016, DOI: 10.4103/0974-8490.175610.
24. Han, X.L., Wang, H., Zhang, Z.H., Tan, Y., Wang, J.H., Study on chemical constituents in seeds of Datura metel from Xinjiang. *Zhong Yao Cai*, 38, 8, 1646–1648, 2015.
25. Sharma, P. and Sharma, R.A., Comparative antimicrobial activity and phytochemical analysis of Datura stramonium L. Plant extracts and callus *in vitro*. *Eur. J. Med. Plants*, 3, 2, 281–287, 2013.
26. Ibiam, O.F.A., Kalu, E.N., Kanayochukwu, L.U., Akpo, S.O., Proximate and phytochemical analysis of the fruits and leaves of Datura stramonium L. and the effect of fungi associated with foliar blight on them. *Austin J. Biotechnol. Bioeng.*, 4, 1, 1073, 2017.
27. Hussain, J., Rehman, N.U., Shinwari, Z.K., Khan, A.L., Al-Harrasi, A., Ali, L., Mabood, F., Preliminary comparative analysis of four botanicals used in the traditional medicines of Pakistan. *Pak. J. Bot.*, 46, 4, 1403–1407, 2014.
28. Ayuba, V.O., Ojobe, T.O., Ayuba, S.A., Phytochemical and proximate composition of Datura innoxia leaf, seed, stem, pod and root. *J. Med. Plants Res.*, 5, 14, 2952–2955, 2011, https://doi.org/10.5897/JMPR.9000995.
29. Hameed, I. and Hussain, F., Proximate and elemental analysis of five selected medicinal plants of family solanaceae. *Pak. J. Pharm. Sci.*, 28, 4, 1203–1215, 2015.
30. Roy, S., Samant, L., Ganjhu, R., Mukherjee, S., Chowdhary, A., Assessment of *in vivo* antiviral potential of Datura metel Linn. extracts against rabies virus. *Pharmacogn. Res.*, 10, 1, 109, 2018.
31. Maheshwari, N.O., Khan, A., Chopade, B.A., Rediscovering the medicinal properties of Datura sp.: A review. *J. Med. Plant Res.*, 7, 39, 2885–2897, 2013, https://doi.org/10.5897/JMPR11.1657.
32. Baloch, I.A.H. and Hanif-ur-Rehman, I.A.B., Review article on poisonous plants of balochistan. *Balochistaniyat*, 6, 01, 36–56, 2017.
33. Chamani, E., Ebrahimi, R., Khorsandi, K., Meshkini, A., Zarban, A., Sharifzadeh, G., *In vitro* cytotoxicity of polyphenols from Datura innoxia

aqueous leaf-extract on human leukemia K562 cells: DNA and nuclear proteins as targets. *Drug Chem. Toxicol.*, 43, 2, 138–148, 2020, https://doi.org/10.1080/01480545.2019.1629588.
34. Belayneh, Y.M., Birhanu, Z., Birru, E.M., Getenet, G., Evaluation of *in vivo* antidiabetic, antidyslipidemic, and *in vitro* antioxidant activities of hydromethanolic root extract of Datura stramonium L. (Solanaceae). *J. Exp. Pharmacol.*, 11, 29, 2019.
35. Dhiman, A., Lal, R., Bhan, M., Dhiman, B., Hooda, A., Plebeian assessment of antimicrobial and *in vitro* antioxidant zest of D. fastuosa L. seeds. *J. Pharm. Sci. Innov.*, 1, 4, 49–53, 2012.
36. Rajesh, and Sharma, G.L., Studies onantimycotic properties of Daturametel. *J. Ethnopharmacol.*, 80, 2-3, 193–197, 2002, https://doi.org/10.1016/S0378-8741(02)00036-3.
37. Anitha, V.S. and Suseela, L., *In vitro* antioxidant activity and wound healing activity of the alcoholic extract of the alcoholic extract of the ariel parts of Datura fastuosa Linn. *J. Pharm. Res.*, 2, 8, 1176 1179, 2010.
38. Pandiarajan, G., Govindaraj, R., Makesh, B.K., Sankarasivaraman, K., Antifertility activity in the acetone extracts of Datura metel, L seeds on female mouse. *J. Pharmacogenomics Pharmacoproteomics*, 3, 4, 111, 2012, http://dx.doi. org/10.4172/2153-0645.1000111.
39. Bania, T.C., Chu, J., Bailes, D., O'Neill, M., Jimson weed extract as a protective agent in severe organophosphate toxicity. *Acad. Emerg. Med.*, 11, 4, 335–338, 2004, https://doi.org/10.1197/j.aem.2003.12.002.
40. Swathi, S., Murugananthan, G., Ghosh, S.K., Pradeep, A.S., Larvicidal and repellent activities of ethanolic extract of Datura stramonium leaves against mosquitoes. *Int. J. Pharmacogn. Phytochem. Res.*, 4, 1, 25–27, 2012.
41. Singh, V. and Singh, R., Effect of Datura metel seed methanol extract and its fractions on the biology and ovipositional behaviour of *helicoverpa armigera*. *J. Med. Aromat. Plant Sci.*, 30, 2, 157–163, 2008.
42. Alarcón, B., González, M.E., Carrasco, L., Antiherpesvirus action of atropine. *Antimicrob. Agents Chemother.*, 26, 5, 702–706, 1984, https://doi.org/10.1128/AAC.26.5.702.
43. Yamazaki, Z. and Tagaya, I., Antiviral effects of atropine and caffeine. *J. Gen. Virol.*, 50, 2, 429–431, 1980, https://doi.org/10.1099/0022-1317-50-2-429.
44. Abena, A.A., Miguel, L.M., Mouanga, A., Assah, T.H., Diatewa, M., Evaluation of analgesic effect of Datura fastuosa leaves and seed extracts. *Fitoterapia*, 74, 5, 486–488, 2003, https://doi.org/10.1016/S0367-326X(03)00124-2.
45. Abena, A.A. and Miguel, L.M., Neuropsychopharmacological effects of leaves and seeds extracts of Datura fastuosa. *Biotechnol. (Pakistan)*, 3, 109–113, 2004, https://scialert.net/abstract/?doi=biotech.2004.109.113.
46. Prabhakar, E. and Kumar, N.N., Spasmogenic effect of Datura metel root extract on rat uterus and rectum smooth muscles. *Phytother. Res.*, 8, 1, 52–54, 1994, https://doi.org/10.1002/ptr.2650080113.

47. Umamaheswari, M., AsokKumar, K., Somasundaram, A., Sivashanmugam, T., Subhadradevi, V., Ravi, T.K., Xanthine oxidase inhibitory activity of some Indian medical plants. *J. Ethnopharmacol.*, 109, 3, 547–551, 2007, https://doi.org/10.1016/j.jep.2006.08.020.
48. Agarwal, R., Gupta, R., Yadav, R., Asati, V., Rathi, J.C., Anti-Inflammatory activity of seeds extract of Datura stramonium against carrageenan induced paw edema on albino wistar rats. *J. Pharm. Biol. Sci.*, 7, 1, 41–46, 2019, https://10.18231/j.jpbs.2019.007.
49. Chandan, G., Kumar, C., Chibber, P., Kumar, A., Singh, G., Satti, N.K., Saini, R.V., Evaluation of analgesic and anti-inflammatory activities and molecular docking analysis of steroidal lactones from Datura stramonium L. *Phytomedicine*, 89, 153621, 2021, https://doi.org/10.1016/j.phymed.2021.153621.
50. Jaafar, F.R., Ajeena, S.J., Mehdy, S.S., Anti-inflammatory impacts and analgesic activity of aqueous extract Datura innoxia leaves against induced pain and inflammation in mice. *J. Entomol. Zool. Stud.*, 6, 1894–1899, 2018.
51. Muthusamy, P., Nivedhitha, M., Jayshree, N., Analgesic and anti-inflammatory activities of Datura metel Linn. Root in experimental animal models. *Res. J. Pharm. Technol.*, 3, 3, 897–899, 2010.
52. Iqbal, S., Sivaraj, C., Gunasekaran, K., Antioxidant and anticancer activities of methanol extract of seeds of Datura stramonium L. *Free Radic. Antioxid.*, 7, 2, 184–189, 2017, https://doi.org/10.5530/fra.2017.2.28.
53. Girmay, S., Preliminary phytochemical screening and *in vitro* antimicrobial activity of Datura stramonium leaves extracts collected from Eastern Ethiopia. *Res. J. Chem. Sci.*, 22, 31, 606X, 2015.
54. Kalim, M., Hussain, F., Ali, H., Ahmad, I., Iqbal, M.N., Antifungal activities of methanolic extracts of Datura inoxia. *PSM Biol. Res.*, 1, 2, 70–73, 2016.
55. Bachheti, R.K., Rai, I., Mishra, V.K., Joshi, A., Antioxidant and antimicrobial properties of seed oil of Datura metel. *J. Environ. Biol.*, 39, 2, 182–188, 2018, DOI:10.22438/jeb/39/2/MRN-341.
56. Usha, K., Singh, B., Praseetha, P., Deepa, N., Agarwal, D.K., Agarwal, R., Nagaraja, A., Antifungal activity of Datura stramonium, calotropis gigantea and azadirachta indica against fusarium mangiferae and floral malformation in mango. *Eur. J. Plant Pathol.*, 124, 4, 637–657, 2009, https://doi.org/10.1007/s10658-009-9450-2.
57. Choudhary, M., Sharma, I., Agrawal, D.C., Dhar, M.K., Kaul, S., Neurotoxic potential of alkaloids from thorn apple (Datura stramonium.): A commonly used indian folk medicinal herb, in: *Medicinal Herbs and Fungi*, pp. 391–420, Springer, Singapore, 2021, https://doi.org/10.1007/978-981-33-4141-8_16.
58. Kuraki, T., Bronchodilators for COPD: At what stage should therapeutic intervention be initiated?, in: *Chronic Obstructive Pulmonary Disease*, pp. 211–243, Springer, Singapore, 2017.

59. Papi, A., Fabbri, L.M., Kerstjens, H.A., Rogliani, P., Watz, H., Singh, D., Inhaled long-acting muscarinic antagonists in asthma–A narrative review. *Eur. J. Intern. Med.*, 85, 14–22, 2021, https://doi.org/10.1016/j.ejim.2021.01.027.
60. Soni, P., Siddiqui, A.A., Dwivedi, J., Soni, V., Pharmacological properties of Datura stramonium L. as a potential medicinal tree: An overview. *Asian Pac. J. Trop. Biomed.*, 2, 12, 1002–1008, 2012, https://doi.org/10.1016/S2221-1691(13)60014-3.
61. Ibrahim, M., Siddique, S., Rehman, K., Husnain, M., Hussain, A., Akash, M.S.H., Azam, F., Comprehensive analysis of phytochemical constituents and ethnopharmacological investigation of genus Datura. *Crit. Rev. Eukaryot. Gene Expr.*, 28, 3, 223–283, 2018, DOI: 10.1615/CritRevEukaryotGeneExpr.2018022531.
62. Al-Snafi, A.E., Arabian medicinal plants with dermatological effects-plant based review (part 1). *IOSR J. Pharm.*, 8, 10, 44–73, 2018.
63. Falcão, H.S., Mariath, I.R., Diniz, M.F.F.M., Batista, L.M., Barbosa-Filho, J.M., Plants of the American continent with antiulcer activity. *Phytomedicine*, 15, 1-2, 132–146, 2008, https://doi.org/10.1016/j.phymed.2007.07.057.
64. Murthy, B.K., Nammi, S., Kota, M.K., Rao, R.K., Rao, N.K., Annapurna, A., Evaluation of hypoglycemic and antihyperglycemic effects of Datura metel (linn.) seeds in normal and alloxan-induced diabetic rats. *J. Ethnopharmacol.*, 91, 1, 95–98, 2004, https://doi.org/10.1016/j.jep.2003.12.010.
65. Dhiman, A.K. and Kumar, A., *Ayurvedic drug plants*, Daya Publishing House, New Delhi, 2006.
66. Kirtikar, K.R. and Basu, B.D., Indian medicinal plants, in: *Indian Medicinal Plants*, 1935.
67. Gulco, P., *Medicinal plants in Folklores of Bihar and Orissa*, p. 25, CCRUM, New Delhi, 2001.
68. Bhattacharjee, S.K., *Handbook of medicinal plants*, Aavishkar Publishers, Jaipur (India), 2000.
69. Kirtikar, K.R. and Basu, B.D., *Indian medicinal plants*, vol. 03, pp. 1783–1792, Lalit Mohan Basu, New Delhi, India, 2012.
70. Dar, G.H., Bhagat, R.C., Khan, M.A., *Biodiversity of the Kashmir Himalaya*, Valley Book House, India, 2002.
71. Bhat, N.A., *Some common wild flowers of Srinagar*, Makoff Printers, Delhi, 2002.
72. Lone, F.A., *The exploration of Uri sector, Kashmir valley*, Shipra, New Delhi, 2005.
73. Freye, E., Toxicity of Datura stramonium, in: *Pharmacology and Abuse of Cocaine, Amphetamines, Ecstasy and Related Designer Drugs*, pp. 217–218, Springer, Dordrecht, 2009, https://doi.org/10.1007/978-90-481-2448-0_34.

74. Roemer, H.C., Both, H.V., Foellmann, W., Golka, K., Angel's trumpet and the eye. *J. R. Soc. Med.*, 93, 6, 319–319, 2000.
75. Burns, M.J., Linden, C.H., Graudins, A., Brown, R.M., Fletcher, K.E., A comparison of physostigmine and benzodiazepines for the treatment of anticholinergic poisoning. *Ann. Emerg. Med.*, 35, 4, 374–381, 2000, https://doi.org/10.1016/ S0196-0644(00)70057-6.
76. Shonle, I. and Bergelson, J., Evolutionary ecology of the tropane alkaloids of Datura stramonium L. (Solanaceae). *Evolution*, 54, 3, 778–788, 2000, https://doi.org/10.1111/j.0014-3820.2000.tb00079.x.
77. List, G.R., Spencer, G.F., Hunt, W.H., Toxic weed seed contaminants in soybean processing. *J. Am. Oil Chem. Soc.*, 56, 8, 706–710, 1979, https://doi.org/10.1007/BF02663046.

10

Bioactive Properties and Health Benefits of *Amaranthus*

Nisha Singhania[1], Rajesh Kumar[1*], Pramila[2], Sunil Bishnoi[1], Aradhita B. Ray[1] and Aastha Diwan[1]

[1]*Department of Food Technology, Guru Jambheshwar University of Science & Technology, Hisar, Haryana, India*
[2]*Pt. Jawahar Lal Nehru Government College, Faridabad, Haryana, India*

Abstract

Amaranthus (Amaranthus viridis L.) or *Amaranth* is generally taken from the word 'Anthos' a Greek word that means evergreen flower, belongs to Amaranthaceae family and Amaranthus genus; it is native to Mesoamerica, Mexico and Central America. Grain of amaranth is a pseudo-cereal with unique nutritional and agronomic attributes. Amaranth contains significant amount of starch (29.36 %), essential protein (14.25 %), fat or lipid (8.18 %), and dietary fiber (4.04 %) these nutritional values varied from species to species. Amaranth also contained a significant amount of mineral especially potassium magnesium calcium and iron; vitamins include riboflavin, vitamin C, folic acid and vitamin E. Amaranth leaves are an ideal source of different antioxidants viz., anthocyanins, betalain, carotenoids, chlorophylls, β-cyanin, β-xanthin, and other phytochemicals such as ascorbic acid, phenolic acids, and flavonoids. Phytochemicals such as betaxanthins, betacyanins, rutin and isoquercetin; apart from these some major phenolic compounds like vanillic acid, flavonoids, gallic acid, salicylic, sinapic acid, ferulic, syringic, ellagic, p-coumaric, quercetin, kaempferol, and epigallocatechin gallates are also present in amaranth species. Phenolic compounds are identified in its seed are caffeoylglucaric acid, coumaroylglucaric acid, feruloylglucaric acid were glucaricisomers, caffeoylquinic acid, coumaroylquinic acid and feruloylquinic acid. These bioactive compounds act as a natural defense system against several diseases such as anti-gastric ulcer, gastroprotective, anti-colorectal cancer activity, anti-inflammation activity, atherosclerosis, arthritis, cardiovascular diseases, cataract, emphysema, retinopathy, and neurodegenerative diseases.

**Corresponding author*: rajeshsheoran92@gmail.com

Prerna Gupta, Navnidhi Chhikara and Anil Panghal (eds.) Harvesting Food from Weeds, (351–384)

Keywords: *Amaranthus*, bioactive compounds, antioxidant, polyphenols, phytochemicals

10.1 Introduction

The word *Amaranthus (Amaranthus viridis L.)* or *Amaranth* is generally taken from the word "Anthos," which is a Greek word that means ever-lasting or unwilting flower. This belongs to Amaranthaceae family and Amaranthus genus. The genes of amaranth contains about 400 species, those are spread all over the world. Species of the amaranth are divided on the basis of utilization, such as vegetable amaranth, grain amaranth, seed amaranth, and weedy amaranth. Amaranth species is native to Mesoamerica, and is extensively found in Mexico and Central America [1]. A wide range of species also cultivated and utilized in Asia, South-eastern and eastern Africa. Amaranth species were consumed as a vegetable form in subtropics and tropical of the Old World during colonial times. Currently, it is used

Table 10.1 Vernacular name of amaranthus from different origin.

Sr. no.	Name of country	Local name of Amaranthus
1	India	Chaulai, Rajgira, Ramdana, Kantachaulai, Tanduliya, kuppacheera
2	China	Een choy, yin choy, in-tsai, hsientsai, xiancai
3	Indonesia and Malaysia	Bayam
4	Laos	Pak hom
5	Philippines	Kulitis
6	Sri Lanka	Thampala
7	Thailand	Pak khom hat, pakkhomsuan
8	Vietnam	Yan yang
9	Peru	Anchita, achos, achis, incajataco, coimi, kiwicha
10	Bolivia	Coimi, millmi
11	Ecuador	Sangoracha, alaco
12	Caribbean	Calaloo

as leafy vegetable, highly consumed in Togo, Leone, Benin, and Sierra, and is a significant part of diet in low-lying area like Tanzania, Nigeria, Kenya, and DR Congo. It is often used as vegetable in many tropical areas other than Africa such as India, Bangladesh, Sri Lanka, and the Caribbean. Bangladesh species has big fleshy stems, which are consumed with leaves. *A. cruentus* is grown as a leaf vegetable throughout South-East Asia and in Indonesia, it is grown in mountain areas, where the climate is too cold for the more common amaranth. Grain amaranth is produced commercially in hot and dry areas of the United States, Argentina, and China. These species are known with their common name according to their growing regions (Table 10.1). The crop can grow on marginal lands that can adapt to a wide range of environments with a tolerance of drought conditions as well as salinity condition. The grain amaranth is a pseudo-cereal with unique nutritional and agronomic attributes [2].

10.2 Species

Amaranthus species are widely spread all over the word and these varies according to the cultivating environment. The following are the identified species of the amaranthus leafy vegetable. *A. acanthobracteatus, A. acanthochiton, A. acutilobus, A. albus* L., *A. anderssonii, A. Arenicola, A. asplundii, A. atropurpureus, A. aureus, A. australis, A. bahiensis, A. bengalense, A. blitoides, A. blitum* L., *A. brandegeei, A. brownie, A. californicus, A. cannabinus* (L.), *A. capensis, A. cardenasianus, A. caudatus* L., *A. celosioides, A. centralis, A. clementii, A. cochleitepalus, A. commutatus, A. congestus, A. crassipes, A. crispus, A. cruentus* L., *A. cuspidifolius, A. deflexus* L., *A. dinteri, A. floridanus, A. furcatus, A. graecizans* L., *A. grandifloras, A. greggii, A. hunzikeri, A. hybridus* L., *A. hypochondriacus* L., *A. induratus, A. kloosianus, A. lepturus, A. lombardoi, A. looseri, A. macrocarpus, A. minimus, A. neei, A. obcordatus, A. palmeri, A. paraganensis, A. pedersenianus, A. persimilis, A. peruvianus, A. polygonoides* L., *A. powellii, A. acanthobracteatus, A. acanthochiton, A. acutilobus, A. albus* L., *A. anderssonii, A. Arenicola, A. asplundii, A. atropurpureus, A. aureus, A. australis, A. bahiensis, A. bengalense, A. blitoides, A. blitum* L., *A. brandegeei, A. brownie, A. californicus, A. cannabinus* (L.), *A. capensis, A. cardenasianus, A. caudatus* L., *A. celosioides, A. centralis, A. clementii, A. cochleitepalus, A. commutatus, A. congestus, A. crassipes, A. crispus, A. cruentus* L., *A. cuspidifolius, A. deflexus* L., *A. dinteri, A. floridanus, A. furcatus, A. graecizans* L., *A. grandifloras, A. greggii, A. hunzikeri, A. hybridus* L., *A. hypochondriacus* L., *A. induratus, A. kloosianus, A. lepturus, A. lombardoi, A. looseri, A. macrocarpus, A. minimus, A. neei,*

A. obcordatus, A. palmeri, A. paraganensis, A. pedersenianus, A. persimilis, A. peruvianus, A. polygonoides L., *A. powellii, A. praetermissus, A. pumilus, A. retroflexus* L., *A. rhombeus, A. rosengurttii, A. saradhiana, A. scariosus, A. schinzianus, A. scleranthoides, A. scleropoides, A. sonoriensis, A. muricatus, A. sparganicephalus, A. spinosus* L., *A. squamulatus, A. standleyanus, A. tamaulipensis, A. thunbergia, A. torreyi, A. tortuosus, A. tricolor* L., *A. tuberculatus, A. tucsonensis, A. tunetanus, A. undulatus, A. urceolatus, A. viridis* L., *A. viscidulus, A. mitchellii, A. vulgatissimus, A. watsonii, A. wrightii* (http://www.theplantlist.org/browse/A/Amaranthaceae/Amaranthus;https://www.newworldencyclopedia.org/entry/Amaranth) [3].

Amaranths genes occupied about 70 species, of which about 40 are native to the Americas. Among these 17 species are classified as edible leaves and 3 species for grain amaranths. *A. cruentus* belongs to both categories (leaves and grain). The three main domestic species of amaranth are, *A. Caudatus* L., *A. cruentus* L., and *A. hypochondriacus* L., which originated in America and mainly grow for grain propose.

10.3 Plant Physiology and Environmental Factors for Growth of Amaranth

Amaranth is a summer-tolerant green leafy vegetable and also a grain very similar to quinoa and couscous as shown in Figure 10.1. Amaranth is adapted to all type of soil but grow best in fertile or well-drained loam soil. Amaranth is basically herbaceous shrub or plant, having three or five stamens and tepals on flowers, whereas its leaves are about 6.5 to 15 cm with an elliptical or oval shape and its grain structure remains consistent across the family. Amaranth's primary roots have a deeper spread along with secondary fibrous root structures. Its flowers are bisexual and function as unisexual having a small, bristly, or prickly texture of perianth and pointy bracts symmetrically [4]. Species in this genus are either belongs to monecious (*A. hybridus*) or dioecious (*A. palmeri*). Fruits are in the form of capsules that split while maturity to release seed. The seeds are circular, shiny and have smooth seed coats with 1 to 1.5 mm in diameter. The panicle is harvested 200 days after cultivation with approximately 1,000 to 3,000 seeds harvested per gram [5]. The soil for growing lettuce evenly suitable for amaranth. Seeds of amaranth should be sowed properly with one fourth inches depth in soil. The temperature preference for cultivating amaranth was a warm climate around 70°F with well-drained soil. Amaranth can successfully grow in the area having low or less than 250 mm of yearly rainfall. Initially, amaranth leaves grow slowly, but afterword it proliferates

Figure 10.1 Various amranthus species.

and the crop having low maintenance during its growth period. It can grow up to 5 feet height. Flowers bloom on long straight stems and enduring the first hard frost. Removing terminal buds of the amaranth plant will help in encouraging branching and also help in the development of tender or young leaves and shoots for a green salad.

10.4 Edible Part and Uses

Several amaranth species are useful as food crops and grown both for their leaves and edible seeds. Amaranth seeds are very nutritious and considered pseudo cereals, which is nongrass seed like cereals grains. Many of species used a common garden ornamental such as love-lies-bleeding (*A. caudatus*), prince's feather (*A. hypochondriacus*), and Joseph's coat (*A. tricolor*). Whereas some species are considered as weed seed. However, there is not often a distinction between vegetable and grain amaranth species because young leaves of grain species are also consumed as vegetables. Leaves and seeds of amaranth are high in nutrition and very good sources of dietary fibre, calcium, and iron; its seeds are also contained a high amount of protein. Amaranth leaves possess great nutritional value along with antioxidative properties [1]. Leaves are consumed like spinach; they can be eaten fresh or cooked and often used in salads or soups. Seeds taste alike nuts that can be popped, cooked into porridge, and can be grounded in a powder that is used in grain flour for baked goods [6]. Red-leaf vegetable amaranth has an attractive color splash with medium green leaves overlay burgundy-red that is a delightful part of summer time salads. Black seeded Amaranth varieties tend to remain quite gritty when cooked whereas golden or lighter colored seeds tend to cook better, and all of them have delectable greens salad. Tampala is one of the tastiest varieties grown for its greens. Amaranth starch is used as a thickening in soups, gravies, and sauces, and it's also found in breakfast cereals, muffins, pastries, snacks, kinds of pasta, and health foods. Cosmetics, biodegradable films, paper coatings, and laundry starch are some of the other current and future commercial uses of amaranth starch [7].

10.5 Nutritional Properties

10.5.1 Carbohydrates

Carbohydrates are the major portion of the amaranth grain. Its constituents about 65% to 72% of the grains composition. Starch is the prime carbohydrate present in amaranth grains, accounting for 48–69% of its total dry weight. The granules of starch range in size from 0.8 to 2.5 m in diameter and amaranth starch was quickly degraded by α-amylases due to its small size [8]. The size of amaranth starch granules and their distribution have an impact on the crop's functional attributes. Compound-starch particles made composed of tiny granules make up the majority of starch raw materials. Particles clump together, reducing surface area and generating distinctive compounds. As the granule diameter shrinks, the specific surface area of starches increases dramatically. However, the amylose concentration of amaranth starch is significantly lower than that of other cereal starches, ranging from 0.1% to 11.1%, depending on genotype [9]. The amaranth starch granule's compact size and high amylopectin percentage comprise the majority of the starch's physical features. Amaranth starch has superior freeze-thaw and retrogradation stability than corn starch, responsible feature is its higher gelatinization temperature, high viscosity, higher water-binding capacity, lower solubility, higher water uptake with higher water activity values, and higher swelling power and enzyme susceptibility.

Resistant starch (RS) is a type of starch that occurs naturally in food, like dietary fibre. This starch is not digested during the digestion process of food and hence, resistant starch makes its way to the colon, where it is fermented by the bacterial biota. RS has various health benefits such as it helps in lowering the blood lipids, decreasing the possibility of colon cancer, etc. Capriles, *et al.* (2008) [10] discovered that raw amaranth seeds had an RS level of 0.05 percent and an RS/total starch proportion of 0.86 percent. The RS content of a food is determined by the starch characteristics (granule morphology, size, and amylose/amylopectin ratio of starch), as well as the testing method and food processing parameters used. Roasted grains of the amaranth exhibited higher RS content than raw grains. This could be because amaranth starch contains roughly 1% amylose, and after cooling, retrogradation of amylose chains occurs, increasing the RS content of roasted seeds. Cooking, on the other hand, tends to reduce RS levels, owing to the gelatinization of starch [11]. Amaranth contains about 3% to 5% of monosaccharides and disaccharides. When compared to wheat grain and non-starch polysaccharide components, the predominant saccharide observed in the carbohydrates profile of amaranth seeds was sucrose, which had a two to three times higher amount. The dominant sugar present in amaranth is sucrose with a range of 0.58 to 0.75 g/100 g, while total sugar content is found in the range of 1.84 to 2.17 g/100 g in *A. cruentus* and *A. caudatus* species according to [11]. Inositol, maltose, raffinose, and stachyose were also found in minor proportions in the amaranth seed. Amaranth seeds have about 1% inositol, with a minor amount of glucose, fructose, and other monosaccharides (0.05–0.67%), as well as disaccharides like raffinose (0.27–2.3%), sucrose (0.4–2%), maltose (0.02–0.36%), and stachyose (0.02–0.36%). Amaranth has a higher raffinose level than wheat, but it is lower than corn. Low-molecular-weight carbohydrates were found in *A. cruentus* and *A. caudatus* in amounts ranging from 0.12 to 0.17 g/100 g for fructose, 0.34 to 0.42 g/100 g for glucose, 0.02 to 0.04 g/100 g for inositol, 0.24 to 0.28 g/100 g for maltose, 0.39 to 0.48 g/100 g for raffinose, and 0.15 to 0.13 g/100 g for stachyose [11].

10.5.2 Dietary Fiber

Amaranth seeds have a crude fiber content ranging from 1.5% to 4.5% (Table 10.2), where the soluble dietary fiber (SDF) portion accounts for 14% of the total fibre content. Soluble dietary fibre containing pectin, uronic acid, and undigested biopolymers such as glucose, arabinose, xylose, mannose, and galactose were also found in the SDF. Whereas, lignin, cellulose, and hemicellulose make up the insoluble dietary fibre portion. Both soluble and insoluble fibres have been shown various human health-improving effects. In amaranth species such as *A. caudatus*, *A. cruentus*, and *A. hypochondriacus* contains total dietary fibre content (includes soluble and insoluble fibre) ranges from 9.8% to 14.5% [12]. The crude fibre content of amaranth grain was about 3.6-4.2% [7], which is significantly higher than that of wheat (2.6%), maize (2.3%), and rice (0.9%) [13]. When these dietary fibres were exposed in the large intestine that undergoes bacterial fermentation and generates acetic acid, butyric acid, lactic acid, and some fatty acid, which promote health beneficial effects.

10.5.3 Protein

The pseudo-cereal's nutritional value is mainly considered due to their protein content and amaranth contains higher protein content than buckwheat or quinoa. Amaranth contains about 16% to 18% of protein (Table 10.2), as compared with other cereals. The main constituents of amaranth protein are globulins, albumins and contain a small proportion of prolamin proteins, which is considered as a major compound of gluten protein fraction [15]. Gluten shows unique properties that as the ability to retain gas, stretchability, coagulability, extensibility, and form a protein-starch matrix in bakery products [16]. Its presence determines the overall appearance and rheological properties of cereal-based products. Amaranth protein has a significant effect on human health due to the presence of an excellent amount of amino acids that a body cannot be synthesized. It is rich in lysine content (three-time

Table 10.2 Quantitative analysis of macro and micro nutrients of different varieties of amaranthus grain.

Parameters	***A. caudatus***	***A. cruentus***	***A. hybrid***	***A. hybridus***	***A. hypochondriacus***
Moisture (%)	10.85	11.17	10.01	9.30	11.54
Ash (%)	3.66	3.14	4.94	4.40	4.20
Protein (%)	14.25	15.81	15.78	17.89	13.82
Starch (%)	29.36	31.21	29.05	38.01	37.65
Sugar (%)	1.63	1.66	1.18	1.62	1.95
Fat (%)	8.18	8.68	5.92	7.46	4.85
Crude Fibre (%)	4.04	3.48	1.89	2.33	3.25
Iron (mg/100 g)	144.14	117.94	102.04	145.94	149.71
Zinc (mg/100 g)	43.19	59.49	47.19	42.20	46.02
Copper (mg/100 g)	4.14	6.62	4.64	4.50	3.37
Sodium (mg/100 g)	8.95	1.39	2.57	Nd	0.97
Manganese (mg/100 g)	63.55	136.44	87.13	94.05	100.84
Potassium (mg/100 g)	5137.57	4281.06	4742.75	5172.61	5362.34
Calcium (mg/100 g)	1287.77	1642.45	1620.52	1115.90	1392.09
Magnesium (mg/100 g)	2219.15	2035.24	2193.16	1760.23	2199.70
Aluminum (mg/100 g)	111.09	47.55	86.59	94.98	110.19
Selenium (mg/100 g)	0.24	0.78	0.82	0.23	0.27

Source: Akin-Idowu *et al.*, 2017 [14].

higher than wheat flour) which is found low in corn, wheat, and rice [8, 17]. Besides lysine, it is also significantly rich in sulfur-containing amino acids that are lower in pulse crops. It was found that according to FAO, amaranth protein contains near to ideal composition for adults. As per FAO/WHO Nutritionist's Protein Value Chart, the score of amaranth is also higher than other seeds such as barley, soybean, wheat, and maize. A score of 100 is considered as ideal score and that is perfect balance of essential amino acids on the nutritionist's scale of protein quality based on the amino acid composition. Amaranth protein scored 75 (highest content of lysine), as compared to soybeans which are 68, and peanut that has 52 scores, according to FAO/WHO/UNU, 2986). The mixture of amaranth flour and corn flour scores almost reach to perfect 100, because one amino acid deficiency is covered by other flour and this combination can use in making bread, tortillas, and other food products. Therefore, it is considered a perfect protein-contained food commodity as compared with other (e.g., egg), that meets virtually all the requirements of the body. The amaranth leaves contain protein content in the range of 17.2% to 32.6% on a dry weight basis (Table 10.3), and its values depend on the amaranth species, their genotype, growing condition, and also the pattern of amino acid present in amaranth. Where major fraction is of albumin that covers about 40% of protein followed by glutelin (25–30%), globulin (20%), and prolamin (2–3%), these values vary because of extraction and fractionation procedures and also genetical. According to Gorinstein *et al.* (2004) [18], alcohol-soluble prolamin protein was found in a lower amount (1.2–1.4%), and another portion of prolamin was 0.48% to 0.79% [19]. Scanning electron microscopy (SEM) and SDS-PAGE shows a very close similarity between the protein fraction of amaranth and soybean. According to the sedimentation coefficient, two main classes of globulins can be differentiated as 7S and 11S globulins. In amaranth, similar 7S (conamaranthin) and 11S (amaranthin) storage globulins are found. Water-soluble protein fractions such as albumins and globulins are responsive to thermal treatment, these could be degraded on heating, on the other hand, alcohol-soluble protein fractions (prolamins) also show a similar tendency towards thermal treatment.

10.5.4 Lipids

Amaranth seeds contain 4.5% to 9% lipids (Tables 10.2 and 10.3), and their fat content is two to three times higher than that of other cereals, however, it varies widely between species. More than 75% of the fats in amaranth oil are unsaturated. It had a higher amount of linoleic acid (approximately 47 % fatty acid), followed by oleic acid (24 %) and palmitic acid (23.4%),

Table 10.3 Quantitative analysis of macro and micro nutrients of different varieties of amaranthus leaves.

Nutritional parameters	***A. cruentus***	***A. hybridus***	***A. lividus***	***A. hypochondriacus***
Moisture (%)	3.15–3.78	1.36–4.46	2.85–3.37	1.60–4.41
Ash (%)	14.08–19.95	14.14–18.37	14.77–16.88	15.15–17.65
Crude Protein (%)	14.69–18.90	14.14–23.28	19.97–20.81	14.48–21.50
Vitamin C (mg/100 g)	44.30–117.79	48.81–93.70	52.33–86.45	30.30–115.57
Iron (mg/100 g)	17.26–27.06	15.51–31.17	23.50–15.63	14.21–22.82
Zinc (mg/100 g)	1.50–3.46	1.05–2.87	1.03–1.36	1.07–2.83
Calcium (mg/100 g)	1883–1984	1512–2332	2028–2354	1647–2381
Potassium (mg/100 g)	1539–1677	1320–1658	1473–1563	1329–1592
Magnesium (mg/100 g)	468.7–490.9	383.4–513.9	478.7–507.2	414.6–484.4

Source: Saker *et al.*, 2020 [20].

and lower levels of stearic acid (4.16%) and linolenic acid (about 3.9 % and 0.7 %, respectively, with a high degree of unsaturation. Amaranth oil has a fatty acid profile that was extremely comparable to cottonseed and sesame oils [21]. Triglycerides, sterols, phospholipids, glycolipids, tocopherols, and hydrocarbons are among the other lipids identified in amaranth [11]. Squalene, a lipid typically derived from the livers of deep-sea dogfish, is found in substantial proportions in amaranth, ranging from 2.4% to 8% of the extractable oil (0.3–0.4% of the total seed mass). Squalene is a biological precursor of cholesterol and other steroids and is a highly unsaturated open-chain triterpene [22]. Shin *et al.* (2004) [23] discovered that squalene lipid from amaranth lowers down the cholesterol level by removing steroids that interfere with cholesterol absorption through the fecal. It is more effective than that of shark-liver squalene. Amaranth lipid also aids in the reduction of serum and hepatic cholesterol, as well as triglycerides. Because of its high squalene concentration, amaranth oil will gain popularity as an alternative plant source of squalene and simple vacuum distillation can easily extract it from amaranth grain.

Phospholipids from amaranth make up a 3.6% oil fraction, or roughly 5% cephalin was 13.3%, lecithin was 16.3%, and phosphoinositol was 8.2%, in addition to phospholipids. Amaranth oil contains 24.6 parts per million sterols, and all of the sterols in amaranth are esterified. The percentage of free (non-esterified) sterols in most vegetable oils is usually substantially higher. Clerosterol (42%) is the most abundant sterol in amaranth oil, and it has antimicrobial properties [24]. Due to the presence of high sterol content in amaranth, it is majorly used in pharmaceutical and medicinal purposes [21].

10.5.5 Minerals

Amaranth has a mineral concentration that is roughly twice that of other grains. It differs by variety and is further influenced by processing. The grain and leafy portion of amaranth contain a significantly higher amount of Ca, Mg, Fe, K, and Zn (Tables 10.2 and 10.3). Sanz-Penella *et al.* (2012) [25] tested the influence of phytates on *in-vitro* iron absorption by supplementing wheat bread with whole amaranth flour (0%, 20%, and 40%), depending on the amount used in bread formation, whole amaranth flour has been observed to limit accessible Fe. However, using up to 20% whole amaranth flour into bread formulations looks to be a promising strategy for increasing bread nutritional value, since it provides a useful vehicle for increasing the concentration of accessible Fe in bread. Manganese (425.2

mg/100 g) and copper (1.25 mg/100 g) are found in significant amounts, along with nickel, chromium, zinc, and selenium are all abundant in amaranth seeds. Iron level in amaranthis 29.35 mg/100 g, which is several times more than in typical grains. 159 mg calcium, 248 mg magnesium, 557 mg phosphorus, 508 mg potassium, 4 mg sodium, 2.87 mg zinc, 0.53 mg copper, 3.33 mg manganese, and 18.7 g selenium were found in 100 g of uncooked amaranth grain [26].

10.5.6 Vitamins

Overall, amaranth isn't a particularly good contributor to vitamins. Amaranth is a good source of riboflavin, vitamin C, and especially folic acid and vitamin E. [11]. Folic acid levels were estimated to be 82 g/100 g, which is twice the amount found in wheat (43 g/100 g). Bruni *et al.* (2002) [27] observed total tocopherol values of 100–129 mg/kg in amaranth seeds using supercritical fluid extraction. Another member of the vitamin E family is tocotrienols, which are essential molecules having anticholesterolemic properties. Amaranth grains possess a significant amount of tocotrienols, ranging from 5.02 to 11.47 mg/kg of seed [28]. Ogrodowska *et al.* (2014) [29] found that *A. Cruentus* seeds and their *products were high in tocopherols. β-tocopherol is the dominating homolog of vitamin E identified in amaranth seeds, representing around 38%* of tocopherols, according to HPLC analysis. About 32% contribution of δ-tocopherol and an 18% contribution of α-tocopherol were revealed. The lowest concentration of γ-tocopherol was estimated to be 1.29 mg/100 g, contributing to 12% of all tocopherols. Ultrasonic and supercritical carbon dioxide method revealed α-tocopherol (17.8 to 20.6 mg/kg d.w), β-tocotrienol (35.4 to 39.8 mg/kg d.w), γ-tocotrienol (2.0 to 4.0 mg/kg d.w), δ-tocotrienol (15.5 to 18.4 mg/kg d.w).γ- and δ-tocotrienols decrease HMG-CoA reductase activity in a dose-dependent manner, whereas α-tocopherol has antagonistic properties with opposing effects by boosting reductase activity [30].

10.5.7 Bioactive Compounds

10.5.7.1 Polyphenolic Compounds

Polyphenolic compounds are considered as the significant source of natural antioxidants in plant food. The amaranth leaves are an ideal source of pigments such as anthocyanins, betalain, carotenoids, chlorophylls, β-cyanin, β-xanthin, and other phytochemicals such as ascorbic acid, phenolic acids, and flavonoids that are responsible for natural antioxidants [31]. Some other active constituents like alkaloids, glycosides, steroids, terpenoids, saponins, and catechuic tannins. The antioxidants present in amaranth act as a natural defence system against several diseases such as atherosclerosis, arthritis, cancer causing properties, cardiovascular diseases, cataract, emphysema, retinopathy, and neurodegenerative diseases [1]. The composition and level of phytochemicals present in amaranth may vary among species due to environmental conditions and also, processing and cooking alter the phytochemical composition. Amaranth possessed high content of squalene that makes it more preferable natural source of squalene over marine sources. Values of phytochemicals such as alkaloid, flavonoid, phenols, saponin, tannin, phytic acid, and hydrocyanic acid are influenced by different processing (sun drying, oven drying, steaming, etc.) [32]. Phytochemical properties in *A. viridis* were investigated that alkaloid, tannins, saponins, and glycosides were present in both leaf and seed of amaranth. Whereas leaves contained a higher percentage of these phytochemicals than seed, leaves contain 13.14% of alkaloids whereas seed contain 11.42%, similarly, tannins in leaves 6.07% and seed 5.96%, saponins in leaves 53% and seed 32%. It was observed that 80% of the methanolic content of leaves and seeds extract has 2.81 mg GAE/g and 3.27 mg GAE/g of phenolic content whereas, 18.4 mg QE/g and 2.51 mg QE/g of flavonoids [33, 34].

A. tricolor and *A. hypochondriacus* leaves are storehouse of phytochemicals such as betaxanthins, betacyanins, rutin, and isoquercetin; apart from these some major phenolic compounds like vanillic acid, flavonoids, gallic acid, salicylic, sinapic acid, ferulic, syringic, ellagic, and p-coumaric were present. *A. viridis* and *A. spinosus* species contained 267.85 to 302.56 μg/g chlorophyll *a* content and 135.26 to 152.42 μg/g chlorophyll *b* content. A significant difference was observed in chlorophyll *a* and *b* content of *A. viridis* (413.61 μg/g) and *A. spinosus* genotypes (445.22 μg/g). It was observed that β-cyanins content variations in *A. spinosus* and *A. viridis* genotypes were 185.52 to 538.51 μg/g. Chlorophyll *a* (302.56 μg/g), chlorophyll *ab* (445.22 μg/g), β-cyanins (287.56 μg/g),and carotenoids content (92.87 mg/100 g) in *A. viridis* genotype, although *A. spinosus* possessed β-xanthins (274.96 μg/g), chlorophyll *b* (152.42 μg/g), betalains content (561.42 μg/g), and β-cyanins (286.46 μg/g) in a significant amount [35, 36]. Similarly, green and red amaranth contains a considerable amount of betalains, carotenoid content, chlorophyll *a*, chlorophyll *b*, chlorophyll *ab*, β-cyanins, β-xanthins. *A. viridis* variety had the highest values of chlorophyll *a*, chlorophyll *ab*, β-cyanins, and carotenoids content, while

A. spinosus exhibited the maximum values of chlorophyll *b*, β-xanthins, and betalains content [37]. β-carotene content in *A. viridis* and *A. spinosus* was ranged from 46.76 to 64.22 mg/100 g, both varieties exhibited a significantly high amount of β-carotene as compared to other green leafy vegetables (Sarker *et al.*, 2018b) [38]. *A. viridis* and *A. spinosus* varieties showed prominent variations in ascorbic acid content with a range of 44.62 to 107.45 mg/100 g, both species exhibited high ascorbic acid as compared to leafy vegetables. Ascorbic acid was the highest in *A. viridis* (107.45–106.64 mg/100 g) and the lowest in *A. spinosus* (44.62 mg/100 g). Significant variations can be observed in total polyphenol content (TPC) of *A. viridis* and *A. spinosus* which ranged from 25.98 to 46.72 µg GAE/g. *A. viridis* showed a higher phenolic content of 46.72 µg GAE/g, whereas *A. spinosus* exhibited a lower phenolic content (24.98 µg GAE/g). *A. viridis* and *A. spinosus*, both species had high flavonoids content (174.58 µg RE/g to 182.46 µg RE/g) [37]. Seeds of *A. hypochondriacus* species were extracted using different solvents (methanol, ethanol, and hexane) by the soxhlet extraction method and obtained total phenolic content (TPC) varied from 16 to 25 mg GAE/100 g (DW) [39]. There was found a significant difference between methanol, ethanol, and hexane extracted values. Similarly, antioxidant activity values varied from 25 to 30 mg TroloxE/100 g [40, 41]. Quercetin, kaempferol, and epigallocatechin gallates; three phenolic compounds were commonly found compounds in amaranth species. Another unbounded phenolic compound was protocatechuic acid, its amount was abundant in *A. cruentus* (13.6 ± 9.4 µmol/100 g) and gallic acid (400–440 µg/mg) predominantly found followed by *p*-hydroxybenzoic acid (8.5–20.7 µg/mg) in two varieties of *A. cruentus* seeds. The methanol and hot water extract contain tannins with a concentration of 0.316 mg TAE/g and 0.511 mg TAE/g [42]. Vanillic acid was found in some varieties, although *p*-coumaric and syringic acid were detected in the sprouted of amaranth, and ferulic acid and caffeic acid were found in leaves but not found in seeds. In *A. cruentus*, ferulic acid, *p*-hydroxybenzoic, and vanillic were observed in higher amounts while the concentration of caffeic, sinapic, and cinnamic acid were found in lower [43, 44].

Other phenolic compounds were caffeic, protocatechuic, *p*-coumaric, 4-hydroxybenzoic, syringic, rutin, ferulic, and salicylic acid found in various species of amaranth. Three major polyphenols were rutin, nicotiflorin, and isoquercitin identified and isolated in different varieties of amaranth [45]. Different amaranth varieties contain different values of rutin such as 70 µg/g for *A. hypochondriacus*, 11 µg/g for *A. retroflexus*, 99 µg/g for *A. hybridus*, 55 µg/g for *A. caudatus*, and 7 µg/g for *A. tricolor*.

The sconcentration of rutin was found maximum (10.1–4.0 μg/g) followed by the concentration of isoquercitrin (0.3–0.5 μg/g) and nicotiflorin (7.2–4.8 μg/g). *A. hypochondriacus* varieties contained 4.8–7.2 μg/g nicotiflorin. Vitexin and isovitexin, flavanones found in a minor amount, while flavonoids are present in abundance amount 1.53 mg catechin equivalents (CE)/g. Vitexin (410 μg/g) and isovitexin (266 μg/g) were detected in the seeds of *A. rawa* [46]. *A. hypochondriacus* and *A. tricolor* contain two main pigment compounds which are classified under the category of anthocyanins, betacyanins (182.7 to 784.8 betanin equivalent μg/g) and betaxanthins (374.7 to 655.4 indicaxanthin equivalent μg/g) [43]. In *A. blitum* genotypes, betacyanin ranged from 10.54 to 53.63 μg/100 g [47, 48], where betalains were found higher in *A. tricolor* (997.9 to 1375.9 ng/g) than *A. hypochondriacus* (557.5 to 777.3 μg/g) cultivars [36].

Amaranth seed bran contains lignans or resinols, which are part of phenolic compounds and form during the dimerization of cinnamic alcohols. Some other resinols derivative present in amaranth bran like hydroxymatairesinol, 7-oxomatairesinol, medioresinol, and secoisolariciresinol [40]. Phenolic content of three different amaranth varieties (*A. caudatus, A. hypochondriacus, and A. cruentus*) seed contains a higher amount than their stalks. These phenolic compounds were identified as caffeoylglucaric acid, coumaroylglucaric acid, feruloylglucaric acid were glucaricisomers, caffeoylquinic acid, coumaroylquinic acid, and feruloylquinic acid were quinic isomers, rutin, kaempferol-3-O-rutinoside, caffeic acid and quercetin glucoside [48, 49]. The prominent hydroxybenzoic acid was salicylic acid, found in red amaranth with a concentration of 34.68 μg/g and the second most abundant hydroxybenzoic acid was vanillic acid found in both, red and green amaranth with a concentration of 9.53 μg/g and 9.29 μg/g. Green amaranth also contained a high concentration of isoquercetin content with hyperoside (quercetin-3-galactoside that is present in both red and green amaranth [50].

Phytic acid interferes with mineral absorption, especially with zinc, and also leads to inhibiting the enzymatic digestion reaction by forming complexes with protein residues. Phytate content can be reduced up to 20% by cooking and heating; and with 48 hours germination, it is reduced by 22% [11]. Rutin and its metabolites are compounds that could imply the prevention of several pathologies by inhibiting the formation of the glycation end-products (AGE) preventing aging by forming a complex with heterogeneous molecules and helping in curing diabetes complications [51]. Quercetin, a polyphenol act as a free radical scavenger and show the protective action against oxidative damages. Nicotiflorinis used in potentially therapeutic and pharmacological treatment for curing cerebral ischemic illness and has a protective effect for reducing memory dysfunction.

10.5.7.2 Alkaloids

Alkaloids are basic compounds having amino acids that contain one or more heterocyclic nitrogen atoms that are derived from plant-based. These plant-based alkaloids are natural antioxidants having pharmacological activity and preventing various diseases such as malaria, cancer, diabetics, and cardiac dysfunction. Some species of amaranth reported the presence of alkaloids in a different part of plant such as leave or root of *A. Spinosus*, leave or seed of *A. Viridis*, leave of *A. Graecizans*, and leave or seed *A. Hybridus* [1, 52].

10.5.7.3 Sterols

Amaranth seed contains about 4% to 8% of lipid content that is low as compared to true cereals but biologically it was high due to its squalene content. Amaranth seeds are rich in unsaturated fatty acids, mostly linolenic acid (43%) followed by oleic acid (22%) with antioxidant properties, while saturated fatty acid was palmitic acid. The saturated fatty acids were detected in a lesser amount up to 26% to 27%, while unsaturated fatty acids were found in a higher amount up to 72% to 73% [53]. Amaranth oil also contained a high content of phospholipids, tocopherols, phytosterols. In *A. kharkov* and *A. andijan* varieties, linoleic acid was found as 40.1% and 47.3%, followed by oleic acid as 31.3% and 24.5% and palmitic acid as 23.19% and 22.2% respectively. Apart from these compounds, stearic, linolenic, arachidic, and myristic acids were identified in amaranth seed oil [54]. In *A. caudatus*, α-linoleic acid content was about 683 μg/g, and also rich source of eicosapentaenoic acid [55]. Mono-galactosyl and di-galactosyl compounds are a form of glycolipids that present 6.4% of total lipids, and phosphatidylcholine, phosphatidylethanolamine, and phosphatidylinositol are a form of phospholipids that present about 3.6% of total lipids. While their percentage varies depending on extraction methods and solvent used during extraction [56]. It was reported that about 20% of sterols are present in free form while 80% of sterols are in esterified form in *A. cruentus*.

Some other sterols identified in amaranth oil were terpenic alcohol and methyl sterols were found in amaranth oil, such as taraxerol, dammaradienol, *b*-amyrin, gramisterol, cycloartenol, citrostadienol [20]. Supercritical fluid extraction of amaranth oil with solvent extraction represents the following sterols: campesterol, 5,24-stigma stadienol, 24-metylenecholesterol, Δ7-stigmastenol, α-spinasterol, + sitosterol, stigmasterol, sitostanol, Δ5-avenastero, Δ7-ergosterol, Δ7-avenasterol, cycloartenol, citrostadienol [57]. α-Spinasterol and sitosterol, these sterols were present in higher concentrations than other sterols whereas, spinasterol content was ranged from 48% to 53%. Amaranth plants are good source of phosphorus, which is produced through the presence of phytic acid. It is mainly found in the hull therefore dehulling or extraction process reduces the amount of phytic acid. Amaranth seeds contain 0.3 to 0.6% of phytic acid which is responsible for lowering the cholesterol level in the human body. Some amaranth species also contain saponins, a seed of *A. cruentus* has 0.09% to 0.1% dry weight of saponins. β-D-glucopyranosyl (1-4)-β-D-glucopyranosyl(1-4)-β-D-glucuronopyranosyl(1-3)-oleanolic acid, β-D-glucopyranosyl(1-2)-β-D-glucopyranosyl(1-2)-β-D-glucopyranosyl(1-3)-α-spinasterol, and β-D-glucopyranosyl(1-4)-β-D-glucopyranosyl (1-3)-α-spinasterol were isolated from the root of *A. spinosus*. The other four saponins were also isolated, these were 3-beta-(O-glucopyranosyl) ester, 3-beta-O-(alpha-L-rhamnopyranosyl(1-3)-beta-glucuronopyranosyl)-2 beta, 3 beta, 23-trihydroxyolean-12-en-28-oic acid 28-O-(beta-D-glucopyranosyl) ester, 3 beta-O-(beta-glucopyranosyl)-2 beta, 3 beta-dihydroxy-30 norolean-12,20(29)-diene-23,28-dioicacid 28-O-(beta-D-glucopyranosyl) ester together with known as chondrillasterol (5 alpha-stigmasta-7,22-dien-3 beta-ol) and its 3-O-glucopyranoside [4].

10.6 Non-Nutritional Compounds

Non-nutritive compounds found in amaranth seeds include trypsin and chymotrypsin inhibitors, some phenol compounds such as saponins and tannins, phytates, and phytohemagglutinins. They differ from cereal grains or other raw materials in terms of content, which is either the same or lower. They frequently have dual-action, both positive and negative action, such as they act as natural antioxidant activity or increasing the body's resistance mechanisms. Saponin is the least desirable non-nutrient component in amaranth seed (saponin glycosides). Saponins irritate the digestive tract and cause hemolysis when they enter the bloodstream [1, 52]. Furthermore, they inhibit acetylcholinesterase, resulting in transmission nerve stimulation problems. Saponins form complexes compounds with minerals such as zinc and iron, and reduce the bioavailability of minerals. Saponins are mainly absorbed in the intestine in a very small amount that leads to interference in mineral absorption. Amaranth seeds contain a non-significant amount of triterpene glycoside approximately 0.1% of dry weight, which may cause toxicity, its lethal dose was about 1500 to 1750 mg/kg body weight [30].

10.7 Medicinal Properties

Amaranth grain is a good source of nutritional and bioactive compounds that exhibits a lot of health-promoting properties therefore there should be more research on the utilization of amaranth in diets [40]. Amaranth species have promising potential in terms of their investigation for achieving health and industrial objectives at the highest level. It was reported that regular consumption of amaranth lowing the cholesterol level, hypertension, and also cardiovascular disease. Amaranth species are a vital source of vitamins such as vitamin C, vitamin A, vitamin B complex—B_6, B_9, etc. [58]. Variation in nutritional and chemical profiles was observed in amaranth species that was due to climatic conditions, genetics, and other factors [59]. It also helps in improving liver function and inhibits cancer cause effects. Caffeic, protocatechuic, ferulic acid, hydroxybenzoic acid, rutin, isoquercitrin, and nicotiflorin were bioactive isolated from amaranth grain. These active bioactive compounds show various properties like they help in reducing the aging process, prevent oxidation, and especially nicotiflorin promotes memory function. Tagwira *et al.* (2006) [59] discovered that consumption of amaranth promotes health benefits such as healing properties of mouth sores, improve appetites, reducing obesity, and other effects among rural communities in Zimbabwe, which made them healthier. Amaranth grain has been discovered to be a viable replacement for wheat and other cereals in the production of food products for celiac patients. Celiac disease is an autoimmune condition of the small intestine characterized by a persistent sensitivity to gluten proteins. The treatment now available for patients with celiac disease is a gluten-free diet [60].

10.7.1 Antioxidant Activity

Antioxidant activity and phytochemicals content of different species of *Amaranthus* was assessed by many authors [33, 40, 61, 62, 63, and 64]. There are approximately 70 *Amaranthus* species and among these only 20 produces edible leaves [36]. There were significant differences in antioxidant activity and total polyphenols content among the varieties as well as in species of the amaranthus. Upper layers of Amaranthus plant leaves are a rich source of biologically active agents with antioxidant activity. The idiosyncrasies of vegetable leaf crops were found: the greatest number of calculated antioxidant metabolites accumulates in the leaves and amaranth inflorescences. A complex of antioxidant metabolites such as amaranthine, ascorbic acid, flavonoids, phenol, and carboxylic acids are found in aqueous extracts of the amaranthus leaves. Dissimilarities were identified by authors, which can be used as basic knowledge for the application of Amaranthus species according to their content. According to Tatiya *et al.* (2007) [64] total polyphenols content (TPC) in two species *A. spinosus* and *A. viridis* were 43.4 gallic acid equivalent (GAE) μg/g and 25.7 GAE μg/g fresh weight and total flavonoids content ranges between 177.6-179.2 rutin equivalent (RE) μg/g dried weight. DPPH and ABTS results presented by Sarker and Oba (2019) [37] of both the species were 23.5, 26.6 μg/g trolox equivalent antioxidant capacities (TEAC), and 48.4 and 50 TEAC μg/g dried weight respectively. Results reported by Li *et al.* (2008) [65] by the method of ferric reducing antioxidant power (FRAP) and oxygen radical absorbance capacity (ORAC) of two species namely *A. hypochondriacus and A. caudatus* revealed lower values of total polyphenols and total flavonoids. Total polyphenols content of *A. Caudatus, A. Cruentus, A. Hybrid, A. Hypochondriacus,* and *A. Hybridus* were 27.5, 30.5, 29.7, 30.0, and 30.8 mg GAE/100 g respectively as reported by Pamela *et al.* (2017) [66].

Although the results of the different studies cannot be compared since their methods of evaluation and standard materials used are not identical, for example, Quercetin and Rutin were used by Tativa *et al.* (2007) [66], catechin in Pamela *et al.* (2017) [66]. Reproducible results of the antioxidant capacity of plant extracts by the standard equipment and protocols can be used for comparative studies. The antioxidant activity of polyphenols depends on the structure of functional groups, as several hydroxyl groups influence free radical scavenging and metal chelation ability [67]. Environmental conditions of plants also cause differences in total polyphenols content and their antioxidant activity [68]. According to Stracke *et al.*, (2009) [69] photosynthesis by sunlight manipulates the polyphenols content. Antioxidant activities may differ according to the time of plantation. The photosynthetic performance of plants was also evaluated by Bang *et al.*, (2021) [41] and suggested that in lower rainfall seasons TPC and TFC were higher because of the proper photosynthesis process. The extracts of the leaves and the flower of amaranthus vegetable varieties are a rich source of antioxidants, phytochemicals that represent the pharmacological use of different species of amaranthus.

10.7.2 Gastroprotective Activity

Amaranthus extracts demonstrated the gastroprotective activity in a number of *in vivo* researches. Gastric or peptic ulcers are a general disorder of the gastrointestinal tract that develops in the first part of the duodenum. Currently, available drugs for the treatment of these ulcers have limited effects and severe side effects. The use of natural products for the treatment of these gastrointestinal disorders of different pathologies is continuously increasing all over the world. This is predominately factual with regards to flavonoids and other antioxidant agents with probably beneficial human health effects that are widely distributed in plants and widely consumed in diets. Amaranthus also contains these secondary metabolites and those exhibit antigastric ulcer activities. Ethanolic extraction of the whole plant of *A. spinosus* at different concentrations from 100 to 400 mg/kg presented antipeptic ulcer activity in all three models conducted by Hussian *et al.* (2009) [70], ethanol, aspirin, and cold restraint stress-induced ulcers in rats through reduction of ulcer index concurrently with containment of lipid peroxidation. The same results were revealed by Panda *et al.*, (2017) [71] with the extract of *A. spinosus* leaves with petroleum ether and chloroform. Ethanol extracts provide better results than petroleum ether, chloroform, and acetate extracts of *A. tricolor* leaves [72].

According to Kumar *et al.*, (2008) [73] in the Ayurvedic system, boiled leaves of *A. spinous* (weed variety) have been used for 2 to 3 days to promote health. Decoction prepared from the roots of these species is also used to treat stomach aches. Recently, four species (*A. spinosus*, *A. viridis*, *A. dubius*, and *A. tricolor*) of Amaranthus were comparatively investigated for their polyphenol profiles in methanolic extracts from the upper parts of leaves [52]. Rutin (quercetin-3-O-rutinoside), myricetin-3-O-rutinoside, ferulic acid, and p-coumaric acid (4-hydroxycinnamic acid) (Table 10.4) were present in all indicated Amaranthus species, while quercetin, astragalin, and caffeoylquinic acid were found in *A. spinosus* and *A. dubius*.

Table 10.4 Pharmacological activities of Amaranthus species.

Activities	Amaranthus species	Extraction solvent	References
Antigastric ulceractivity	*A. spinosus* (whole plant)	50% Ethanol	[70]
	A. tricolor (leaf)	Ethanol and ethyl acetate	[72]
	A. spinosus (leaf)	Ethanol	[71]
Antilipid peroxidation activity	*A. spinosus* (whole plant)	50% Ethanol	[70]
Antitrypsin activity	*A. hypochondriacus* (leaf)	Deionized water	[75]
Antiulcerative colitis activity	*A. roxburghianus* (root)	Ethanol	[74]
Anti–*H. pylori* activity; gastric mucosa healing	*A. cruentus* (seed oil)	CO_2 extraction	[76]
Anti-α-amylase activity	*A. spinosus* (leaf))	Methanol	[78]
	A. viridis (leaf, stem, seed, and root)	Methanol, ethanol, ethyl acetate, and water	[79]
	A. caudatus (whole plant)	Methanol	[80]
	A. hypochondriacus (leaf)	Deionized water	[75]
	A. viridis *A. tricolor* (whole plant)	Water	[81]

(*Continued*)

Table 10.4 Pharmacological activities of Amaranthus species. (*Continued*)

Activities	Amaranthus species	Extraction solvent	References
	A. tricolor (leaf and stem)	Distilled water	[82]
Glucose entrapment activity	*A. viridis* (leaf, stem, seed, and root)	Methanol, ethanol, ethyl acetate, and water	[79]
Anti-α-glucosidase activity	*A. tricolor* (leaf and stem)	Distilled water	[82]
Antilipase activity	*A. tricolor* (leaf and stem)	Distilled	[82]
Antibacterial activity	*A. viridis* (leaf and seed)	Methanol	[83]
	A. tricolor (leaf)	Ethyl acetate and methanol	[83]
	A. tricolor (leaf)	Methanol	[84]
Antidiarrheal activity	*A. spinosus* (leaf)	Methanol	[85]
	A. spinosus (whole plant)	50% Ethanol	[70]
	A. tricolor (leaf)	Ethanol and ethyl acetate	[86]

(*Continued*)

Table 10.4 Pharmacological activities of Amaranthus species. (*Continued*)

Activities	**Amaranthus species**	**Extraction solvent**	**References**
Anticolorectal cancer activity	*A. tricolor* (leaf)	Water and 90% ethanol	[77]
	A. caudatus (seed)	Protein isolated (defatting with hexane) and digested peptides	[87]
Laxative activity	*A. spinosus* (leaf)	Distilled water	[76]
	A. spinosus (whole plant)	Methanol	[88]
	A. spinosus (whole plant)	Methanol, ethyl acetate, and water	[88]
	A. caudatus (leaf)	Ethanol	[89]

All these acids and flavonoids are anticipated to take part in a gastroprotective activity. Similar results were interpreted by Paucar-Menacho *et al.*, 2018 [45], that quercetin p-coumaric acid, ferulic acid, rutin, caffeoylquinic acid, and caffeic acid have antigastric ulcer activity. Furthermore, the aqueous leaf extract of *A. hypochondriacus* has antitrypsin activity and ethanol extract of roots of *A. roxburghianus* has antiulcerative colitis activity [74, 75]. The seed oil of *A. cruentus* exerts anti-*Helicobacter pylori* activity gastric mucosa restitution in patients with duodenal peptic ulcers [76]. Active components present in Amaranthus species have gastroprotective activity and controlled use of leaf extracts and roots decoction of this may be beneficial for stomach disorders as well as for gastrointestinal ulcers.

10.7.3 Anticolorectal Cancer Activity

Colorectal cancer is a threat to the world's public health system. This is the second most diagnosed cancer in the whole world with more than 1.9 million cases and 935000 casualties in the last year only [77]. Food or diet is deciding factor in exposing humans to carcinogenic factors of the environment and anticarcinogenic factors, accordingly mitigating or aiding on the progress of different types of cancers. The recently researched capability of *Amaranthus* species to mitigate the growth of cancer cells provides the basis for the use of these leafy vegetables' active phytochemicals as anticarcinogenic agents [71]. Amaranthus peptides are released after simulated gastrointestinal digestion to reduce proliferation and to promote decomposition and apoptosis of HT-29 colon cancer cells. Aqueous solutions from the *Amaranthus tricolor* reveal IC_{50} value was more than 100 µg/mL in a human intestinal cell line derived from colorectal adenocarcinoma (Caco-2 cells) and results were repeated in ethanol extraction of active components [77]. Human colorectal adenocarcinoma cell line, showing that the protein isolated from A. caudatus seeds significantly inhibits cell proliferation and is cytotoxic to cancer cells. Based on the previous work done and shreds of evidence, Amaranths plants are one of the promising possibilities of investigation for plant-derived components (Table 10.5) with anticolorectal cancer potential. Squalene is also effective against colon cancer and functions as a protective and preventive agent in cancer treatment, it helps in reducing the negative effects of chemotherapy. Squalene, in modest amounts, acts as a cancer-fighting agent, skin cancer, breast cancer, lowering the risk of bowel, and also slowing the growth of tumor cells [30].

Table 10.5 Anticolorectal cancer properties of *Amaranthus* sp.

Phytochemicals	Anticolorectal cancer properties
Myricetin	Cytotoxic, mutagenic, pro-apoptotic, autophagic effects [77]
Catechins	Antitumorigenic, antimetastatic effects [72]
Apigenins	Antiproliferative, antimetastatic, apoptotic, antichemosensitive effects [36]
Naringenin	Pro proliferative effects [44]
Rutin	Antimetastatic and antiproliferative effects, induced cell cycle arrest [65]

10.7.4 Anti-Inflammatory Activity

The amaranth protein inhibited inflammation by activating bioactive peptides that lowered the expression of various pro-inflammatory markers, therefore the consumption of amaranth could assist to prevent diseases caused by inflammation. In this respect, it is suggested that amaranth seeds be included in the diet to help reduce inflammation and avoid chronic diseases caused by inflammation. Antiallergic and anti-inflammatory characteristics are found in the fatty acid composition, which also encourages cell growth. It is indeed exploited in cosmetics as a skin-care ingredient, as high-grade machine oil, and in pharmaceutical formulations as a medication carrier.

10.8 Conclusion

Amaranth leaves of many species are multicolour and their coloring components are the ideal source of polyphenolic compounds, such as anthocynins, betalins, carotenoids, chlorophylls, β cyanin, β xanthin. *A. spinosus* exhibits the higher values of chlorophylls b, β xanthins and betalins. Antioxidant activities of polyphenols depends on the structure of the active functional groups and environmental conditions of plants also affects the total polyphenols content of the leaves. Ethanolic extracts of the active component of leaves of amaranthus exhibited the antipeptic ulcer activity. Amaranthus species have gastroprotective activity and controlled use of leaf extracts or roots decoction of this may be beneficial for stomach disorders. Amaranthus peptides are released after stimulated gastrointestinal digestion to reduce proliferation and to promote decomposition and apoptosis of HT-29 colon cancer cells. According to future prospective, there is a need to evaluate the amaranth species individually for the active component and activities against particular diseases and disorders. This will enhance the usage of natural antioxidants for the gastrointestinal disorders and cancer cells apoptosis.

References

1. Adegbola, P.I., Adetutu, A., Olaniyi, T.D., Antioxidant activity of *Amaranthus* species from the Amaranthaceae family–A review. *S. Afr. J. Bot.*, 133, 111-117, 2020.
2. Maurya, N.K. and Arya, P., *Amaranthus* grain nutritional benefits: A review. *J. Pharmacogn. Phytochem.*, 7, 2, 2258–2262, 2018.
3. http://www.theplantlist.org/browse/A/Amaranthaceae/Amaranthus;https://www.newworldencyclopedia.org/entry/Amaranth.
4. Sellers, B.A., Smeda, R.J., Johnson, W.G., Kendig, J.A., Ellersieck, M.R., Comparative growth of six Amaranthus species in Missouri. *Weed Sci.*, 51, 3, 329–333, 2003.
5. Rastogi, A. and Shukla, S., Amaranth: A new millennium crop of nutraceutical values. *Crit. Rev. Food Sci. Nutr.*, 53, 2, 109–125, 2013.
6. Emmanuel, O.C., Amaranth production and consumption in South Africa: The challenges of sustainability for food and nutrition security. *Int. J. Agric. Sustain.*, 20, 1–12, 2021.
7. Choi, H., Kim, W., Shin, M., Properties of Korean amaranth starch compared to waxy millet and waxy sorghum starches. *Starch-stärke*, 56, 10, 469–477, 2004.
8. Arendt-Elke, K. and Zannini, E., Amaranth, in: *Cereal Grains for the Food and Beverage Industries*, vol. 248, pp. 439–473, Elsevier, Amsterdam, The Netherlands, 2013.
9. Chandra, S., Dwivedi, P., Baig, M.M.V., Shinde, L.P., Importance of quinoa and amaranth in food security. *J. Agric. Ecol.*, 5, 26–37, 2018.
10. Capriles, V.D., Coelho, K.D., Guerra-Matias, A.C., Arêas, J.A.G., Effects of processing methods on amaranth starch digestibility and predicted glycemic index. *J. Food Sci.*, 73, 7, 160–164, 2008.
11. Gamel, T.H., Linssen, J.P., Mesallam, A.S., Damir, A.A., Shekib, L.A., Effect of seed treatments on the chemical composition of two amaranth species: Oil, sugars, fibres, minerals and vitamins. *J. Sci. Food Agric.*, 86, 1, 82–89, 2006.
12. Mustafa, A.F., Seguin, P., Gélinas, B., Chemical composition, dietary fibre, tannins and minerals of grain amaranth genotypes. *Int. J. Food Sci. Nutr.*, 62, 7, 750–754, 2011.

13. Ayo, J.A., The effect of amaranth grain flour on the quality of bread. *Int. J. Food Prop.*, 4, 2, 341–351, 2001.
14. Akin-Idowu, P.E., Ademoyegun, O.T., Olagunju, Y.O., Aduloju, A.O., Adebo, U.G., Phytochemical content and antioxidant activity of five grain amaranth species. *Am. J. Food Sci. Technol.*, 5, 6, 249–255, 2017.
15. Janssen, F., Pauly, A., Rombouts, I., Jansens, K.J., Deleu, L.J., Delcour, J.A., Proteins of amaranth (*Amaranthus spp.*), buckwheat (*Fagopyrum spp.*), and quinoa (*Chenopodium spp.*): A food science and technology perspective. *Compr. Rev. Food Sci. Food Saf.*, 16, 1, 39–58, 2017.
16. Shewry, P.R., Corol, D.I., Jones, H.D., Beale, M.H., Ward, J.L., Defining genetic and chemical diversity in wheat grain by 1H-NMR spectroscopy of polar metabolites. *Mol. Nutr. Food Res.*, 61, 7, 1600807, 2017.
17. Amare, E., Mouquet-Rivier, C., Servent, A., Morel, G., Adish, A., Haki, G.D., Protein quality of amaranth grains cultivated in Ethiopia as affected by popping and fermentation. *Food Nutr. Sci.*, 6, 01, 38, 2015.
18. Gorinstein, S., Zachwieja, Z., Katrich, E., Pawelzik, E., Haruenkit, R., Trakhtenberg, S., Martin-Belloso, O., Comparison of the contents of the main antioxidant compounds and the antioxidant activity of white grapefruit and his new hybrid. *LWT-Food Sci. Technol.*, 37, 3, 337–343, 2004.
19. Muchová, Z., Čuková, L., Mucha, R., Seed protein fractions of amaranth (*Amaranthus sp.*). *Rostlinná Výroba*, 46, 7, 331–336, 2000.
20. Sarker, U. and Oba, S., Nutrients, minerals, pigments, phytochemicals, and radical scavenging activity in *Amaranthus blitum* leafy vegetables. *Sci. Rep.*, 10, 1, 1–9, 2020.
21. León-Camacho, M., García-González, D.L., Aparicio, R., A detailed and comprehensive study of amaranth (*Amaranthus cruentus L.)* oil fatty profile. *Eur. Food Res. Technol.*, 213, 4-5, 349–355, 2001.
22. Abdelrahman, M. and Jogaiah, S., Production of plant bioactive triterpenoid and steroidal saponins, in: *Bioactive Molecules in Plant Defence*, pp. 5–13, 2020.
23. Shin, D.H., Heo, H.J., Lee, Y.J., Kim, H.K., Amaranth squalene reduces serum and liver lipid levels in rats fed a cholesterol diet. *Br. J. Biomed. Sci.*, 61, 1, 11–14, 2004.
24. Zhang, Z.S., Kang, Y.J., Che, L., Composition and thermal characteristics of seed oil obtained from Chinese amaranth. *LWT*, 111, 39–45, 2019.
25. Sanz-Penella, J.M., Laparra, J.M., Sanz, Y., Haros, M., Bread supplemented with amaranth (Amaranthus cruentus): Effect of phytates on *in vitro* iron absorption. *Plant Foods Hum. Nutr.*, 67, 1, 50–56, 2012.
26. Palombini, S.V., Claus, T., Maruyama, S.A., Gohara, A.K., Souza, A.H.P., Souza, N.E.D., Visentainer, J.V., Gomes, S.T.M., Matsushita, M., Evaluation of nutritional compounds in new amaranth and quinoa cultivars. *Food Sci. Technol.*, 33, 2, 339–344, 2013.
27. Bruni, R., Guerrini, A., Scalia, S., Romagnoli, C., Sacchetti, G., Rapid techniques for the extraction of vitamin E isomers from *Amaranthus caudatus*

seeds: Ultrasonic and supercritical fluid extraction. *Phytochem. Anal.: An International Journal of Plant Chemical and Biochemical Techniques*, 13, 5, 257–261, 2002.

28. Lehmann, J.W., Putnam, D.H., Qureshi, A.A., Vitamin E isomers in grain amaranths (Amaranthus spp.). *Lipids*, 29, 3, 177–181, 1994.
29. Ogrodowska, D., Zadernowski, R., Czaplicki, S., Derewiaka, D., Wronowska, B., Amaranth seeds and products–the source of bioactive compounds. *Pol. J. Food Nutr. Sci.*, 64, 3, 165–170, 2014.
30. Wolosik, K. and Markowska, A., *Amaranthus* Cruentus taxonomy, botanical description, and review of its seed chemical composition. *Nat. Prod. Commun.*, 14, 5, 1934578X19844141, 2019.
31. Sarker, U.I., Islam, T., Rabbani, G., Oba, S., Antioxidant leaf pigments and variability in vegetable amaranth. *Genetika*, 50, 1, 209–220, 2018 a.
32. Akubugwo, I.E., Obasi, N.A., Chinyere, G.C., Ugbogu, A.E., Mineral and phytochemical contents in leaves of *Amaranthus hybridus L* and *Solanum nigrum L.* subjected to different processing methods. *Afr. J. Biochem. Res.*, 2, 2, 40–44, 2008.
33. Ahmed, S.A., Hanif, S., Iftkhar, T., Phytochemical profiling with antioxidant and antimicrobial screening of *Amaranthus viridis L.* leaf and seed extracts. *Open J. Med. Microbiol.*, 3, 3, 164–171, 2013.
34. Muriuki, E.N., Sila, D.N., Onyango, A., Nutritional diversity of leafy amaranth species grown in Kenya. *J. Appl. Biosci.*, 79, 6818–6825, 2014.
35. Khanam, U.K. and Oba, S., Bioactive substances in leaves of two amaranth species, *Amaranthus tricolor* and *A. hypochondriacus*. *Can. J. Plant Sci.*, 93, 1, 47–58, 2013.
36. Sarker, U. and Oba, S., Antioxidant constituents of three selected red and green color Amaranthus leafy vegetable. *Sci. Rep.*, 9, 1, 1–11, 2019 a.
37. Sarker, U. and Oba, S., Nutraceuticals, antioxidant pigments, and phytochemicals in the leaves of *Amaranthus spinosus* and *Amaranthus viridis* weedy species. *Sci. Rep.*, 9, 1, 1–10, 2019 b.
38. Sarker, U.I., Islam, M.T., Rabbani, M.G., Oba, S., Variability in total antioxidant capacity, antioxidant leaf pigments and foliage yield of vegetable amaranth. *J. Integr. Agric.*, 17, 5, 1145–1153, 2018 b.
39. Pulipati, S.B., Babu, P.S., Naveena, U., Parveen, S.R., Nausheen, S.S., Sai, M.T.N., Determination of total phenolic, tannin, flavonoid contents and evaluation of antioxidant property of *Amaranthus tricolor (L). Int. J. Pharmacogn. Phytochem. Res.*, 9, 6, 814–19, 2017.
40. Karamać, M., Gai, F., Longato, E., Meineri, G., Janiak, M.A., Amarowicz, R., Peiretti, P.G., Antioxidant activity and phenolic composition of Amaranth (*Amaranthus caudatus*) during plant growth. *Antioxidants*, 8, 173, 2019.
41. Bang, J.H., Lee, K.J., Jeong, W.T., Han, S., Jo, I.H., Choi, S.H., Cho, H., Hyun, T.K., Sung, J., Lee, J., So, Y.S., Antioxidant activity and phytochemical content of nine *amaranthus* species. *Agronomy*, 11, 6, 1032, 2021.

42. Peiretti, P.G., Meineri, G., Gai, F., Longato, E., Amarowicz, R., Antioxidative activities and phenolic compounds of pumpkin (*Cucurbita pepo*) seeds and amaranth (*Amaranthus caudatus*) grain extracts. *Nat. Prod. Res.*, 31, 8, 2178–2182, 2017.
43. Nana, F.W., Hilou, A., Millogo, J.F., Nacoulma, O.G., Phytochemical composition, antioxidant and xanthine oxidase inhibitory activities of *Amaranthus cruentus L.* and *Amaranthus hybridus L.* extracts. *Pharmaceuticals*, 5, 6, 613–628, 2012.
44. Venskutonis, P.R. and Kraujalis, P., Nutritional components of amaranth seeds and vegetables: A review on composition, properties, and uses. *Compr. Rev. Food Sci. Food Saf.*, 12, 4, 381–412, 2013.
45. Paśko, P.S., Sajewicz, M., Gorinstein, S., Zachwieja, Z., Analysis of selected phenolic acids and flavonoids in *Amaranthus cruentus* and *Chenopodium quinoa* seeds and sprouts by HPLC. *Acta Chromatogr.*, 20, 4, 661–672, 2008.
46. Sarker, U. and Oba, S., Leaf pigmentation, its profiles and radical scavenging activity in selected *Amaranthus tricolor* leafy vegetables. *Sci. Rep.*, 10, 1, 1–10, 2020.
47. Li, H.D., Deng, Z., Liu, R., Zhu, H., Draves, J., Marcone, M., Sun, Y., Tsao, R., Characterization of phenolics, betacyanins and antioxidant activities of the seed, leaf, sprout, flower and stalk extracts of three *Amaranthus* species. *J. Food Compos. Anal.*, 37, 75–81, 2015.
48. Paucar-Menacho, L.M.V., Dueñas, M., Peñas, E., Frias, J., Martínez Villaluenga, C., Effect of dry heat puffing on nutritional composition, fatty acid, amino acid and phenolic profiles of pseudocereals grains. *Pol. J. Food Nutr. Sci.*, 68, 4, 0–0, 2018.
49. Khanam, U.K.S., Oba, S., Yanase, E., Murakami, Y., Phenolic acids, flavonoids and total antioxidant capacity of selected leafy vegetables. *J. Funct. Foods*, 4, 4, 979–987, 2012.
50. Pashikanti, S., De Alba, D.R., Boissonneault, G.A., Cervantes-Laurean, D., Rutin metabolites: Novel inhibitors of nonoxidative advanced glycation end products. *Free Radical Biol. Med.*, 48, 5, 656–663, 2010.
51. Islam, M.I., Islam, Z., Kochi, A.M., Azad, A.K., Phytochemical screening and pharmacological evaluation of methanol extract of *Amaranthus tricolor L.*(stem). *Int. J. Adv. Res. Rev.*, 6, 9, 01–06, 2021.
52. Mondor, M., Melgar-Lalanne, G., Hernández-Álvarez, A.J., Cold pressed amaranth (*Amaranthus tricolor*) oil, in: *Cold Pressed Oils, Green Technology, Bioactive Compounds, Functionality, and Applications*, pp. 113–127, Academic Press, Cambridge, Massachusetts, United States, 2020.
53. Salikhov, S.I., Chemical composition and biological activity of seed oil of amaranth varieties. *Noval Biotechnol. Chim.*, 17, 1, 66–73, 2018.
54. Yelisyeyeva, O., Semen, K., Bielawska, K., Biernacki, M., Kaminskyy, D., Yavorskyi, O., Skrzydlewska, E., Amaranth oil in prevalent pulmonary arterial hypertension: Changes in fatty acid panel and products of lipid peroxidation. *Free Radical Biol. Med.*, 124, 576–577, 2018.

55. Mwaurah, P.W., Kumar, S., Kumar, N., Attkan, A.K., Panghal, A., Singh, V.K., Garg, M.K., Novel oil extraction technologies: Process conditions, quality parameters, and optimization. *Compr. Rev. Food Sci. Food Saf.*, 19, 1, 3–20, 2020.
56. Czaplicki, S., Ogrodowska, D., Zadernowski, R., Derewiaka, Characteristics of biologically-active substances of amaranth oil obtained by various techniques. *Pol. J. Food Nutr. Sci.*, 62, 4, 235–239, 2012.
57. Musa, A. and Ogbadoyi, E.O., Determination of anti-nutrients and toxic substances of selected fresh leafy vegetables obtained from Minna Town, Nigeria. *Nig. J. Basic Appl. Sci.*, 22, 3-4, 79–83, 2014.
58. Alvarez-Jubete, L., Auty, M., Arendt, E.K., Gallagher, E., Baking properties and microstructure of pseudocereal flours in gluten-free bread formulations. *Eur. Food Res. Technol.*, 230, 3, 437–445, 2010.
59. Tagwira, M., Tagwira, F., Dugger, R., Okum, B., Using grain amaranth to fight malnutrition. *RUFORUM Working Document*, 1, 201–206, 2006.
60. Aderibigbe, O.R., Ezekiel, O.O., Owolade, S.O., Korese, J.K., Sturm, B., Hensel, O., Exploring the potentials of underutilized grain amaranth (*Amaranthus spp.)* along the value chain for food and nutrition security: A review. *Crit. Rev. Food Sci. Nutr.*, 62, 3, 656–669, 2022.
61. Lopez-Mejía, O.A., Lopez-Malo, A., Palou, E., Antioxidant capacity of extracts from amaranth *Amaranthus hypochondriacus L.* seeds or leaves. *Ind. Crops Prod.*, 53, 55–59, 2014.
62. Ishtiaq, S., Ahmad, M., Hanif, U., Akbar, S., Kamran, S.H., Phytochemical and *in-vitro* antioxidant evaluation of different fractions of *Amaranthus graecizan subsp. SilvestrisVill. Brenan. Asian Pac. J. Trop. Med.*, 412, 965–971, 2014.
63. Maiyo, Z.C., Ngure, R.N., Matasyoh, J.C., Chepkorir, R., Phytochemical constituents and antimicrobial activity of leaf extract of three *Amaranthus* plant species. *Afr. J. Biotechnol.*, 9, 3178–3182, 2010.
64. Tatiya, A.U., Surana, S.J., Khope, S.D., Gokhale, S.B., Sutar, M.P., Phytochemical investigation and immunomodulatory activity of *Amaranthus spinosuslinn. Indian J. Pharm. Educ. Res.*, 444, 337–341, 2007.
65. Li, H.B., Wong, C.C., Cheng, K.W., Chen, F., Antioxidant properties *in vitro* and total phenolic contents in methanol extracts from medicinal plants. *LWT-Food Sci. Technol.*, 41, 385–390, 2008.
66. Pamela, E.A.I., Olufemi, T.A., Yemisi, O.O., Aduloju, O.A., Usifo, G.A., Phytochemical content and antioxidant activity of five grain amaranth species. *Am. J. Food Sci. Technol.*, 5, 249–255, 2017.
67. Heim, K.E., Tagliaferro, A.R., Bobilya, D.J., Flavonoid antioxidants: Chemistry, metabolism and structure-activity relationships. *J. Nutr. Biochem.*, 13, 572–584, 2002.
68. Lee, K.J., Baek, D.Y., Lee, G.A., Cho, G.T., So, Y.S., Lee, J.R., Ma, K.H., Chung, J.W., Hyun, D.Y., Phytochemicals and antioxidant activity of korean black soybean (*Glycine max L.)* landraces. *Antioxidants*, 9, 213, 2020.

69. Stracke, B.A., Rüfer, C.E., Weibel, F.P., Bub, A., Watzl, B., Three-year comparison of the polyphenol contents and antioxidant capacities in organically and conventionally produced apples (Malus domestica bork. cultivar 'golden delicious'). *J. Agric. Food Chem.*, 57, 4598–4605, 2009.
70. Hussain, Z., Amresh, G., Singh, S., Rao, C.V., Antidiarrheal and antiulcer activity of *Amaranthus spinosus* in experimental animals. *Pharm. Biol.*, 47, 10, 932–939, 2009.
71. Panda, S.K., Sarkar, G., Acharjya, M., Panda, P., Antiulcer activity of *Amaranthus spinosus* leaf extract and its comparision with famotidine in shay rats. *J. Drug Deliv. Ther.*, 7, 2, 96–98, 2017.
72. Devaraj, V.C. and Krishna, B.G., Gastric antisecretory and cytoprotective effects of leaf extracts of *Amaranthus tricolor Linn.* in rats. *J. Chin. Integr. Med.*, 9, 9, 1031–1038, 2011.
73. Kumar, A., Lakshman, K., Jayaveera, K., Effect of *Amaranthus spinosus* leaf extract on gastrointestinal tract. *Pharmacologyonline*, 1, 233–238, 2008.
74. Nirmal, S.A., Ingale, J.M., Pattan, S.R., Bhawar, S.B., Amaranthus roxburghianus root extract in combination with piperine as a potential treatment of ulcerative colitis in mice. *J. Integr. Med.*, 11, 3, 206–212, 2013.
75. Sánchez-Hernández, C., Martínez-Gallardo, N., Guerrero-Rangel, A., Valdés-Rodríguez, S., Délano-Frier, J., Trypsin and α-amylase inhibitors are differentially induced in leaves of amaranth (*Amaranthus hypochondriacus*) in response to biotic and abiotic stress. *Physiol. Plant*, 122, 2, 254–264, 2004.
76. Cherkas, A., Zarkovic, K., CipakGasparovic, A., Jaganjac, M., Milkovic, L., Abrahamovych, O., Yatskevych, O., Waeg, G., Yelisyeyeva, O., Zarkovic, N., Amaranth oil reduces accumulation of 4-hydroxynonenal-histidine adducts in gastric mucosa and improves heart rate variability in duodenal peptic ulcer patients undergoing *Helicobacter pylori* eradication. *Free Radical Res.*, 52, 2, 135–149, 2018.
77. Sani, H.A., Rahmat, A., Ismail, M., Rosli, R., Endrini, S., Potential anticancer effect of red spinach (*Amaranthus gangeticus*) extract. *Asia Pac. J. Clin. Nutr.*, 13, 4, 396–400, 2004.
78. Ashok Kumar, B.S., Lakshman, K., Nandeesh, R., Arun Kumar, P.A., Manoj, B., Kumar, V., Sheshadri Shekar, D., *In vitro* alpha-amylase inhibition and *in vivo* antioxidant potential of *Amaranthus* spinosusin alloxan-induced oxidative stress in diabetic rats. *Saudi J. Biol. Sci.*, 18, 1, 1–5, 2011.
79. Helen, P. and Bency, B.J., Inhibitory potential of Amaranthus viridison α-amylase and glucose entrapment efficacy *in vitro*. *Res. J. Pharm. Technol.*, 12, 5, 2089–2092, 2019.
80. Kumar, A., Lakshman, K., Jayaveera, K.N., Sheshadri Shekar, D., Narayan Swamy, V.B., Khan, S., Velumurga, C., *In vitro* alpha-amylase inhibition and antioxidant activities of methanolic extract of *Amaranthus caudatuslinn.* *Oman Med. J.*, 26, 3, 166–170, 2011a.

81. Kumar, R., Kumar, S., Shashidhara, S., Anitha, S., *In-vitro* anti-oxidant, anti-amylase, anti-arthritic and cytotoxic activity of important commonly used green leafy vegetables. *Int. J. Pharmtech. Res.*, 3, 4, 2096–2103, 2011b.
82. Yang, Y.C., Mong, M.C., Wu, W.T., Wang, Z.H., Yin, M.C., Phytochemical profiles and anti-diabetic benefits of two edible *Amaranthus* species. *CyTA-J. Food*, 18, 1, 94–101, 2020.
83. Pulipati, S., Babu, P.S., Narasu, M.L., Quantitative determination of tannin content and evaluation of antibacterial activity of *Amaranthus tricolor(L).* *Int. J. Biol. Pharm. Res.*, 5, 623–626, 2014.
84. Pulipati, S., Phytochemical analysis and antibacterial efficacy of Amaranthus tricolor(L) methanolic leaf extract against clinical isolates of urinary tract pathogens. *Afr. J. Microbiol. Res.*, 9, 20, 1381–1385, 2015.
85. Olajide, O.A., Ogunleye, B.R., Erinle, T.O., Anti-inflammatory properties of *Amaranthus spinosus* leaf extract. *Pharm. Biol.*, 42, 7, 521–525, 2004.
86. Alam, S., Evaluation of anti-diarrhoeal activity of *Amaranthus tricolor Linn* in experimental animals. *Innoriginal: Int. J. Sci.*, 2, 1, 5–8, 2015.
87. Sabbione, A.C., Suárez, S., Añón, M.C., Scilingo, A., Amaranth functional cookies exert potential antithrombotic and antihypertensive activities. *Int. J. Food Sci. Technol.*, 54, 5, 1506–1513, 2019.
88. Chaudhary, M.A., Imran, I., Bashir, S., Mehmood, M.H., Rehman, N.U., Gilani, A.H., Evaluation of gut modulatory and bronchodilator activities of *Amaranthus spinosus* Linn. *BMC Complementary Med. Ther.*, 12, 1, 166, 2012.
89. Saba, A.B. and Oridupa, O.A., Pharmacological reactivity of isolated Guinea pig ileum to ethanol leaf extracts of *Amaranthus caudatus* and *Solanum melongena. Niger. J. Physiol.*, 27, 1, 73–78, 2012.

11

Corchorus Species: Health Benefits and Industrial Importance

Kavya Ganthal, Nehal Sharma and Narinder Kaur*

Department of Food Technology and Nutrition, Lovely Professional University, Jalandhar, India

Abstract
The genus Corchorus is the member of family Malvaceae (formerly Tiliaceae) and is widely distributed in the tropical as well as subtropical regions of the world. The primary centre of origin of wild taxa of Corchorus was in Africa. The genus is basically dealt with jute fibers and vegetative parts for their applications in textile industry. In this chapter, a study of some of the species under this genus, *C. capsularis, C. olitorious, C. cunninghami, C. erodioides, C. siliquosi, C. walcotti,* and *C. tridens* have been discussed. Apart from this, wherever available details of their taxation, origin, history, botanical structure have also been discussed. Since these species possess various health benefits and are industrially important also, the species have been discussed in the mentioned context also.

Keywords: *Corchorus, C. capsularis, C. olitorious, C. cunninghami,* jute

11.1 Introduction

Corchorus is a Malvaceae family genus of flowering plants indigenous to tropical and subtropical climates of the world. Jute is the name for the plant's fiber, while jute mallow leaves are the name for the leaves. The family Malvaceae includes the subfamily Grewioideae, which includes the genus *Corchorus*. It has between 40 and 100 species. Oceanopapaver, a previously ambiguous genus, has recently been synonymized with Corchorus. Guillaumin coined the name for the sole species Oceanopapaver neocaledonicum in 1932. The genus

**Corresponding author*: kaurnarinder88@gmail.com

Prerna Gupta, Navnidhi Chhikara and Anil Panghal (eds.) Harvesting Food from Weeds, (385–406)

Capparaceae has been divided into four families: Capparaceae, Cistaceae, Papaveraceae, and Tiliaceae [1].

11.1.1 Botanical Description and Taxonomy of the Corchorus

Annual herbs that grow to a height of 2 to 4 m and are either unbranched or have a few side branches. The leaves are simple, alternate, lanceolate, which are 5 to 15 cm in length, with a finely serrated or lobed margin and an acuminate tip. The flowers are yellow in color and are small (2–3 cm in diameter) with five petals, whereas the fruit is a seed capsule with several seeds. It grows well just about anywhere and can be grown all year [2].

The sub-family Grewioideae, which includes the genus Corchorus, is part of the Malvaceae family. It has between 40 and 100 different species. The different species of Corchorus have been shown in Figure 11.1. Corchorus has recently been synonymized with Oceanopapaver, a

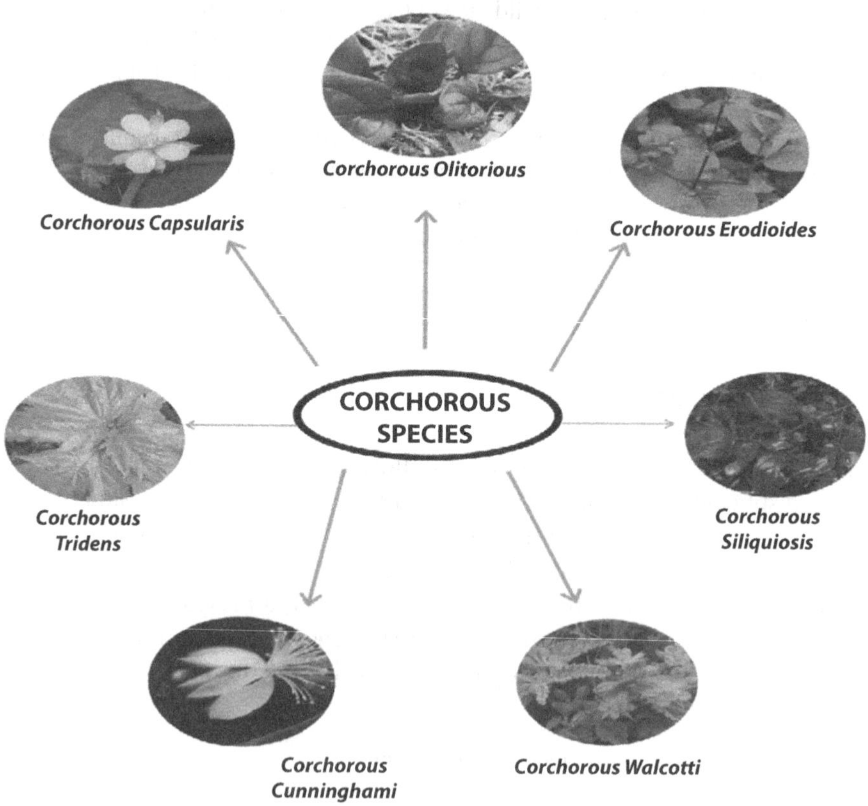

Figure 11.1 Different species of *Corchorus*.

previously ill-defined genus. Guillaumin named the sole New Caledonian species *Oceanopapaver neocaledonicum* Guillaumin in 1932. Capparaceae, Tiliaceae, Papaveraceae, and Cistaceae, are the four families that make up the genus Capparaceae. The imaginary family name "Oceanopapaveraceae" has shown in print and on the Internet on a few occasions; however, it is a nomen nudum, which has never been properly published or recognized by any plant taxonomic system [3].

11.1.2 Uses of the Corchorus

Corchorus plants are not only beneficial for the food industry, but also to the textile industry by providing jute fiber. The leaves and seeds of different species of corchorus genus are used as cuisines in different countries. For example, the leaves of corchorus *Capsularis* and corchorus *Olitorius* are used mainly in parts of Asia and Africa giving a similar texture that of okra. Also, the seeds of these species are used as a flavoring agent. The leaves are rich in beta-carotene, iron, calcium, vitamin C, and α-tocopherol [3].

11.1.3 Origin and History of the Corchorus

The majority of species can be found in the region of Africa and Australia, with minor population in tropical America and southern Asia region. The cultivated species are only *C. Olitorius* and *C. Capsularis*, and the tender leaves are eaten as leafy vegetables, while the bark of stem is used to make jute. The two cultivated species found in African region and later spread to Asia. Previously, corchorus was categorized as a Tiliaceae genus. Tiliaceae was discovered to be paraphyletic, thus numerous species, including corchorus, were reallocated to the Grewioideae subfamily, having formed a monophyletic group within the Malvaceaes family [4–10].

11.1.4 Corchorus as Cuisines in Various Countries and Regions

Corchorus *Olitorius* leaves are mostly utilized in southern Asian, Middle Eastern, North African, and West African cuisines, but corchorus *capsularis* leaves are used in Japan and China. It has a mucilaginous texture similar to okra when cooked. The seeds are used as additive like flavoring agent, while the dried leaves are used to make herbal tea.

Malukhiyah is the Arabic name for the young leaves of Corchorus species. It is a popular dish in Egypt and is regarded as the country's national dish. In turkey and Cyprus, plant is called as molocha or molohiya, and is used to make one kind of chicken stew. Since the time of the Pharaohs, in

Egypt corchorus leaves have been a staple food, and it is because of this that it has gained recognition and popularity. Middle Eastern cuisine is known for its various mallow-leaf stews served with rice.

It is also commonly used in Nigerian cuisine, particularly among the Yoruba, ewedu is a type of stew, as a condiment for many other starch-based meals like amala, or in gbegiri, a traditional Nigerian soup. In Northern Nigeria, it is called as Ayoyo. They use it for making (Miyan Ayoyo), a type of sauce, which is usually served along with Tuwon Allebo or Tuwon Masara.

Corchorus species is mostly eaten in its vegetative form by the northern people of Ghana. It is generally consumed with *Tuo Zaafi*, i.e., a food prepared with corn flour, thereby increasing its acceptability. It is usually called as *Ayoyo* in this region. It is known as krain krain (or crain crain) in Sierra Leone and is prepared as a stew. Typically, the stew is served with rice or foo foo (a traditional food made from cassava).

The people of Luhya from Western Kenya consume jute leaves, they refer to as mrenda or murere. It is usually eaten along with starchy meals that are starchy like ugali, a staple in most Kenyan homes. Khudra, which means "green" in Sudanese Arabic, is the name for it in Northern Sudan. It is referred to as fakohoy by the Songhai people of Mali. It is referred to as nalta sag in India. During the summer, especially in Sambalpur and the western part of Odisha, it is a popular dish. It is typically lightly sautéed and served with rice or rice gruel. In the Philippines, *C. Olitorius* is known as saluyot. It is commonly consumed as a leafy vegetable together with bamboo shoots. The leaves of the *Corchorus Olitorius* are blanched and served with plain rice congee in Thai cuisine. The flavour is similar to spinach and samphire [15].

11.2 Various Species of Corchorus

In the following section, health benefits and industrial importance of various species of corchorus, their origin, history, botanical structure have been discussed.

11.2.1 *Corchorus capsularis*

The genus corchorus includes the shrub corchorus *capsularis*, which belongs to the Malvaceae family. It is also known as White Jute, and it is one of the greatest sources of jute fiber. The roots, unripe fruit, and leaves are all utilized in traditional medicine. It has an upright structure with sharp leaves and yellow five-petaled blooms that can grow to be 2 m tall or more (Table 11.1). It is an annual shrub with globular fruits that grows in

Table 11.1 Classified study of structures of different species of *Corchorus*.

S. no.	Species	Structure	Leaf	Flower	Fruit	Ref.
1.	*C. Capsularis*	Erect	Acute	5-petaled	-	[11]
2.	*C. Cunninghami*	Herbaceous Sparsely Hairy	Alternate Three-veined	Yellow, Narrow, Oval-shaped Petals	Dark brown to black ellipsoid-shaped capsules	[12]
3.	*C. Erodioides*	Tall Unbranched	Alternate Lanceolate	Yellow 5-petaled	Many seeded capsule	[13]
4.	*C. Walcotti*	Erect Brown colored branched shrub	Possess sinuate margins and indumentums	Yellow colored flowers Corolla is obvious and prominent	-	[14]
5.	*C. Olitorious*	Erect and Fairly Branched	Alternate and Acute	Solitary, 5-petaled and sepaled	Spindle-shaped greyish-blue to green or brownish-black	[15]
6.	*C. Tridens*	Simple plant with upright structure	Simple and Alternate Linear Lanceolate lamina	Yellow colored Flowers Numerous stamens	Capsule shaped Linear with triangular cross-section	[16]
7.	*C. Sillquosus*	Grows as a small shrub	Alternate Acute Ovate Crenate Leaf Margin	Actinomorphic Unfused green-colored sepals Yellow-colored petals	Elongated Capsule like Fruit (from a superior ovary)	[17]

a spherical shape. It is supposed to have originated in China, although it is currently widely produced in Bangladesh, India, and tropical Africa. It is also cultivated in the Amazonian area of Brazil.

It is one of the major sources of jute for producing a white and high-quality fiber. The water is made to retain so as to extract fiber from the cut stems after removing tissue. The fiber is obtained after drying process. It is used to make sacks, bags, carpets, curtains, textiles, and paper. India and Bangladesh are the primary producers, as they thrive in the Ganga and Brahmaputra river deltas. When young, the shoots and leaves of this plant are commonly consumed with salads, and when older, they are cooked as a leafy vegetable. The dried and powdered leaves are used to thicken soups or make the tea. The immature fruits can be eaten raw or cooked [18].

11.2.1.1 Health Benefits

C. capsularis shows outstanding performance in different ailments and has been used traditionally.

Gastrointestinal Problems

C. capsularis leaves have been used as a stimulant, appetite stimulant, digestive aid, laxative, and stimulant. Dysentery has been treated with the leaves and roots, and fever has been treated with a leaf infusion. Both animals and insects are poisoned with the seeds, which has a digoxin-like substance in it [19].

For skin and scalp

A smooth and soft skin can be obtained by grinding few leaves of White Jute to make out a paste. The plant leaves also helps to get rid of dandruff by applying its paste on the scalp thereby helps in attaining a healthy scalp. Also, induces the growth of hair and increases its strength.

Anticancer effect

The presence of active components, phytol (3,7,11,15-tetramethyl-2-hexadecen-1-ol) and mono-galacto-syl-di-acyl-glycerol (1,2-di-O-α-linolenoyl-3-O-β-D-galactopyranosyl-sn-glycerol) makes it as anti-tumours or anti-cancer agents [20].

Antioxidant effect

The extracts of its leaves shows antioxidant activity in both assays with the percentage of inhibition nearly 90% (37) due to radical scavenging activity against DPPH [21].

Anti-inflammatory, analgesic, and antipyretic effects

The extract of this plant, if taken in the dose of 20, 100 and 200 mg/kg, exhibits antinociceptive, anti-inflammatory and antipyretic properties respective [22].

Antimicrobial effects

The extracts of *corchorus capsularis* possess antimicrobial, antifungal and anti-yeast activity. N-hexane fraction of methanolic extract of leaves of *corchorus capsularis* showed the highest activities against Gram positive bacteria (*Bacillus subtilis, Staphylococcus aureus, Beta hemolytic streptococcus, Bacillus cereus* and *Streptococcus pyrogen*) and Gram-negative bacteria (*Shigella boydii, Salmonella typhi E. coli, Klebsiella*, and *Vibrio mimicus*). This extract also exhibited antimicrobial behavior for yeast and fungi (*Candida albicans, Saccharomyces cerevisiae* and *Bacillus megaterium*) with a zone of inhibition 0.9 to 1.5 mm [23].

Insecticidal effect

The plant extract acts against common malarial vector, *Anopheles stephensi* and a dengue vector *Aedes aegypti* [24].

Industrial Importance

C. Capsularis have huge industrial potential as food and fiber source. As the jute fiber obtained from *C. Capsularis* are whiter in color and also the quality is higher than other jute fibers, it is widely used in the textile industry to make sacks, curtains, bags, carpets, textile, and paper. The leaves are also consumed as salad and the powdered form of these leaves can be used as thickening agents in foods such as soups and tea. Due to numerous health benefits, this can be used for the manufacturing of herbal medicines [36].

11.2.2 *Corchorus olitorius*

Jute mallow, or nalta jute, is common name for corchorus *olitorius*. Other than *C. capsularis*, it is also a major source of jute fiber. The seeds are edible, as are the leaves and young fruits. The dried leaves are used for tea and as a soup thickener, and the seeds are edible. In other local languages, it is known as "tossa jute," "bush okra," "krinkrin," "etinyung," "molokhia," and "West African sorrel." It is a tall, erect herbaceous plant with a lot of branches that reaches a height of 1.5 m. When grown for production of fiber, it can grow up to heights of 4 m. The taproot gives way to a tough, hairless stem that is green with a subtle red-brownish tinge and turns woody on the ground. The alternating serrate acute leaves are 6 to 10 cm long and 2 to 4 cm broad. Flowers are borne by the plant alone or in two-flowered cymes opposing the leaf. The blooms are at the end of a short stalk and feature 5 sepals, 5 petals, 10 yellow stamina, and are free. The fruit is spindle-shaped, dehiscent, and divided into transverse pieces by five valves. The fruit can be grayish-blue, green, or brownish-black in color and can be 2 to 8 cm long. Each seed chamber contains 25 to 40 seeds, bringing the total number of seeds per fruit to 125 to 200. It is one of the most widely cultivated and traded leafy vegetables in many countries. There are no statistics on production or marketing available. Jew's mallow is sold in Europe as a powdered Libanese product under the Arabic name 'meloukhia'.

Origin and History

It is assumed that this species is originated from Africa and the area of Indo- Burmese in India. It is believed that this species had been under cultivation for a long time from where it originated and as such had grown as a cash crop or wild crop in every country of tropical Africa.

Growth Conditions

Corchorus olitorius is a lowland tropics annual crop that thrives in warm zone of temperate, tropical deserts, and wet forest living zones. They can survive in 400 to 4290 mm of yearly precipitation (optimum 1000 mm per year) and temperature ranging from 16.8 to 27.5 °C. Soil pH should be between 4.5 and 8.2. The plant flourishes well in humus rich, fertile, well-drained alluvial soil, but it will grow in less-than-ideal conditions. Before sowing, the land is thoroughly prepared through ploughing, and the seeds are disseminated or dribbled behind the plough during the wet season. To break dormancy, the seeds must be pre-soaked for 10 seconds in hot water (about 93 °C) 24 hours prior sowing. It is easier to sow the small seeds if they are mixed with sand. Germination occurs 2 to 3 days after sowing if the soil is wet. Seedlings are transplanted at a height of 10 cm in some systems. Plants are planted in lines with a gap of 20 to 50 cm between them. When the seedlings reach a height of 8 to 25 cm, they are harrowed three to four times with a rake and weeded two to three times. Cow dung, wood ashes, or decaying water hyacinth or its ashes are used as manure [25–27].

Storage Conditions

For fresh eating, the leaves should be kept at a temperature of 8 to 15 °C. Leaf browning is caused by low temperatures of 1°C to 8°C, whereas leaf yellowing is caused by extremely high storage temperatures. Six weeks after flowering, the fruits can be picked to yield seeds. The dry capsules are husked and kept for eight to twelve months in hermetically sealed jars. For storage, the moisture content should be about 9%.

Chemical Composition

The leaves of *corchorus olitorius* are well-known for their high concentrations of a range of chemical substances. The constituents of C. *olitorious* have been discussed in Table 11.2. Among the 17 active nutritional molecules present in Jute leaves are protein, fat, carbohydrate, fiber, ash, calcium, potassium, iron, salt, phosphorus, beta-carotene, thiamine, riboflavin, niacin, and ascorbic acid (Table 11.3). The composition is quite comparative with other dark green leafy vegetables, however fresh Jew's

Table 11.2 Constituents of C. *olitorious.*

Constituents	Quantity (g/100g)
Water	80.4 g
Energy	243 kJ
Protein	4.5 g
Fat	0.3 g
Carbohydrate	12.4 g
Fiber	2.0 g

Table 11.3 Chemical composition of C. *olitorius.*

Constituents	Quantity
Calcium	360 mg
Phosphorous	122mg
Iron	7 mg
β-carotene	6410 μg
Thiamine	0.15mg
Riboflavin	0.53mg
Niacin	1.2mg
Ascorbic acid	80mg

mallow leaves have a greater dry matter content than typical. External variables like as soil fertility and fertilization have a significant impact on the composition, particularly the micronutrient content. Micronutrient content, such as Fe, P, Ca, carotene, and vitamin C, is considerably improved by nitrogen fertilizer.

The mucilaginous polysaccharide in the leaves is rich in uronic acid (65%) and consists of rhamnose, galactose, glucose, galacturonic acid and glucuronic acid in a molar ratio of 1.0:0.2:0.2:0.9:1.7 in addition to 11.7% acetyl groups.

Health Benefits

Anti-oxidant Property

The leaves of corchorus *olitorius* contain anti-oxidative phenol compounds, of which 5-caffeoylquinic acid is the most important. Some ionone glycosides have also been isolated from these leaves. As such they show inhibitory activity on histamine release from rat peritoneal exudate cells induced by antigen-antibody reaction [28].

Fulfils Nutritional Requirements

The leaves are rich in potassium, vitamin B_6, iron, vitamin A, and vitamin C, which match the high energy needs of micronutrient-deficient staple food crops. This vegetable is mostly consumed in Africa and Asia. Mulukhiyah is a Syrian, Lebanese, Tunisian, Turkish Cypriot, Jordanian, Palestinian, and Egyptian dish with a long history [28].

Acts as a Tonic

When ingested, the leaves are claimed to serve as a demulcent, deobstruent, diuretic, lactagogue, purgative, and tonic. Aside from that, it is utilized to treat aches and pains, diarrhea, enteritis, fever, pectoral pains, and tumors. Ayurvedic medicine uses the leaves to treat ascites, discomfort, piles, and tumors. In various regions of the world, the leaves are used to cure cystitis, dysuria, fever, and gonorrhoea. The cold infusion is claimed to help you restore your appetite and vigour. Anti-inflammatory, gastro protective and antifertility effects are all found in it (Table 11.4) [28].

Miscellaneous

In Kenya, the root scrapings of this plant are used to alleviate toothache. In Congo, the leafy twigs are used to treat cardiac problems. In Tanzania, an infusion made from the leaves is used to treat constipation and in Nigeria, the seeds are used as a purgative and febrifuge.

Table 11.4 Medicinal uses of *Corchorus* species.

Properties	**Species**	**Medicinal Property**
Gastro-intestinal properties	*C. capsularis*	i) Increases appetite. ii) Aids Dysentery.
	C. olotorious	i) Fulfill energy requirements. ii) Acts as digestion tonic. iii) Infusion work against constipation.
Dermatological properties	*C. capsularis*	i) Helps to attain dry skin. ii) Acts against dandruff.
	C. olotorious	-----------------
Anti-Cancer Properties	*C. capsularis*	Presence of compounds of phytol and glycerol acts as anti-cancer agents.
	C. olotorious	----------------
Anti-Oxidant Properties	*C. capsularis*	Assays of *C. capsularis* work as anti-oxidant agents.
	C. olotorious	Phenolic and gylocosidic compounds.
Anti-Inflammatory Properties	*C. capsularis*	Extracts work as anti-inflammatory agents.
	C. olotorious	-
Anti-Microbial and Insecticidal Effects	*C. capsularis*	Extracts work as anti-microbial agents and posses some insecticidal effects.
	C. olotorious	-

Industrial Importance

Textile Industry

Jute is used to manufacture yarn, twine, sacking, carpet backing fabric, and various mixed textiles, among other things. It is also utilized in the creation of cords and strings as a raw material. The bark tissue of *C. olitorius*, which is mostly cultivated in South Asian nations, is used to make jute fiber, albeit *C. olitorius* fiber is of poorer quality. The completed fibers have a length of up to 3 m and a diameter of 2.4 m, and they seem golden and smooth. The fiber obtained are fine well separated from unwanted part by means of pulling up, rippling, partly retting, breaking, spinning, and combing the plant stalk. After that, the fibers are treated and dried [35].

Cuisines at various places

C. *Olitorius* is cultivated as a potherb in Syria, Lebanon, Palestine, and Egypt, and it has been used in Egyptian cuisine since ancient times. It is a prominent leafy vegetable in Côte d'Ivoire, Benin, Nigeria, Ghana, Cameroon, Sudan, Uganda, Kenya, and Zimbabwe. It is cultivated and consumed all over the world, including the Caribbean, Brazil, the Middle East, India, Bangladesh, Japan, and China. Its leaves are particularly popular among the Boros of northeast India, who use dried leaves to prepare a mucilaginous delicacy called narji with fatty pork and lye. In Nigeria, the leaves are cooked to form a sticky, mucilaginous sauce that is served with otherwise dry cassava balls.

In the production of Sauces

Jew's mallow is a mucilaginous leafy vegetable. Cooked leaves produce a slimy, sticky sauce similar to okra. This sauce is found to be suitable for consuming starchy balls made from cassava, yam, or millet in Nigeria. During the dry season, this sauce is made with a powder made from dried leaves. For the slimy sauce, the immature fruits, known as bush okra, are also dried and ground to a powder. Jew's mallow can be cooked with cowpeas, pumpkin, cocoyam leaves, sweet potato, milk and butter, meat, and peppers and lemon in several East African recipes.

As a source of packaging

For more than a century, jute has been the most widely used packaging fiber due to its strength and durability, low production costs, ease of manufacturing, and availability in large, consistent quantities. In Africa, however, jute production is negligible. *Corchorus olitorius* used as a leaf vegetable and used for jute production are different in nature. Jute stems in their entirety can be used to make paper pulp. When jute is used for pulping, however, it is usually in the form of burlap cuttings, old sugar bags, and wrappings. The pulp is then turned into a hard, thick paper that can be used for cards and labels (Table 11.5).

11.2.3 *Corchorus Cunninghami*

C. cunninghammi is another species under genus corchorus. It is an herbaceous, perennial shrub, which grows to a height of 1.5 to 2 m. The stems are generally covered with hair sparsely. The leaves of this plant are 5 to 15 cm long, 1.5 to 5 cm wide, attached to stalks, which are 1 to 2.5 cm long and are three-veined from near their base and arranged in an alternate pattern around the stem. This plant's blooms are tiny, with four yellow, slender, oval-shaped petals that are 9 to 11 mm long and 3 to 5 mm broad. They can form a cluster of two to seven blooms from a single stalk on the side of an upper stem opposite a leaf, or they can grow singly at leaf or flower nodes. Buds are pear-shaped and 3 to 4 mm in diameter. At the base of the flower or bud, four pale yellow to green pointed sepals measuring 7 to 11 mm are visible. Flowers have 60 to 80 stamens and a weak, 3 or 4 ellipsoid ovary (1.5–3 mm long and 4–6 mm broad) with three or four cells each containing 18 to 21 ovules (1.5–3 mm long and 4–6 mm wide). November to May is the peak flowering period for *C. cunninghammi*. Between December and May, fruiting bodies form on the plant, but they can also be found at other times of the year. They have ellipsoid-shaped capsules that are 1.5 to 11.5 cm long and 4 to 6 mm wide and are dark brown to black in color. The capsules contain three or four chambers each with 2 to 22, elliptical or rounded matt brown to black seeds (2–3 mm in length). The species is distributed from Brisbane, South East Queensland to Lismore, New South Wales. Found in the ecotone between rainforest and eucalypt communities.

Corchorus cunninghammi as an Endangered Species

C. cunninghammi is a naturally rare plant species that is endangered due to habitat loss, genetic isolation, and competition with invasive weed species, as well as ineffective fire and land management regimes and forestry operations. Other significant populations for the species include one in Wongawallan, which contains more than 85% of the total number of plants in southeast Queensland, and another at Mount Cotton, which is genetically distinct from others and may be required to maintain the species' genetic diversity in the long run. As a result, Marion Saunders and the Rainforest Ecotone Recovery Team drafted a recovery plan, which was approved by the Commonwealth Government in 2001. *C. cunninghammi* is found in the thin ecotone that divides subtropical rainforest from open eucalypt woodland, according to them. The species may be found on higher hill-slopes or hill-crests facing south-east or east, at low to mid altitudes (110–430 m). Although it prefers upper hill slopes, the species may grow anywhere between the ridge and the gully depending on where the open forest-rainforest ecotone is found. As the species occurs on both metamorphic and igneous substrates, as well as loam or clay soil, there does not appear to be a specific geology or soil type associated with it. As a result, there does not appear to be a specific habitat that is critical to the species' survival. *Eucalyptus propinqua* (grey gum), *Lophostemon confertus* (brush box), and *Eucalyptus siderophloia* are some of the common canopy species found alongside *C. cunninghammi* (grey ironbark). Introduced weed species such as *Lantana camara* (lantana), *Rivina humilis* (coral berry), and

Table 11.5 Classified study of industrial and economic importance of *Corchorus* genus.

S. no	Species	Origin	Vegetative and medicinal parts	Economical parts	Useful products	References
1.	*C. capsularis*	China	Leaves as food Unripe Fruits and Roots used in traditional medicines	Cut stems after removing tissues	i.) Sacks ii) Carpets iii) Bags iv) Curtains v) Bags vi) Papers vii) Thickeners for soups and salads. viii) Herbal Medicines	[1]
2.	*C. cunninghami*	From Brisbane to Queensland (then became extinct)	-	-	-	[11]
3.	*C. erodiodes*	Yemen	-	-	Jute fiber	[14]
4.	*C. olitorious*	Africa and Indo-Burmese region of India	Leaves Young Fruits Seeds	Bark Tissue	i) Leafy vegetables ii) Sauces iii) Fiber iv) Packaging materials	[19]

(*Continued*)

Table 11.5 Classified study of industrial and economic importance of *Corchorus* genus. (*Continued*)

S. no	Species	Origin	Vegetative and medicinal parts	Economical parts	Useful products	References
5.	C. tridens	Tropics	Leaves	Leaves and stem	i) Medicine ii) Jute fibers	[23]
6.	*C. silIquosus*	Central America	Leaves	Leaves and stems	i) Tea ii) Medicine iii) Directly consumed as spinach	[29]
7.	*C. walcotti*	Australia	Leaves	Leaves and stem	i) Medicine ii) Jute fibers	[34]

Ageratina adenophora (crofton weed) frequently occur in the shrub layers, with density and composition varying between sites.

Need for Recovery

C. cunninghammi communities in Queensland have high nature and conservation importance because of the presence of numerous rare or vulnerable plant species. Some of the species are *Choricarpia subargentea* (giant ironwood), *Endiandra floydii, Macadamia integrifolia* (macadamia nut), *Pouteria eerwah* (black plum), *Sophora fraseri* (brush sophora), and *Randia moorei* (spiny gardenia). Another advantage of executing the *C. cunninghamii* recovery plan in Queensland is the conservation and maintenance of biodiversity within the ecotonal zones where *C. cunninghamii* naturally exists. Exotic species, such as *Lantana camara* (lantana), are particularly susceptible to invasion in these ecotonal areas, and a better understanding of the role of fire and disturbance in these habitats will greatly aid in maintaining native species diversity and controlling invasive species [30].

11.2.4 *Corchorus Erodioides*

Corchorus erodioides is a species of flowering plant in the family Malvaceae senslato or Tiliaceae or Sparrmanniaceae family. It is found only in Yemen. Annual herbs that attain a height of 2 to 4 m and are unbranched or have only a few side branches are the most common. The leaves are alternating, simple, lanceolate, 5 to 15 cm long, with a finely serrated or lobed edge and an acuminate tip, and are alternate, simple, lanceolate, 5 to 15 cm long. The blooms are little yellow flowers with five petals (2–3 cm in diameter), and the fruit is a seed capsule with numerous seeds.

11.2.5 *Corchorus Siliquosus*

Its common name is Slippery Bur. It is a densely branched shrub with a taproot, growing up to 1 m tall. The plant is harvested from the wild for its leaves, which are used to make a tea. It is found in Central America from Panama north to Mexico, southeastern N. America and the Caribbean. The plant is a common weed in some parts of Central America. It grows in moist or dry thickets, often in waste ground, at elevations of 1000 m or lower.

Corchorus siliquosus is a small shrub that can reach a height of 2 m. The leaves are ovate with a crenate leaf margin and acute leaf apex, arranged alternately to 4 cm in length and 1.5 cm in width. The surface of all vegetative matter is glabrous.

The actinomorphic flowers are arranged in umbels that emerge from the nodes. Five unfused greenish sepals make up the calyx. Five unfused, bright yellow petals make up the corolla. There are a lot of stamens in this flower. With five locules and numerous seeds, the ovary is superior. The fruit is an elongate capsule with a diameter of 2 mm and a length of 8 cm.

It thrives in areas that have been altered by humans, such as abandoned fields and roadside ditches. It can also be found on the outskirts of Sabal palmetto woodlands and Dry Broadleaf Evergreen Formation—Shrublands (scrublands). It can be found on all of the Lucayan Archipelago's island groups, as well as the southern United States, the Caribbean region, and all of the New World's tropical and subtropical areas.

Health Benefits and Industrial Importance

- Leaves are used as a cooling medicine, demulcent, gonorrhea (macerated in water) Pot-herb (raw).
- The leaves are sometimes cooked and eaten like spinach.
- The leaves are used as a substitute for China tea [31].

11.2.6 *Corchorus Walcotti*

Woolly corchorus is a common name for it. It belongs to the Malvaceae family of shrubs. It is only found in Australia. In the native range of the species, plants reach a height of 1.2 m and produce yellow flowers from June to November. In 1862, Victorian Government Botanist Ferdinand von Mueller published the first formal description of the species in the third edition of his Fragmenta Phytographiae Australiae. For Mueller's description, Pemberton Walcott gathered plant samples from Hearson Island (possibly Dampier Island), Nickol Bay (near present-day Karratha).

This species may be found in Western Australia, the Northern Territory, and South Australia's northwest. This species is often an upright, spreading, heavily branched shrub with a height of 1.2 m and high. Its blossoms are generally yellow in color. The months of June through November are the best for growing this plant. It grows well in light brown or red sand, sandy loam, and brown loamy clay.

Scientific Description

Hairy stems are typical of these plants and herbs. Sinuate edges and indumentums are found on the 15- to 70-mm long, 12- to 40-mm broad leaves. The calyx and corolla of the perianth are distinctly divided into two whorls (the corolla obvious and prominent). There is a pedicel with a length of 5 to

6.5 mm. The golden Corolla is 6.5 to 8 mm in length and is glabrous. There are several free stamens inserted at the ovary's base. There are filaments with a length of 11.5 to 5 mm and anthers with a length of 0.7 to 1.5 mm.

Health Benefits and Industrial Importance

- Dysentery, worm infestation, and constipation can all be controlled or prevented (Table 11.4).
- The leaves are used in Ayurveda for ascites, piles, pain, and tumors.
- Cystitis, dysuria, fever, and gonorrhea even cold can be cured.
- Used in textile industry [32].

11.2.7 *Corchorus Tridens*

C. tridens is a plant having simple, alternating, stalked, and stipulated leaves that grows upright. The blade has a dentate border and is linear. A filament bends back and extends the first tooth of the lamina's base. Flowers are grouped in bundles on the opposite side of the petioles. They are big and golden, with free sepals and petals and a lot of stamens. The fruit is a linear capsule with a beak on top that splits into three valves in numerous polyhedral seeds.

The Cotyledons of this plant are tiny, 2 to 3 mm long and 1.5 to 2 mm wide, glabrous, quickly obsolete. Also, they have a round shaped lamina with short stalk. The leaves are alternating and simple. A lengthy petiole (5–15 mm) supports leaves. The petiole is coarsely pubescent and is framed at the base by two linear stipules that are lengthy (5–7 mm) and quickly deciduous. The lamina is straight lanceolate with a sharp top edge and a broad or truncated base. It is 6 to 8 cm long and seldom surpasses 15 mm in width. It has a toothed edge. A filament as long as 10 mm in length and purple in color bends back and extends the initial tooth at the blade's base. The base of the limb is trinerved, with several pairs of secondary veins. The upper surface is smooth, with a few extremely small hairs running along the veins on the underside. The leaves have a vibrant green color. The blooms are carried on 1-mm-long stalks. The calyx is made up of five sepals that are linear in form and aciculate at the apex. They measure 5 mm in length. The corolla is made up of five free petals that are 5-mm long and have a rounded tip and thin base. Yellow is the color of the corolla. The stamens are abundant. The ovary is oval and has a short style on top. They are grouped in pairs of two to six opposite the petiole's beams.

The fruit is a dehiscent capsule, linear, with triangular cross-section, with three valves. The capsule is striated longitudinally. It measures 3 to 5 cm long and 2 to 3 mm in diameter. It has a short beak splitting into three points at the top. It is smooth. Each capsule contains a large number of seeds. The shape of seeds is polyhedral. They are 1.5 mm long and 1 mm wide. The seed coat is smooth and dark brown. This species does not have a preference for any specific soil type. It grows in a wide range of soils, from heavy clay vertical soils to ferruginous soils; however, it becomes less frequent as the soil degrades. It is frequently seen in combination with *C. olitorius* in heavy soils. It is, nevertheless, more tolerant of drier, less fruitful environments. The tropics are home to this species.

Health Benefits and Industrial Importance

- Depress the central nervous system, lower the blood pressure (Table 11.4).
- Stimulate smooth muscles (ayurvedic formulations).
- Used in textile industry for jute production [33, 34].

11.3 Future Scope

The study of different species of genus *Corchorus* is an important aspect for upcoming times. The species are an important part of both textile and food industry. The edible part of the plant is consumed as vegetable similar to okra and spinach in different regions as different cuisines, whereas the fibers are used for manufacturing of jute items, such as bags, carpets, sacks, etc. The species can be widely consumed for its antioxidant and other health properties.

References

1. Wilson, T.C., Elkan, L., Henwood, M., Murray, L., Renner, M., Wardrop, C., Prostanthera conniana (lamiaceae, westringieae), a new species from the Southern Tablelands, New South Wales, Australia. *Telopea*, 18, 519–526, 2015.
2. Azuma, K., Nakayama, M., Koshioka, M., Ippoushi, K., Yamaguchi, Y., Kohata, K., Higashio, H., Phenolic antioxidants from the leaves of Corchorus olitorius L. *J. Agric. Food Chem.*, 47, 10, 3963–3966, 1999.
3. Al-Snafi, A.E., The contents and pharmacological importance of Corchorus capsularis-a review. *IOSR J. Pharm.*, 6, 6, 58–611, 2016.

4. Oboh, G., Raddatz, H., Henle, T., Characterization of the antioxidant properties of hydrophilic and lipophilic extracts of jute (corchorus olitorius) leaf. *Int. J. Food Sci. Nutr.*, 60 Suppl 2, 124–34, Biochemistry Department, Federal University of Technology, Akure, Nigeria, (Epub Apr. 23, 2009), 2009.
5. The Editors of Encyclopaedia Britannica, Encyclopedia Britannica Inc., 2014.
6. Calleja, D.O. and Danny, O., Saluyot now a popular vegetable worldwide. *Inquirer*, Agri green, 2010Retrieved August, 7, 2011.
7. Choudhary, S.B., Sharma, H.K., Kumar, A.A., Maruthi, R.T., Karmakar, P.G., The genus corchorus L. (malvaceae) in India: Species distribution and ethnobotany. *Genet. Resour. Crop Evol.*, 64, 7, 1675–1686, 2017.
8. Cumo, C. (Ed.), *Encyclopedia of Cultivated Plants: From Acacia to Zinnia [3 Volumes]*, 1236 pages, ABC-CLIO, Nature, 25-Apr-2013.
9. Duke, J.A., Ecosystematic data on economic plants. *Quart. J. Crude Drug Res.*, 17, 3-4, 91–110, 1979.
10. Fern, K., *Plants For A Future: Edible & Useful Plants For A Healthier World*, Permanent Publications, UK, (https://www.permanentpublications.co.uk/about/) 1997.
11. Gledhill, D., *The Names of Plants*, Cambridge University Press, Cambridge, 2008.
12. Brummitt, N.A., Bachman, S.P., Griffiths-Lee, J., Lutz, M., Moat, J.F., Farjon, A., Lughadha, E.M.N., Green plants in the red: A baseline global assessment for the IUCN sampled red list index for plants. *PloS One*, 10, 8, e0135152, 2015.
13. Halford, D., Conservation statement and draft recovery plan for corchorus cunninghamii f. Muell, NSW Department of Environment and Conservation, Hurstville, NSW, 1995.
14. Heywood, V.H., Brummitt, R.K., Culham, A., Seberg, O., *Flowering Plant Families of the World*, vol. 88, Firefly Books, Ontario, 2007.
15. Islam, M.M., Biochemistry, medicinal and food values of jute (corchorus capsularis L. and c. olitorius L.) leaf: A review. *Int. J. Enhanc. Res. Sci. Technol. Eng.*, 2, 11, 135–44, 2013.
16. Nyadanu, D., Adu Amoah, R., Kwarteng, A.O., Akromah, R., Aboagye, L.M., Adu-Dapaah, H., Tsama, A., Domestication of jute mallow (Corchorus olitorius L.): Ethnobotany, production constraints and phenomics of local cultivars in Ghana. *Genet. Resour. Crop Evol.*, 64, 6, 1313–1329, 2017.
17. Simmonds, M. and Playford, J., Genetics, ecology and convervation of the endangered plant corchorus cunninghamii f. Muell. *Proc. R. Soc. Qld.*, 110, 61–711, 2002.
18. Harden, G.J., *Flora of New South Wales, Vols. 1–4*, New South Wales University Press, Kensington, 1990.
19. Halford, D.A., *Notes on tiliaceae in Australia, 2*, A revision of the simple-haired species of the genus corchorus L. Austrobaileya, 297–320, 1995.

20. Hazra, S.K., Mahapatra, A.K., Saha, A., Saha, D., Gupta, D., *Therapeutic and medicinal potential of Indian corchorus and their conservation*, Sci. Cult., 70(07/08), 260-264, 2004.
21. Islam, M.T., de Freitas, R.M., Sultana, I., Mahmood, A., Hossain, J.A., Homa, Z., Uddin, M.M., A comprehensive review of Corchorus capsularis: A source of nutrition, essential phytoconstituents and biological activities. *J. Biomed. Pharm. Res.*, 2, 1, 01–08, 2013.
22. Mueller, F., II. A record of the plants collected by Mr Pemberton Walcott and Mr Maitland Brown, in the year 1861, during Mr F. Gregory's exploring expedition into North-West Australia, in: *Transactions of the Botanical Society of Edinburgh*, vol. 7, pp. 479–500, Taylor & Francis Group, Permanent Publications, January 1863.
23. Muthoni, J. and Nyamongo, D.O., Traditional food crops and their role in food and nutritional security in Kenya. *J. Agric. Food Inf.*, 11, 1, 36–50, 2010.
24. Owoyele, B.V., Oyewole, A.L., Alimi, M.L., Sanni, S.A., Oyeleke, S.A., Anti-inflammatory and antipyretic properties of Corchorus olitorius aqueous root extract in wistar rats. *J. Basic Clin. Physiol. Pharmacol.*, 26, 363–8, 2015.
25. Perret, F., Marnotte, P., Le Bourgeois, T., Carrara, A., Détermination pratique de quelques espèces de convolvulacées, adventices de l'Afrique du Centre et de l'Ouest, 1997.
26. Sattler, P.S., *The Conservation Status of Queensland's Bioregional Ecosystems*, Environmental Protection Agency, Queensland Government, 1999.
27. Saunders, M., *Recovery Plan for the Endangered Native Jute Species, Corchorus Cunninghamii F. Muell. in Queensland (2001-2006)*, Environmental Protection Agency, It is the recovery plan made by Natural Heritage Trust in Australia 2001.
28. Simmonds, M. and Playford, J., Genetics, ecology and convervation of the endangered plant corchorus cunninghamii f. Muell. *Proc. R. Soc. Qld.*, 110, 61–711, Natural Heritage Trust in Australia, 2002.
29. Slatkin, M., Gene flow and the geographic structure of natural populations. *Science*, 236, 4803, 787–792, 1987.
30. Stanley, T.D. and Ross, E.M., *Flora of South-Eastern Queensland, Volume 11*, Queensland Department of Primary Industries, Queensland, Australia, 1989.
31. Stewart, R.H., *The Corchorus (Jute) Pages. Malvaceae Info*, Queensland, Australia, 2011, http://www. malvaceae. info/Genera/Corchorus/Corchorus. html.
32. Whitlock, B.A., Karol, K.G., Alverson, W.S., Chloroplast DNA sequences confirm the placement of the enigmatic oceanopapaver within corchorus (grewioideae: Malvaceae sl, formerly tiliaceae). *Int. J. Plant Sci.*, 164, 1, 35–41, 2003.
33. Wunderlin, R.P., Hansen, B.F., Franck, A.R., Essig, F.B., *Atlas of Florida Plants*, Retrieved April 13, University of South Florida, Tampa, USA, 2017, 2017.

34. Zeghichi, S., Kallithraka, S., Simopoulos, A.P., Nutritional composition of molokhia (corchorus olitorius) and stamnagathi (cichorium spinosum). *World Rev. Nutr. Diet.*, 91, 1–21, University of South Florida, Tampa, USA, 2003.
35. Biswas, A., Dey, S., Li, D., Liu, Y., Zhang, J., Huang, S., Deng, Y., Comparison of phytochemical profile, mineral content, and *in vitro* antioxidant activities of corchorus capsularis and corchorus olitorius leaf extracts from different populations. *J. Food Qual.*, 2020, 1–14, 2020.

Index

Also of Interest

Other Books in the series, "Bioprocessing in Food Science"

MICROBES IN THE FOOD INDUSTRY, Edited by Navnidhi Chhikara, Anil Panghal, and Gaurav Chaudhary, ISBN: 9781119775584. This newest volume in the groundbreaking new series, "Bioprocessing in Food Science," focuses on the latest processes, industrial applications, and leading research on microbes in the food industry, for engineers, scientists, students, and other industry professionals.

NOVEL TECHNOLOGIES IN FOOD SCIENCE, Edited by Navnidhi Chhikara, Anil Panghal, and Gaurav Chaudhary, ISBN: 9781119775584. The third book in the series, "Bioprocessing in Food Science," covers the latest and greatest new technologies in food science.

FUNCTIONAL FOODS, Edited by Navnidhi Chhikara, Anil Panghal, and Gaurav Chaudhary, ISBN: 9781119775560. Presenting cutting-edge information on new and emerging food engineering processes, *Functional Foods*, the second volume in the groundbreaking new series, "Bioprocessing in Food Science," is an essential reference on the modeling, quality, safety, and technologies associated with food processing operations today.

THERMAL FOOD ENGINEERING OPERATIONS, Edited by Nitin Kumar, Anil Panghal, and M. K. Garg, ISBN: 9781119775591. Presenting cutting-edge information on new and emerging food engineering processes, Thermal Food Engineering Operations, the first volume in the new series, "Bioprocessing in Food Science," is an essential reference on the modeling, quality, safety, and technologies associated with food processing operations today.

Other related titles from Scrivener Publishing

FUNCTIONAL POLYMERS IN FOOD SCIENCE, Part 1: Food Packaging, Edited by Giuseppe Cirillo, Umile Gianfranco Spizzirri, and Francesca Iemma, ISBN: 9781118594896. Provides an extensive overview of the link between polymer properties and food quality. *NOW AVAILABLE!*

FUNCTIONAL POLYMERS IN FOOD SCIENCE, Part 2: Food Processing, Edited by Giuseppe Cirillo, Umile Gianfranco Spizzirri, and Francesca Iemma, ISBN: 9781118595183. Provides an extensive overview of the link between polymer properties and food quality. *NOW AVAILABLE!*

ADVANCES IN FOOD SCIENCE AND TECHNOLOGY Volume 1, Edited by Visakh, P.M., Sabu Thomas, Laura B. Iturriaga, and Pablo Daniel Ribotta, ISBN: 9781118121023. Written in a systematic and comprehensive manner the book reports recent advances in the development of food science and food technology areas. *NOW AVAILABLE!*

ADVANCES IN FOOD SCIENCE AND NUTRITION Volume 2, Edited by Visakh P.M., Laura B. Iturriaga, and Pablo Daniel Ribotta, ISBN: 9781118137093. This important book comprehensively reviews research on new developments in all areas of food chemistry/science and nutrition. *NOW AVAILABLE!*

Applied Water Science Volume 1: Fundamentals and Applications, Edited by Inamuddin, Mohd Imran Ahamed, Rajender Boddula and Tauseef Ahmad Rangreez, ISBN 9781119724766. Edited by one of the most well-respected and prolific engineers in the world and his team, this is the first volume in a two-volume set that is the most thorough, up-to-date, and comprehensive volume on applied water science available today.

Sustainable Water Purification, by Safiur Rahman and M. R. Islam, ISBN 9781119650997. This is the only book that takes a zero-waste approach to propose 100% sustainable water purification techniques. *NOW AVAILABLE!*

Printed in the USA/Agawam, MA
September 16, 2026

910661.136